# Impact of Information Society Research in the Global South

Arul Chib • Julian May • Roxana Barrantes
Editors

# Impact of Information Society Research in the Global South

Springer Open

*Editors*
Arul Chib
Wee Kim Wee School for Communication
  and Information
Nanyang Technological University
Singapore

Julian May
Institute for Social Development
University of the Western Cape
Bellville, Western Cape
South Africa

Roxana Barrantes
Instituto de Estudios Peruanos
Lima, Peru

This work was carried out with the aid of a grant from the International Development Research Centre, Ottawa, Canada. The views expressed herein do not necessarily represent those of IDRC or its Board of Governors.

ISBN 978-981-287-380-4    ISBN 978-981-287-381-1   (eBook)
DOI 10.1007/978-981-287-381-1

Library of Congress Control Number: 2014958028

Springer Singapore Heidelberg New York Dordrecht London
© The Editor(s) (if applicable) and The Author(s) 2015. The book is published with open access at SpringerLink.com.
**Open Access** This book is distributed under the terms of the Creative Commons Attribution Noncommercial License, which permits any noncommercial use, distribution, and reproduction in any medium, provided the original author(s) and source are credited.
All commercial rights are reserved by the Publisher, whether the whole or part of the material is concerned, specifically the rights of translation, reprinting, reuse of illustrations, recitation, broadcasting, reproduction on microfilms or in any other physical way, and transmission or information storage and retrieval, electronic adaptation, computer software, or by similar or dissimilar methodology now known or hereafter developed.
The use of general descriptive names, registered names, trademarks, service marks, etc. in this publication does not imply, even in the absence of a specific statement, that such names are exempt from the relevant protective laws and regulations and therefore free for general use.
The publisher, the authors and the editors are safe to assume that the advice and information in this book are believed to be true and accurate at the date of publication. Neither the publisher nor the authors or the editors give a warranty, express or implied, with respect to the material contained herein or for any errors or omissions that may have been made.

Printed on acid-free paper

Springer Science+Business Media Singapore Pte Ltd. is part of Springer Science+Business Media (www.springer.com)

# Foreword

The contributors to this volume make a crucial and forceful point. There are numerous theories and methodologies that can be used to yield research findings about the potential of information and communication technologies (ICTs) to make a positive difference in people's lives. Whether the findings in any particular research project actually contribute in this way depends on multiple factors, some affecting researchers themselves, others on whether actors beyond university-based research communities are interested and have the resources that are necessary for learning. It is not only researchers for whom capacity building is essential. It is essential that all the actors who have an interest in whether digital ICTs are produced and used in beneficial ways have the capacity to learn about how they can make a difference in people's lives. The need to build capacity for learning from a range of types of evidence developed by both researchers and practitioners applies as much to government actors as it does to those in the private sector and representatives of civil society. This volume illustrates this extremely well.

The SIRCA II (Strengthening Information Societies Research Capacity Alliance) programme involved researchers in research capacity building focusing on ICTs in contemporary information societies in the African, Asian and Latin American regions. The first part of the volume is concerned with research on 'impact'; the second sets out how research in the global South is contributing to our understanding of the information societies in these regions. The results of this second SIRCA programme offer varied reflections on how learning has accumulated within this community of researchers. In the opening chapter, Arul Chib ('Research on impact of the information society in the Global South: An introduction to SIRCA') says that 'impact, even during the process of evaluation of the proposals [for SIRCA II], was difficult to define, describe, or agree upon, leading to contested debate'. There is a multiplicity of voices, methodologies and theoretical traditions in the interdisciplinary fields of research that investigate the role of ICTs 'in', 'for', and sometimes 'and' development. This volume includes many illustrations of the richness of approaches within the social sciences. In this case, there is also an emphasis on critically evaluating what development means for those with an interest in the social, economic, political or cultural outcomes that may be expected when ICTs are involved.

The coherence of this volume (and of the research programme) lies not in any specific theory or method that is privileged but in the way research questions are posed and the rationale for undertaking the research programme in the first place. We find a strong commitment to several core principles. The first is that understanding the impact of ICTs within information societies requires a commitment to the analysis of developmental change in a way that extends beyond the economic dynamics of the marketplace. The second is a commitment, regardless of theoretical or methodological stance, to participatory research and especially to participatory action research that insists on local stakeholders being able to engage with choices that are made in the process of implementing ICTs. Linked to this, is a third principle. This is ensuring that research is as much about discovery (i.e. the causes of things) as it is about making change in the world in a way that is inclusive and consistent with values of justice and equity.

Debates in universities about what should be understood as the 'impact of research' are unlikely to abate. As is well emphasized in this volume, when 'impact' is set as a criterion for judging the excellence of research, it inevitably creates incentives that shape both the topics that are researched and the way researchers undertake their work. Because there is so much controversy about how to measure the impact of academic research, measures of impact are in constant flux. This suggests that the highest priority for researchers themselves is to work out how their own commitments to enabling positive change through research can be enhanced by exploiting opportunities within the 'impact' agendas of others. When they learn how to proceed in this way, they have the potential to make an even bigger difference than they might otherwise have made. Building capacity within research communities for the strategic exploitation of an always shifting impact agenda is crucial. This is especially so in regions of the world with relatively fewer resources for the conduct of research than is the case in the United Kingdom, for example.

The second part of this volume displays a rich research evidence base. The SIRCA scholars address the implications of the development and implementation of ICTs for poverty reduction, mobile financial and education services development, the reduction of cybercrime, strengthening women's entrepreneurship, and creating new online spaces for public consultation and the expression of public opinion. The results illustrate the value of a collaborative effort that provides mentoring for researchers and supports their initiatives through dialogue with each other and, crucially, with interested others. Each of the chapters in this part succeeds in advancing both conceptual and applied knowledge.

The question remains nevertheless: Does this work demonstrate 'impact'? It is worth pausing to reflect on what researchers and various stakeholder communities mean by this term. For some, it may mean that there is a demonstrable strong effect of the research on someone or something. For others, it may mean that it can be claimed that the research has had an influence, one that may be perceived as being positive or negative depending on where the assessor who makes this judgement is situated. Different actors will have contradictory views about what can be claimed as 'impact' even in cases when there is agreement about the value of participatory research. This is unavoidable because there is often profound disagreement among different communities of actors about what development means and, therefore, about what kind of change researchers investigating the impact of ICTs should be encouraging.

Even if impact assessment tools detect that change has happened and that ICTs are implicated in that change, it matters whether that change is consensual or achieved through external enforcement. We might imagine a world in which dialogue among all the stakeholders leads to a consensus about what would constitute positive change in information societies in the global South or, indeed, in all parts of the world. However, in reality, the process of achieving a dialogue about how ICTs can be introduced in a way that is inclusive, just and equitable is a struggle because it involves competing normative frameworks. The result is that 'impact' – however it is measured – will be regarded as positive, negative or mixed by those with different interests in the developmental process and its outcomes.

The contributors to this volume avoid the common pitfall of opposing curiosity-driven academic research to social problem-driven research. Instead, they acknowledge that learning thrives on both. When considering the 'impact' of research, it is helpful to recall Latour's (2013) point that inquiry and change require an open and negotiated sharing of new knowledge, however it is produced. Albagli and Maciel (2010) argue, similarly, that in considering how information societies are developing in the global South, it is important to understand that there is no single model of change that applies universally. Diverse ICT applications are likely to be welcomed when local actors see their normative commitments being translated into arrangements for the design and implementation of these technologies in a way that they regard as just and equitable. Aiming to achieve participatory involvement in ICTs 'for' development is as good a guide as any for fostering research that demonstrates how ICTs become embedded in societies in ways that are empowering not for the few but for the many. This volume is exemplary in making explicit the need for research that can empower the widest possible number of stakeholders. The SIRCA II programme has been a vital step along a pathway towards ensuring that there is a productive dialogue about the assumptions that underpin research on the role of ICTs in the process of building information societies.

The London School of Economics and Political Science     Robin Mansell
London, UK

# References

Albagli, S., & Maciel, M. L. (Eds.). (2010). *Information, power and politics: Technological and institutional mediations.* Lexington Books.
Latour, B. (2013). *An inquiry into modes of existence.* Cambridge: Harvard University Press.

# Contents

**Research on the Impact of the Information Society in the Global South: An Introduction to SIRCA** ................................. 1
Arul Chib

**Part I  Impact of Research**

**The Impact of Research on Development Policy and Practice: This Much We Know** .................................................. 21
Roger Harris

**Constructing Theories of Change for Information Society Impact Research** ................................................. 45
Alexander Flor

**A New Set of Questions: ICT4D Research and Policy** ..................... 63
Andrea Ordóñez

**Progress Towards Resolving the Measurement Link Between ICT and Poverty Reduction** ................................................. 83
Julian May and Kathleen Diga

**The Impact of mFinance Initiatives in the Global South: A Review of the Literature** ................................................. 105
Arul Chib, Laura León, and Fouziah Rahim

**An Analytical Framework to Incorporate ICT as an Independent Variable** ................................................. 125
Matías Dodel

ix

## Part II Research on Impact

**(Un)Balanced Conversations: Participatory Action Research in Technology Development in Peruvian Primary Schools** .................. 147
Paz Olivera, Komathi Ale, and Arul Chib

**The Institutional Dynamics Perspective of ICT for Health Initiatives in India** ............................................................... 167
Rajesh Chandwani and Rahul De

**Cybersex as Affective Labour: Critical Interrogations of the Philippine ICT Framework and the Cybercrime Prevention Act of 2012** ............................................................ 187
Elinor May Cruz and Trina Joyce Sajo

**The Internet and Indonesian Women Entrepreneurs: Examining the Impact of Social Media on Women Empowerment** ......... 203
Ezmieralda Melissa, Anis Hamidati, Muninggar Sri Saraswati, and Alexander Flor

**The Use of Mobile Communication in the Marketing of Foodstuffs in Côte d'Ivoire** .................................................. 223
Kabran Aristide Djane and Richard Ling

**Designing Web 2.0 Tools for Online Public Consultation** .................... 243
Fabro Steibel and Elsa Estevez

**ICTs and Opinion Expression: An Empirical Study of New-Generation Migrant Workers in Shanghai** .......................... 265
Baohua Zhou

**Impact of Research or Research on Impact: More Than a Matter of Semantics and Sequence** ......................................... 283
Julian May and Roxana Barrantes

# Research on the Impact of the Information Society in the Global South: An Introduction to SIRCA

Arul Chib

The age of globalisation has been defined in terms of access to modern information and communication technologies (ICTs) by some scholars (Hutton and Giddens 2001; Castells 2000; Rantanen 2001). Scholarly debate about the role of ICTs as an agent of social organisation and transformation has raged on before and since, from discussions about the networked information society (Bell 1999; Castells 1996) and consideration about the commercial potential of the technology (Gandy 2002; Shapiro and Varian 1999) to critiques of the systemic divides in organisation, access, use, adaptation and impact (Mansell 1999; Norris 2001; Warschauer 2003). Since these initial considerations, we find ourselves living in a world where ICTs have diffused widely to far-flung corners of the globe and are being deployed to confront some of the world's most complex problems. Scholarly debates in domains such as the global digital divide continue, in which some argue that technologies such as mobile phones have led to the expansion of socio-economic opportunity for the developing world (Donner 2008; Waverman et al. 2005), to those who claim that inequalities remain, with resultant limitations on their societal impact (Carmody 2013; Hilbert 2014). We focus here on notions of the impact of ICTs on international development, going beyond issues of access and use, well documented elsewhere.

As the current Millennium Development Goals (MDGs) approach the initial deadline of 2015, it is timely to take stock of the impact these technologies have had on key development problems. This moment is simultaneously the culmination of the second round of SIRCA II (the Strengthening Information Society Research Capacity Alliance), a capacity-building programme for information society research in the context of development in the Global South. It is then worthwhile too to interrogate the impact of research endeavours on the development process.

A. Chib, Ph.D. (✉)
Wee Kim Wee School for Communication and Information, Nanyang Technological University, Singapore
e-mail: ArulChib@ntu.edu.sg

# 1 History

As this volume is the second in a series resulting from the SIRCA programme (see Chib and Harris 2012), it is probably worthwhile to introduce our new readers to the historical trajectory and source for some of the arguments presented here. The IDRC team of Sinha et al. (2012) notes that the programme was borne out of a need for offering institutional support for interdisciplinary and methodologically sound projects led by emerging researchers with mentorship of senior researchers. Among the myriad challenges faced by the discipline were difficulties in measuring development outcomes (and impact), accompanied by an excessive reliance of anecdotal evidence (Gomez and Pather 2012). The lack of adequate scientific documentation and analysis of failures and successes culminated in limited relevance of the research to both practitioners and policymakers (Harris and Chib 2012). A further systemic problem identified was the lack of voice in the research dialogue of researchers from developing countries, possibly as a result of insufficient resources and training (both in quantity and quality) and unsupportive local research environment due to political, structural and other reasons.

These issues were identified as a result of a programme for 'informed and systematic' research based in Asia which began in 2006 and continued in a workshop at Manila in 2007, culminating in among other endeavours, the launch of the Strengthening ICTD Research Capacity in Asia programme (SIRCA I) in 2008. SIRCA I aimed to address issues related to the limitations in rigour, interdisciplinary research and collaboration in ICTD research, via a research capacity-building programme for emerging scholars in Asia. The programme provided research grants and training opportunities, with a key component being the mentorship of senior researchers provided for 15 emerging principal investigators from Bangladesh (2 projects), Cambodia (2), China, India (4), the Philippines (2), Singapore, Sri Lanka, Thailand and Vietnam. SIRCA was conceived, managed and continues to be implemented by the Singapore Internet Research Centre (SiRC), based at Nanyang Technological University in Singapore. For more details see Chib and Harris (2012)—*Linking research to practice*: *Strengthening ICT for Development research capacity in Asia*.

At the end of the three-year SIRCA I programme, emerging researchers produced research outputs such as publications in internationally recognised forums such as journals (including *Media Asia Journal* and *International Journal on Advances in ICT for Emerging Regions*), books and conferences (including IFIP WG9.4 Conference 2011, the 5th Annual ACRON-REDECOM conference and the 4th International Conference on Theory and Practice of Electronic Governance (ICEGOV2010)), admission and scholarships to prestigious doctoral programmes and assumption of leadership positions at universities and organisations related to individual disciplines (see SIRCA Technical Report available at www.sirca.org.sg). A key question that troubled us at the time concerned whether the impact of the programme related to the personal academic achievements and outputs of the beneficiaries (a measurable output for the programme) or whether there was any influence on the broader practices and policies of the day, a view echoed by others (Elder et al. 2013). One conclusion was that the 'SIRCA programme is moving towards a roadmap for navigating the tortuous route from research to impact via practice' (Harris and Chib 2012, p. 9). The attempt to improve methodological rigour via programme activities and mentoring nonetheless left us short of demonstrable compelling evidence of socio-economic impacts, despite providing a range of lessons and research findings shared with the broader community, i.e., academic impact. A key barrier in bridging these two sets of impact, i.e., socio-economic and academic, identified by Harris and Chib (2012) was in the differential processes, skill sets and motivations required of researchers trained in academic investigation. A series of measures were proposed, including greater engagement with a wider range of stakeholders, development of fresh skill sets, a reorientation of internal systems and incentives and production of a different range of outputs.

While I recognise the inherent contradiction of reporting on the impact of ICTs on broad measures of development in yet another (hardly a different output) academic publication, this volume is nevertheless a result of the exhortations to ourselves in the previous avatar of the programme. The SIRCA II (the Strengthening Information Society Research Capacity Alliance) programme was born in 2011, continuing research capacity-building efforts focused on the information society with an expanded global scope, establishing connections between African, Asian and Latin American researchers. The mentorship model evolved from the original hierarchical knowledge delivery model to emphasise the aspect of bidirectional collaborative learning and experience-sharing for both established and emerging researchers. Finally, to manage the complexities of a programme with global reach, SiRC partnered with the University of Western Cape (UWC),[1] South Africa, and Instituto de Estudios Peruanos (Institute of Peruvian Studies-IEP), Peru, allowing collaboration with my co-editors, Julian May and Roxana Barrantes, based at these respective institutions.

## 2 Conceptual Focus

Beyond the administrative management of an enlarged global scale, the most important aspect of the evolution of the SIRCA II programme has been the focus on an investigation of impact, defined more broadly than economic advancement. The notion of impact emerged from discussions among a variety of stakeholders in the SIRCA process—the donors, reviewers, advisors and the scholars themselves. For example, one key difference versus the SIRCA I programme was that selected investigators were selected not just in terms of their potential for scholarship, but

---

[1] The initial institution of collaboration was University of Kwazulu-Natal (UKZN) based in Durban, South Africa. With the regional collaborator's transfer to UWC, the partnering organisation changed accordingly.

the potential impact of the research on the communities studied was considered as well. We note that impact, even during the process of evaluation of the proposals, was difficult to define, describe or agree upon, leading to contested debates among an interdisciplinary group of senior scholars.

Government rhetoric and media opinion (and more often than not, scholarly tomes) frequently evoke the transformational nature of ICT adoption and use to achieve development goals, with the result that social implications of technology use in the developing world are believed to be enormous. But research indicates that interventions such as the impacts of mobile phone use on marginal communities (Butt and Sarker 2009) and in health care (Chib et al. 2014), e-governance on transparency and openness (Bhatnagar and Singh 2010), etc. all show wide fluctuations in impact. Variations in approach are rife, which means we don't really have the answers we think we have (Heeks 2010).

The first volume of the SIRCA book series (Chib and Harris 2012) questioned whether diffusing the boundaries of academic research into the realm of policy formulation and implementation led to broader impact of information society initiatives. We discussed how to translate theoretical approaches from the ivory tower of academia to real-world practice more consistently. This volume extends and focuses the enquiry into impact—looking closely at the nature of this link between research, practice and policy.

Linking Information Society research to issues of practice and policy first requires the definition of impact within the context of the information society and development. The question of impact has been contentious, and this volume hopes to contribute to that discourse by examining two aspects of the debate:

1. Impact of research: how is the research on ICTs in the Global South playing a role in achieving an information society, through implementation in practice, influence on policy formulation and media coverage for shaping public opinion?
2. Research on impact: what is the evidence of the impact of ICTs on society (i.e. the end objectives of socio-economic development)?

This volume is organised along these two major investigative fault lines. The first section on impact of research addresses dominant and alternative frameworks to evaluate and measure real-world impacts, while the second section on research on impact provides empirical evidence from SIRCA II principal investigators. There are valid reasons behind making these questions salient at this historical juncture. ICTD research has faced criticism for under-reporting of negative results, as well as harbouring a techno-centric or techno-determinist bias (Papert 1987). Although the pre-2015 MDG emphasise the need to leverage on the development potential of ICTs, the impact question has time and again begged for theoretical resolution. After several waves of developmental impact of ICTs which have been equally contested at each stage, this chapter next undertakes a review of the core literature related to trends, linkages and disconnects in impact of ICTD research on policy, public opinion and practice and the research on impact of ICTs on developmental objectives itself.

## 3 SIRCA Projects Situated vis-à-vis Theoretical Literature

It's vital to disaggregate the term impact prior to examining individual contributions from SIRCA authors. Various scholars, policymakers, international development practitioners and others such as journalists have wrestled with questions of how to consider impact when discussing and studying the role of ICTs in the Global South. Flor, in chapter 'Constructing theories of change for information society impact research', raises the question whether digital access and usage have helped achieve the MDGs. He concludes that despite a decade of research, scholars are still divided over identifying the type of impact ICTs have had on development and poverty. Flor argues that while ICTD programmes have tried to address this question by strengthening capacities of information society researchers, the information society discourse still lacks a comprehensive, universally accepted framework that establishes clear causal links between ICTD interventions and achievement of the MDGs.

Information society scholars have suggested diverse frameworks to provide analytical perspectives as to the what, why and how of ICT utilisation and projects in the Global South. Several frameworks, often grounded in development literature, but also drawing from disciplines such as information science, systems theory, sociology, etc., offer useful insights into how technology may improve human lives and bring social change. The conceptual journey[2] of impact is illuminated next with evidence on the second question in the form of findings from the research projects of SIRCA II researchers. As my co-editors May and Barrantes ponder in chapter 'Impact of research or research on impact: More than a matter of semantics and sequence', it is worth considering the theoretical contribution of the research that follows.

### 3.1 Capabilities Approach

Sen (1999) proposed the since-mainstreamed framework in response to prevailing discussions on development focused on the bottom line of economic growth. Improvement in human well-being had previously been equated principally with economic growth. Scholars such as Chib (2009) and Kleine (2010) argue that the mainstream impact argument for ICTD had been heavily focused on economic development due to a need to legitimise projects for a donor base. Sen challenged the prevailing notion of development by prioritising individual ability to make choices, layering the baseline of economic growth with broader, more humanistic

---

[2]Note that this chapter is not intended to provide a rigorous review of the impact literature. The objective instead is to discuss relevant conceptual frameworks informing the issue of the impact of ICTs and situate the contributions of SIRCA participants contained within this volume relative to the broader scholarly discussion.

conceptions of capabilities and freedoms. While the framework has allowed more in-depth, diverse theoretical discussions, there has been considerable effort to operationalise the concept in order to make it applicable as a framework for structuring and evaluating ICTD projects.

## 3.2 Extensions

Sen's capability approach has subsequently been modified and expanded. Hatakka and De (2011) presented an ICTD evaluation framework by adopting a participatory evaluation approach (House 1980). These scholars argue that Sen's work does not directly address the issue of ICT usage and development. Instead, technology needs to be understood as a means to development goals (or as an intervention) that, together with supportive aspects, enable or restrict conversion factors that translate into outcomes. What's important here is that it's not the technology that is enabling but whether its use contributes to enabling choices for various stakeholders.

This conceptualisation departs from techno-deterministic models of impact. Technology introduction in itself is meaningless unless non-technological factors are taken into consideration, for example, the role of teachers in an educational technology project. Conversion factors influence both the enablement of teaching and learning as well as the ability of teachers and students to make choices. Olivera et al. (2015), in their chapter '(Un)Balanced conversations: Participatory action research in technology development in peruvian primary schools', examine the value of utilising participatory action research, involving teachers in the technological design process for educational applications developed for Peruvian primary schools. The authors acknowledge that there is a constant challenge to design successful technical solutions that fully deliver intended capabilities (Rodriguez et al. 2012) as these programmes are operating in resource-constrained environments that are further complicated by social, psychological, geographic and cultural differences (Chib and Zhao 2009). To bridge this disparity, they recommend a participatory action research approach that supports iterative data gathering and ongoing technology development that would, ultimately, positively improve teaching and learning in schools in developing counties. Hence, this shifts the project focus to building capacities of key users to bolster the educational infrastructure in which the new technologies are deployed.

There are increasing calls for focusing on noneconomic measures of development, with Gomez and Pather (2012) stating that the ICTD evaluation field has often neglected the intangible aspects of technology projects. May and Diga (2015) review the debate surrounding the connection of ICTs to poverty in chapter 'Progress towards resolving the measurement link between ICT and poverty reduction'. They argue that the measures of poverty in the current literature are not sophisticated enough to capture the complexity of development problems. These scholars propose incorporating participatory approaches and subjective well-being measures to enhance our understanding of the link between ICTs and poverty reduction.

Gomez and Pather (2012) observe that the corporate sector has focused far better on 'intangible aspects of business benefits [...], such as trust, loyalty and brand improvement in evaluation frameworks,' (p. 4) creating greater awareness of organisational identities and capabilities. The argument is that a greater focus on indirect, intangible benefits (rather than the tangible mentioned above) is necessary to incorporate the 'elusive and ubiquitous nature of ICTD impact measurement' (p. 9). Obviously, it is much more challenging to observe and analyse the intangible. Djane and Ling (2015) in the chapter 'The use of mobile communication in the marketing of foodstuffs in Cote d'Ivoire' find that mobile phones facilitate organisation and communication among large-scale and petite wholesalers embedded within a scattered distribution chain. Nonetheless, the analysis also unearthed the traditional 'Chain of the Grandmother' distribution network and 'African taboos' that limit the role of mobile phones in producing greater efficiencies in the foodstuff distribution system.

## 3.3 Empowerment Framework

Alsop and Heinsohn (2005) focus on individual agency and opportunity structure in achieving development outcomes. For these authors, the existence of choice and the achievement of choice are key measures of the degree of empowerment. Aspects such as empowerment and cohesiveness (McNamara 2003) directly affect people's behaviours and responses (Reimer 2002). Melissa et al. (2015), in their chapter 'The Internet and Indonesian women entrepreneurs: Examining the impact of social media on women empowerment', investigate how going online can boost female entrepreneurship by allowing them to establish and conduct businesses from their homes. These scholars use several indicators to measure empowerment, including those related to both existence and achievement of choice, such as domestic decision-making, access to or control over resources, freedom of movement, economic contribution to the household, appreciation within the household and sense of self-worth. Melissa et al. found that the prevalence of social media boosted gender-preneurship in Indonesia, bringing a wide range of business opportunities and flexibilities in terms of both self-actualization possibilities for women as well as improvement in their socio-economic status.

We note however, with a note of caution, that it is necessary to apply a critical lens when examining impact—while there are intangible benefits, there are also intangible negative consequences that ICT researchers should be aware of. Cruz and Sajo (2015) in the chapter 'Cybersex as affective labour: Critical interrogation of the Philippine ICT framework and the Cybercrime Prevention Act of 2012' explore negative consequences of ICT development— cybersex as an anomalous offshoot of ICT development in the Philippines. These researchers examine the operations of cybersex as a business and its uses of ICT and documented the life histories of cybersex workers to make sense of their work, sexuality and identity. The authors argue that instead of legislating the problem away, there are opportunities to understand the creative uses of technology appropriation that are practical and meaningful to cybersex workers. This study provides a deeper understanding of how ICT can both be a tool for exploitation as well as broaden choices and empower the marginalised to be self-reliant in truly trying circumstances.

Gomez and Pather (2012) highlight several other voices in ICTD research, from Parthasaranthy and Srinivasan (2006) and Qureshi (2005), who make strong cases against excessive quantifiable data and modelling that may miss out on noneconomic, micro-level impact, to Taylor and Zhang (2007), who argue that transformation depends not on the presence of infrastructure per se but on the design and implementation contextualised to social, economic and technological environments. Chandwani and De's (2015) chapter 'The institutional dynamics perspective of ICT for health initiatives in India' exemplifies the importance of institutional context in impact assessment of ICT4D initiatives. Chandwani's research on telemedicine interventions in India suggests that one of the major behavioural reasons hindering the diffusion of telemedicine (and, by extension, similar ICTD projects) is that these interventions focus on modern medical systems, while the rural population, at large, relies on alternate systems of medicine for primary health care.

## 3.4 Sustainable Livelihood Framework

Developed between the 1980s to the 1990s, the framework adopts a multidimensional perspective towards analysing and measuring the socio-economic impact of developmental projects (van Rijn et al. 2012). Specifically, it argues that the livelihood of people is a function of interrelations between tangible and intangible assets (Newton and Franklin 2011). Scoones (1998) argued that sustainable livelihoods could withstand and recover from shocks, maintain and grow in the future while guarding against depletion of natural resources. The framework proposes five capital assets (Carney 1998) that can be observed and measured for the impact of ICTD projects—human, natural, financial, physical and social—as cornerstones to be observed and measured for the impact of ICTD projects.

## 3.5 Choice Framework

Agency is a key concept in Kleine's (2013) work, which is then related to a resource portfolio. The argument is that social context influences the development of one's personal characteristics and how one gets to exercise agency. The resources listed are material, financial, natural, geographical, human, psychological, information, cultural and social (as social capital); together these allow a systemic analysis of ICTD projects.

Structure, along with agency, forms an essential element for the choice framework. Structural factors include various formal and informal elements of norms such as laws, regulations, culture, policy, institution, processes and culture. These form and are embedded in the discourse which contribute to the framework. Structure contributes to the shaping of the individual resource portfolio and can either enhance or limit one's exercise of agency.

The two-pronged approach of 'agency' and 'structure' sounds fine in an abstract conceptual sphere, but how does the approach pan out in practice? Kleine suggests that the choice framework can help development priorities be more participatory, engaging the recipients' interest and considering local and cultural elements into programme formation and implementation. It operationalises Sen's capability approach, linking ICTs as the means which are imbedded in agency and structure which eventually lead to individual choices. The choice framework requires greater work in terms of measurability, as the author admits that it is more easily applicable to individual level than at aggregate meso- or macro-levels.

Bhatnagar and Singh's (2010) assessment of Indian e-Government initiatives examines national-level data, accompanied with fuzzy measurability and myriad complex relationships. The framework identifies key stakeholders and client value by measuring cost to the client of assessing services and their perception of the quality of service and governance. The research question investigates values that can both be monetised and those that are intangible, across multiple stakeholders (beneficiaries, implementation agencies and society), thus requiring a range of methodologies to capture the overall rating for a project's impact.

In the chapter 'An analytical framework to incorporate ICT as an independent variable', Dodel presents an analytical framework based on Selwyn's (2010) 'Impact Assessment Framework' that includes hierarchical stages of ICT involvement—access, usage, appreciation and outcomes—on development. These scholars argue that ICT research would gain greater relevance to non-ICT researchers by including these variables in addition to measures of individual well-being. The model was tested on a sample of young Uruguayans, finding that digital skills were a significant influence on their occupational achievements. Supporting such an analysis, Chew et al. (2011) claim that quantitative analysis is more likely to provide rigorous tests of causality. In their study on ICT use in women-run microenterprises, these scholars find a statistically significant, but limited, causal relationship between access to ICTs and business growth. The authors claim that at the macro level, ICTs have had an overall positive impact in terms of economic growth; however, at the micro-level, it is difficult to find such evidence and, in this case, for micro-entrepreneurs, mostly, due to the cost issues in adopting computers. These examples underscore the challenges of validating higher-level abstractions with measurable empirical evidence—conceptual frames that can bridge local and national-level projects.

## 3.6 RAPID (Research and Policy in Development) Programme/Framework

Young and Court's framework (2004), developed by the Overseas Development Institute, emphasises the link among research, policy and practice. It identifies three overlapping areas—the political context, the scientific evidence and the links between policy and research communities—and external context as the factors which will need to be accounted for if research is contribution to the way that policymakers and practitioners work.

Echoing the importance of political context, Ordóñez (2015) in the chapter 'A new set of questions: ICT4D research and policy' suggests that there is an overemphasis on demonstrating the link between ICT and development. She claims that researchers have neglected other important areas such as understanding the relationship between policies, politics and research. She proposes that three distinct streams of academic inquiry—policy studies, the interface between research and policy and the conceptualization of ICT4D research—are required to understand the motivations of policymakers and the complexity of the political context. By doing so, researchers will be in a better position to set new agendas and offer effective solutions in the context of ICT4D.

The RAPID framework suggests that the influence of policy on research is dependent upon the quality of empirical evidence offered, and as such, quality research can potentially lead to solutions to policy problems. Harris (2015) in the chapter 'The impact of research on development policy and practice: This much we know'. A Literature Review and the Implications for ICT4D' makes a counter-argument that since most of research content revolves solely around academic impact, quality rarely leads to policy changes. This review of the impact of research on development policy and practice reports that information society impact research in the Global South has almost exclusively focused on the impact of ICTs without taking into account the socio-economic impact of research itself. Harris proposes that in order to reduce the disparity between research and practice, researchers ought to interact with stakeholders at varying levels to ensure that the research addresses real-world problems with the goal of producing tangible outputs, an argument echoed by others (Datta 2012).

Datta (2012) disaggregates some of RAPID's elements, finding that traditional approaches to communicating research to policymakers are inadequate. For better deliberative engagement, Datta emphasises that clarification is required during selection of stakeholders, in choosing when to engage in either downstream (top-down) or upstream (bottom-up) engagement and during the selection of appropriate methods of engagement. One example of a public engagement processes using ICTs is provided by Steibel and Estevez (2015) in their chapter 'Designing Web 2.0 tools for online public consultation'. These scholars found that the amalgamation of the different functions of Web 2.0 tools influences certain attributes of the political communication environment. These scholars studied two virtual public consultation spaces in Brazil using three theoretical models of online democratic communication and concluded that the design of the ICT platform would have a great influence on the public's political deliberation and discourse.

As the policy formation actors have diversified and newer channels a created, thereby making bottom-up approaches more possible, Datta suggests that the traditional linear model of research where the results are targeted at a specific, narrow range of the audience has proved to be inadequate in modern policy engagement. Intermediaries such as the media (especially online media) have become essential in influencing people's understanding of policies and issues and can foster better understanding among stakeholders of diverse backgrounds and capacities. Zhou (2015) in his chapter 'ICTs and opinion expression: An empirical study of new-generation migrant workers in Shanghai' examines the basic pattern and antecedents of Chinese rural-to-urban migrant workers' intentions to express themselves, finding that new media channels have become an important space for discussing problems. Zhou concludes that since the availability of online platforms and mobile phones have the ability to empower migrant workers, policymakers and social activists may get inspirations from these findings to consider designing effective campaigns via online platforms to facilitate the active opinion expression among new-generation migrant workers in China. The study also provides cautionary evidence against overdue emphasis on online media, since personal networks continue to exert the most influence among this group at the lack of a link between online behaviour and offline expressive intentions.

## 3.7 Input-Mechanism-Outputs Pathway

Chib, van Velthoven and Car (2014) review mobile health-care studies conducted in the developing world (53 studies represented by 63 papers), pointing out a lack of dominant theory or measures of outputs that are relevant to making policy decisions. They propose a categorisation through an input-mechanism-outputs pathway where *inputs* are issues of technology access and use, *mechanisms* are psychosocial influences and individual preferences, and *outputs* are efficiency measures such as health-care process factors or effectiveness measures such as health indicators within the beneficiary population. The analysis revealed a plethora of pilot studies, implementation evaluations or studies with undefined design or interviews; most of the studies lacked explicit theoretical support and largely failed to address impact. Chib, Leon and Rahim use the pathways model to delve into the literature on the impact of mFinance initiatives in the chapter 'The impact of mFinance initiatives: A review of the literature'. These scholars identify the notions and evidence of impact of mFinance initiatives, broadly including m-banking, m-payments and m-finance (Donner 2007). They examine issues such as how impact is conceptualised in the mFinance literature, what evidence exists for this impact and what alternative definitions of impact can be proposed, beyond traditional notions of economic development such as income and savings. The review of 51 research papers found that studies focused largely on explanation of mechanisms influencing adoption of mFinance. The authors however suggest that lack of the linkage between adoption processes and substantive impact on the poor might lack sufficient value for policymakers keen to support the growth of mFinance initiatives.

## 3.8 Process-Based Impact Frameworks

Heeks and Molla (2009) offer a longitudinal framework examining the shift in research focus from readiness (issues of awareness, digital divide and supply) to availability (questions of infrastructure and capacity) to uptake (issues of demand and usage) and finally to impacts (questions of efficiency, effectiveness and equity). The Heeks and Molla's compendium (2009) proposes that the discipline moves on from talking about potential to how the actuality of 'downstream' development impact is like (not just the 'upstream' elements of accessibility and infrastructure).

Similarly, Batchelor and Norrish present a framework for ICT pilot projects which seek to go 'beyond monitoring and evaluation to applied research' (2005, p. 7). Martin Hilbert suggests a 'Cube Framework' based on observations of Latin American cases where ICTs play a significant role in gradual socio-economic evolution, while technology, policy and social change dimensions are 'interplaying' to bring such evolution in the Global South. The 'Real Access Framework' (brides.org 2005 adapted by Gomez 2008) talks about resource portfolios and the importance of larger institutional and social structures within which the research is situated.

The review of studies providing a basis for policymaking might suggest that the ICTD field has a robust theoretical and rigorous methodological tradition. There are those, however, who question the quality of impact assessments in the ICTD, critiquing both the conceptual foundations and the emphasis on descriptive case studies (Heeks 2010). In spite of the profusion of frameworks and approaches, there are common problems that crop up repeatedly. Almost all process models seem to suggest a clear dividing line between 'inputs' and 'outputs', often treating them in separate silos. There is little determination of causality, leading to some methodological issues. There's an inadequate understanding of the mechanisms of usage and interaction between the inputs and outputs mechanisms.

We may also find ourselves questioning the basic premises and assumption of development studies: Who are the marginalised and poor? Which society do we refer to when discussing the information society? Which types of data collection and analytical methods can claim superiority over others? Despite these misgivings, the data is rich and comes from far afield, often illuminating subjects and regions that are hard to come by. The material contained here helps us validate or questions existing theory, proposes new models for conceptualization and measurement and provides varied evidence for the beneficial, the constrained and the negative impacts of ICTs.

# 4 Conclusion

This chapter has introduced various aspects of the SIRCA programme—the focus on capacity building brought about by support for mentorship in the research process; the multiplicity of voices; methodologies and traditions that constitute development research; the role of underlying theory as a basis for investigation, validation, extension and refutation; and number of lessons we draw from the empirical evidence offered. While many of these aspects may arguably be examining the impact of research upon research itself, this volume may not only be seen as a research output. The partial value of this compendium is in allowing scholars from the Global South to learn about, interrogate and present work gleaned from their communities to a global audience, including policymakers and practitioners. One might argue that the biggest challenge faced in imagining, designing and running the SIRCA II programme was in identifying an over-arching theme to bind the disparate voices to present a coherent and comprehensive whole. I would argue instead that the value of the programme was in trusting in the process—fostering a community of emerging scholars to draw upon the expertise of established scholars and scholarship, to challenge the assumptions and findings in interdisciplinary forums, and encouraging impact beyond traditional academic output. One of the resultant outputs is in your hands, though this is not the only one.

The personal achievements of SIRCA scholars, too numerous to detail here, range from academic publications in international outlets and presentations at global conferences, appointments and promotions, admissions and employment at prestigious institutions. It is worthwhile within the context of impact on policy and practice, possibly adding a dimension of influencing public opinion via media dissemination of research results, to consider other dimensions of impact.

A number of SIRCA II PIs took their projects beyond the academic level—engaging at the policy level with local communities where their fieldwork was based, at the university level with regard to research focus and funding and with national media and policymakers at the international level, representing the research perspective in global public consultations on ICT issues. Nikos Dacanay's research helped obtain national funding for the Burma Women's Union's (BWU) citizen journalism project, which documents political and current events/stories in several parts of Burma. Ezmieralda Melissa's research was replicated at the university level (Swiss-German University, Indonesia), with the incorporation of ICT as a key research theme under the university's umbrella. Fabro Steibel's research led to collaboration with the 'Marco Civil da Internet' or the Brazilian Civil Rights Framework for the Internet. He was invited to design public consultation portals for the Ministry of Justice as well as for the human rights agency of Mercosul, culminating in selection as an independent researcher from Brazil for the global Open Government Partnership. Matías Dodel's work led to the creation of the Research Group on Uruguay, Society and Internet (GIUSI), whose mandate is to create a general analytic framework on the Internet's impact. The group is also the national chapter for the World Internet Project. Sebastián Benítez Larghi discussed his findings with policymakers at the Programa Conectar Igualdad (PCI), an institution focused on digital inclusion in Argentina, and was also invited several times to TV and radio programmes and interviewed by different news agencies and newspapers.

There is one final measure of impact—yours. Enjoy the read!

**Open Access** This chapter is distributed under the terms of the Creative Commons Attribution Noncommercial License, which permits any noncommercial use, distribution, and reproduction in any medium, provided the original author(s) and source are credited.

# References

Alsop, R., & Heinsohn, N. (2005). *Measuring empowerment in practice—Structuring analysis and framing indicators*. Washington, DC: World Bank.
Batchelor, S., & Norrish, P. (2005). *Framework for the assessment of ICT pilot projects: Beyond monitoring and evaluation to applied research*. Washington, DC: World Bank.
Bell, D. (1999). *The coming of post-industrial society*. New York: Basic Books.
Bhatnagar, S. C., & Singh, N. (2010). Assessing the impact of E-government: A study of projects in India. *Information Technologies and International Development, 6*(2), 109–127.
Butt, D., & Sarker, P. P. (2009). ICT for development in Asia Pacific: Emerging themes in a diverse region. In *Digital review of Asia Pacific 2009–2010*. Montreal: Sage Publications.
Carmody, P. (2013). A knowledge economy or an information society in Africa? Thintegration and the mobile phone revolution. *Information Technology for Development, 19*(1), 24–39. doi:10.1080/02681102.2012.719859.
Carney, D. (1998). *Sustainable livelihoods: What contribution can we make?* London: Department for International Development.
Castells, M. (1996). *The rise of the network society*. Oxford: Blackwell Publishing.
Castells, M. (2000). *The information age: Economy society and culture, vol. 1: The rise of the network society*. Oxford: Blackwell.
Chandwani, R., & De, R. (2015). The institutional dynamics perspective of ICT for health initiatives in India. In A. Chib, J. May, & R. Barrantes (Eds.), *Impact of information society research in the global south*. New York: Springer.
Chew, H. E., Levy, M., & Ilavarasan, P. V. (2011). The limited impact of ICTs on microenterprise growth: A study of businesses owned by women in urban India. *Information Technologies & International Development, 7*(4), 1–16.
Chib, A. (2009). The role of ICTs in nurturing the field of information and communication technologies for development. *International Communication Association Newsletter, 37*(2), 13–16. Available at http://www.icahdq.org/pubs/publicnewsletter/2009/3/MAR09FULLPF.asp
Chib, A., & Harris, R. (2012). *Linking research to practice: Strengthening ICT for development research capacity in Asia*. Singapore: Institute of Southeast Asian Studies.
Chib, A., & Zhao, J. (2009). Sustainability of ICT interventions: Lessons from rural projects in China and India. In L. Harter, M. J. Dutta, & C. E. Cole (Eds.), *Communicating for social impact: Engaging communication theory, research, and pedagogy* (pp. 145–159). Cresskill: Hampton Press.
Chib, A., Leon, L., & Rahim, F. (2015). The impact of mFinance initiatives: A review of the literature. In A. Chib, J. May, & R. Barrantes (Eds.), *Impact of information society research in the global south*. New York: Springer.
Chib, A., van Velthoven, M., & Car, J. (2014). mHealth adoption in low-resource environments: A review of the use of mobile healthcare in developing countries. *Journal of Health Communication*. doi:10.1080/10810730.2013.864735.
Cruz, E. M., & Sajo, T. J. (2015). Cybersex as affective labour: Critical interrogation of the Philippine ICT framework and the Cybercrime Prevention Act of 2012. In A. Chib, J. May, & R. Barrantes (Eds.), *Impact of information society research in the global south*. New York: Springer.
Datta, A. (2012). Deliberation, dialogue and debate: Why researchers need to engage with others to address complex issues. *IDS Bulletin, 43*(5), 9–16. doi:10.1111/j.1759-5436.2012.00357.x.

Djane, K. A., & Ling, R. (2015). The use of mobile communication in the marketing of foodstuffs in Cote d'Ivoire. In A. Chib, J. May, & R. Barrantes (Eds.), *Impact of information society research in the global south*. New York: Springer.

Dodel, M. (2015). An analytical framework to incorporate ICT as an independent variable. In A. Chib, J. May, & R. Barrantes (Eds.), *Impact of information society research in the global south*. New York: Springer.

Donner, J. (2007). M-banking. Extending financial services to poor people. *id21 insights*, IDS, University of Sussex. Available at http://r4d.dfid.gov.uk/PDF/Articles/insights69.pdf

Donner, J. (2008). Research approaches to mobile phone use in the developing world: A review of literature. *The Information Society, 24*(3), 140–159. doi:10.1080/01972240802019970.

Elder, L., Emdon, H., Fuchs, R., & Petrazzini, B. (Eds.). (2013). *Connecting ICTs to development: The IDRC experience*. London: Anthem Press.

Gandy, O. H., Jr. (2002). The real digital divide: Citizens versus consumers. In L. A. Lievrouw & S. M. Livingstone (Eds.), *The handbook of new media: Social shaping and consequences of ICTs*. London: Sage.

Gomez, R. (2008). *The quest for intangibles: Understanding ICTs for digital inclusion beyond socio-economic impact.* Paper presented at the Prato CIRN 2008 Community Informatics conference. Prato.

Gomez, R., & Pather, S. (2012). ICT evaluation: Are we asking the right questions? *The Electronic Journal of Information Systems in Developing Countries, 50*(5), 1–14.

Harris, A., & Chib, A. (2012). Perspectives on ICTD research and practice. In A. Chib & R. Harris (Eds.), *Linking research to practice: Strengthening ICT for development research capacity in Asia* (pp. 3–11). Singapore: Institute of Southeast Asian Studies.

Harris, R. (2015). The impact of research on development policy and practice: This much we know. In A. Chib, J. May, & R. Barrantes (Eds.), *Impact of information society research in the global south*. New York: Springer.

Hatakka, M., & De, R. (2011). *Development, capabilities and technology: An evaluative framework*. Proceedings from IFIP WG9.4: 11th international conference on Social Implications of Computers in Developing Countries. Kathmandu.

Heeks, R. (2010). Do information and communication technologies (ICTs) contribute to development? *Journal of International Development, 22*(5), 625–640. doi:10.1002/jid.1716.

Heeks, R., & Molla, A. (2009). *Compendium on impact assessment of ICT-for-development projects*. Retrieved from https://digital.lib.washington.edu/researchworks/handle/1773/25541

Hilbert, M. (2014). Technological information inequality as an incessantly moving target: The redistribution of information and communication capacities between 1986 and 2010. *Journal of the Association for Information Science and Technology, 65*(4), 821–835. doi:10.1002/jid.1716.

House, E. R. (1980). *Evaluating with validity*. Beverly Hills: Sage.

Hutton, W., & Giddens, A. (Eds.). (2001). *On the edge: Living with global capitalism*. London: Vintage.

Kleine, D. (2010). ICT4WHAT? Using the choice framework to operationalise the capability approach to development. *Journal of International Development, 22*(5), 674–692. doi:10.1002/jid.1719.

Kleine, D. (2013). *Technologies of choice? ICTs, development, and the capabilities approach*. Cambridge: MIT Press.

Mansell, R. (1999). New media competition and access: The scarcity-abundance dialectic. *New Media and Society, 1*(2), 155–182. doi:10.1177/14614449922225546.

May, J., & Diga, K. (2015). Progress towards resolving the measurement link between ICT and poverty reduction. In A. Chib, J. May, & R. Barrantes (Eds.), *Impact of information society research in the global south*. New York: Springer.

McNamara, K. S. (2003). *Information and communication technologies, poverty and development: Learning from experience*. Washington, DC: InfoDev, World Bank.

Melissa, E., Hamidati, A., Saraswati, M. S., & Flor, A. (2015). The Internet and Indonesian women entrepreneurs: Examining the impact of social media on women empowerment. In A. Chib, J. May, & R. Barrantes (Eds.), *Impact of information society research in the global south*. New York: Springer.

Newton, J., & Franklin, A. (2011). Delivering sustainable communities in China: Using a sustainable livelihoods framework for reviewing the promotion of "ecotourism" in Anji. *Local Environment, 16*(8), 789–806. doi:10.1080/13549839.2011.569536.

Norris, P. (2001). *Digital divide: Civic engagement, information poverty, and the internet worldwide*. Cambridge: Cambridge University Press.

Olivera, P., Ale, K., & Chib, A. (2015). (Un)Balanced conversations: Participatory action research in technology development in peruvian primary schools. In A. Chib, J. May, & R. Barrantes (Eds.), *Impact of information society research in the global south*. New York: Springer.

Ordóñez, A. (2015). A new set of questions: ICT4D research and policy. In A. Chib, J. May, & R. Barrantes (Eds.), *Impact of information society research in the global south*. New York: Springer.

Papert, S. (1987). Information technology and education: Computer criticism vs. technocentric thinking. *Educational Researcher, 16*(1), 22–30.

Parthasarathy, B., & Srinivasan, J. (2006, October). Innovation and its social impacts: *The role of ethnography in the evaluation and assessment of ICTD projects*. Paper presented at the Global Network for Economics of Learning, Innovation, and Competence Building Systems conference, Trivandrum.

Qureshi, S. (2005). *How does information technology affect development? Integrating theory and practice into a process model*. Proceedings from the 11th Americas conference on Information Systems. Omaha.

Rantanen, T. (2001). The old and the new: Communication technology and globalization in Russia. *New Media and Society, 3*(2), 85–105.

Reimer, B. (2002). *Understanding and measuring social capital and social cohesion*. Montreal: Concordia University, Canadian Rural Restructuring Foundation.

Rodriguez, P. P., Nussbaum, M. M., & Dombrovskaia, L. L. (2012). Evolutionary development: A model for the design, implementation, and evaluation of ICT for education programmes. *Journal of Computer Assisted Learning, 28*(2), 81–98.

Scoones, I. (1998). *Sustainable rural livelihoods: A framework for analysis* (unpublished paper). Brighton: University of Sussex Institute of Development Studies.

Selwyn, N. (2010). Degrees of digital division: Reconsidering digital inequalities and contemporary higher education. *RUSC. Universities and Knowledge Society Journal, 7*(1). http://dx.doi.org/10.7238/rusc.v7i1.660.

Sen, A. (1999). *Development as freedom*. New York: Oxford University Press.

Shapiro, C., & Varian, H. R. (1999). *Information rules: A strategic guide to the network economy*. Boston: Harvard Business School Press.

Sinha, C., Elder, L., & Smith, M. (2012). SIRCA: An opportunity to build and improve the field of ICT4D. In A. Chib & R. Harris (Eds.), *Linking research to practice: Strengthening ICT for development research capacity in Asia* (pp. 12–24). Singapore: Institute of Southeast Asian Studies.

Steibel, F., & Estevez, E. (2015). Designing Web 2.0 tools for online public consultation. In A. Chib, J. May, & R. Barrantes (Eds.), *Impact of information society research in the global south*. New York: Springer.

Taylor, R., & Zhang, B. (2007, September). *Measuring the impact of ICT: Theories of information and development*. Paper presented at the Telecommunications Policy Research conference, Washington, DC.

van Rijn, F., Burger, K., & den Belder, E. (2012). Impact assessment in the sustainable livelihood framework. *Development in Practice, 22*(7), 1019–1035. doi:10.1080/096/4524.2012.696586.

Warschauer, M. (2003). Dissecting the 'digital divide': A case study in Egypt. *The Information Society, 19*(4), 297–304. doi:10.1080/01972240309490.

Waverman, L., Meschi, M., & Fuss, M. (2005). The impact of telecoms on economic growth in developing countries. *The Vodafone Policy Paper Series, 2*(03), 10–24.

Young, J., & Court, J. (2004). *Bridging research and policy in international development: An analytical and practical framework.* Retrieved from http://www.odi.org/publications/159-bridging-research-policy-international-development-analytical-practical-framework

Zhou, B. (2015). ICTs and opinion expression: An empirical study of new-generation migrant workers in Shanghai. In A. Chib, J. May, & R. Barrantes (Eds.), *Impact of information society research in the global south.* New York: Springer.

# Part I
# Impact of Research

# The Impact of Research on Development Policy and Practice: This Much We Know

Roger Harris

This chapter highlights the near absence of research into the nonacademic impact of ICT4D research within the ICT4D literature. It draws on studies in international development to review the literature on the impact of research on development policy and practice and reflects on the implications for ICT4D research. Noting the cultural and professional differences between researchers and practitioners as well as their differing perspectives of impact, it goes on to describe the dominant themes in the literature. ICT4D research is characterised as lacking in certain respects, which would tend to inhibit its capacity for policy impact, but having overcome these, further adjustments to research conduct and culture are implied for such impact to emerge. Consequential recommendations include revised incentive structures for academic institutions as well as closer engagement between researchers and practitioners.

## 1 Introduction

The first phase of the SIRCA programme categorised the impact of research into academic impact and socio-economic impact, with the latter comprising impacts on socio-economic benefits, policy and capacity building. It was argued, moreover, that achieving academic impact does not automatically lead to achieving socio-economic impact. To date, it is evident that information society impact research in the global south has focused almost exclusively on the impact of ICTs, largely ignoring the socio-economic impact of the research itself. Where the impact of research has been addressed, this has deliberated almost entirely on academic impact, with discussions

---

R. Harris (✉)
CEO of Roger Harris Associates, Tai Po, New Territories, Hong Kong
e-mail: roger.harris@rogharris.org

© The Author(s) 2015
A. Chib et al. (eds.), *Impact of Information Society Research in the Global South*,
DOI 10.1007/978-981-287-381-1_2

about where to publish research findings and how to maximise the citations that such publications receive. However, policy research programmes do not normally use traditional academic citations in peer-reviewed journals as a principal monitoring and evaluation tool. While there is considerable reflection on the socio-economic impact of *ICTs*, there is a paucity of research on the socio-economic impact of *research*. What does all this say about the value of information society research in the global south? That its main purpose is to support academic careers?

In SIRCA I, we pointed out that achieving socio-economic impacts from research required different skills, roles and processes from those that are used to achieve academic impact. Given the near absence of research into the nonacademic impact of ICT4D research within the ICT4D literature, we need to look elsewhere for a better understanding of these skills, roles and processes in order to establish some sense of how they might operate effectively for achieving socio-economic impact with information society research.

The related but wider field of international development has shown greater interest in the impact of research, with governments and major agencies calling for a clearer articulation of socio-economic benefits from the research that they fund. There are also calls for social policy formulations to be based more on evidence, and this implies a heightened role for the research that will be capable of delivering such evidence. For example, the UK government, which invests around £3 billion annually in research, requires funding applicants to demonstrate the contribution of their research to society and the economy. The UK's Department for International Development (DFID) has stated that without a greater focus on getting research into use, the potential for improving lives through research and innovation will not be fully realised.

An examination of the impact of information society research in the global south therefore needs to address all dimensions of impact, and in the light of the foregoing, this should feature an assessment of its impact on policy and practice. In this chapter, we review the literature on the impact of development research on policy and practice and reflect on the implications for ICT4D research, drawing on the SIRCA programme experiences. We begin with some important observed differences between the two worlds of research and policymaking that have emerged as consistent themes in the literature and which give shape to much of the advisory outcomes of the review. This is followed by a summary of the other themes. Finally, we present an assessment of what they might mean for ICT4D research and for the SIRCA programme.

## 2 Opposing Perspectives

### 2.1 Two Communities

The two worlds of academic research and policy formulation are characterised in the literature as being very different. Some observers comment that researchers, practitioners and policymakers live in parallel universes (Court and Maxwell 2005;

Court and Young 2006; Stone 2009). Stone (2009) argues that researchers and policymakers operate with different values, languages, timeframes, reward systems and professional ties to such an extent that they live in separate worlds. Moreover, for some, researchers cannot understand why there is resistance to policy change despite clear and convincing evidence, while policymakers bemoan the inability of many researchers to make their findings accessible and digestible in time for policy decisions (Court and Young 2006).

According to Grejin (2008), researchers often live in very separate worlds from policymakers, civil society organisations and practitioners. As a result, research-based evidence is often only a minor factor when policies for development are formulated and practices shaped. Too often new public policies are rolled out nationally with little trialling or evaluation. In effect, governments experiment on the whole population at once. Even where there is plenty of evidence, there may be a failure to ensure that the evidence being collected and analysed is made relevant to the needs of decision-makers and is acted upon (Mulgan and Puttick 2013). Additionally, as Datta (2012) suggests, researchers in any one field tend not to speak with one voice, and not all researchers see policy engagement as part of their role. Shanley and López (2009) go further by claiming that strong organisational disincentives dissuade researchers from engaging in outreach beyond the scientific community. Others indicate that researchers working in universities and other publicly funded institutions report structural barriers to engaging in knowledge translation activities, suggesting that a failure to transfer knowledge has been attributed to the "two communities" problem—an explanation that points to cultural differences between researchers and users as barriers to such engagement (Jacobson et al. 2004). As a result, says Carden (2009), policymakers lack confidence in their own researchers.

Despite these misgivings relating to the prospects for development research having an influence over development practice and policymaking, there is room for optimism. As de Vibe et al. (2002) put it, notwithstanding the assumption that there is a clear divide between researchers and policymakers (the two communities model), which underpins the traditional view of the link between research and policy, literature on the research-policy link is now shifting away from these assumptions, towards a more dynamic and complex view that emphasises a two-way process between research and policy, shaped by multiple relations and reservoirs of knowledge. Accordingly, much of what emerges from the literature review presented here by way of recommendations for bringing these two communities closer together argues the case for overcoming these perceived gaps, and offers prescriptions for doing so.

## 2.2 What Is Impact?

An understanding of the impact of research on policy and practice requires agreement on what "impact" means for researchers and for policymakers. For academics, the impact of their research is usually reflected by the impact factor that is assigned to the journal in which the research report is published. The impact factor of an academic journal is a measure of the average number of citations that have been made to its recently published articles. It is frequently used as a proxy for the relative importance of a journal within its field; journals with higher impact factors are considered to be more important (influential) than those with a lower impact factor. Thomson Reuters, the academic publisher, computes the impact factor of a journal by dividing the number of current year citations to the source items published in that journal during the previous 2 years.[1]

In contrast to the academic perspective of research impact, practitioners hold a very different view. For example, Young (2008) claims that for research to have any impact, the results must inform and shape policies and programmes and be adopted into practice. Researchers wishing to maximise the impact of their work have to attract the interest of policymakers and practitioners and then convince them that a new policy or different approach is valuable and then foster the behavioural changes that are necessary to put them into practice (Young 2008). For Sumner et al. (2009), impact is multilayered and refers to use (i.e. consideration) or actual outcome(s) of social change. It can be visible or invisible, progressive or regressive, intended or unintended and immediate or long term. The Research Council of the UK acknowledges academic impact—as the demonstrable contribution that excellent research makes to academic advances—but it also emphasises the need for economic and societal impacts as the demonstrable contribution that excellent research makes to society and the economy by, among other things, increasing the effectiveness of public services and policy.

The difference between these contrasting interpretations of what is meant by impact has serious implications for the discourse surrounding the research-policy nexus. According to Shanley and López (2009), appropriation of the word "impact" to designate a journal's ranking constitutes a potential misrepresentation of what impact is. The effect of this can be seen from their survey of 268 researchers in 29 countries which revealed that the largest percentage (34 %) ranked scientists as the most important audience for their work and that engagement with the media, production of training and educational materials and popular publications as outlets for scientific findings was perceived as inconsequential in measuring scientific performance. They conclude that directly and inadvertently, academic and nonacademic research institutions discourage impact-oriented research by prioritising the number and frequency of publications in peer-reviewed journals (Shanley and López 2009).

Chief among the barriers between research and policy impact is the reward and incentive system of the academy, i.e. promotion and tenure. This is seen as a system that, in general, continues to value traditional types of within-group activity, e.g. publication in peer-reviewed journals, presentations at disciplinary

---

[1] http://thomsonreuters.com/products_services/science/free/essays/impact_factor/. Accessed 6 March 2013.

conferences and receipt of research grants from government agencies, over the more broadly directed outreach and production activities associated with the transfer of knowledge. While the importance of knowledge transfer may be endorsed in rhetoric, the rewards, resources and priorities reflect the enduring value accorded to the more traditional academic activities. In many disciplines, knowledge transfer—the exchange, synthesis and application of knowledge—is noted to pose risks to an academic career. This is because the activities that make up much of the work of knowledge transfer are not widely accepted as legitimate forms of scholarship (Jacobson et al. 2004).

Gendron (2008) goes even further in developing a critique of the excessive spread of performance measurement practices in academia, whereby productivity is measured through performance indicators predicated on hard data such as grants, citations and the number of publications. This has given rise to an identity representation of academics as performers. Journal rankings and performance measurement schemes are becoming increasingly influential within many fields of research, thereby consolidating the prevalence of *performativity* on the life and research endeavours of many academics. The influence of journal rankings leads to researchers being assessed on the basis of their "hits" instead of on the substance of their work. Thus, the mania surrounding the practice of performance measurement stifles innovation while engendering and/or reinforcing pressures of superficiality and conformity (Gendron 2008).

Perhaps as a consequence of the serious shortcomings within peer-reviewed journals and the academic reward system, in the world of policy research, the mechanisms of academic peer review and conventional citation counting are regarded as too limited. Although rankings and rating systems applying to both journals and individual academics are acknowledged to provide a useful proxy guide to the quality of a research study, the validity of such rankings for such purposes is noted to be subject to considerable debate (DFID 2013a, b, c). Moreover, not all well-designed and robustly applied research is to be found in peer-reviewed journals, and not all studies in peer-reviewed journals are of high quality. Journal rankings do not always include publications from southern academic organisations or in online journals. Accordingly, policy research programmes will not usually use conventional academic citations in peer-reviewed journals as a primary monitoring and evaluation tool (Hovland 2007). Potentially, this robs the policy arena of a hugely valuable resource because, as Shanley and López (2009) put it, until communication and impact are seriously integrated into (academic) performance measurement systems, it is likely that only a limited number of independently motivated scientists will engage in the time-consuming processes needed to disseminate research effectively.

Despite the foregoing observations, there is again room for optimism when contemplating the possibility of stronger links between research academics and policymakers. Firstly, it can be seen that the two perspectives of impact held by each are not mutually exclusive; research that is highly regarded in peer-reviewing processes and published in high-ranking journals retains its potential for influencing policy. Indeed, high-quality research is a prerequisite for policy influence, although its publication alone seems insufficient for it to do so. This is despite the observation that there is an assumption among some actors that research communication is often an unnecessary add-on, or a dispensable luxury (Harvey et al. 2012). There is also a less polarised perspective of research that it exists as a continuum between research that is used for more conceptual purposes of raising awareness and increasing understanding and knowledge at one end and the more instrumental uses of research such as changes to policy and practice, at the other end (Nutley et al. 2007).

Secondly, pressures on higher education funding mean that academics are increasingly being asked to demonstrate the public benefit of their work. For example, the UK government's 2014 Research Excellence Framework will for the first time explicitly assess the impact of research beyond academia. The framework defines impact as "any effect on, change or benefit to the economy, society, culture, public policy or services, health, the environment or quality of life, beyond academia". Submissions for funding will be assessed under this category through impact case studies and details of the strategy for achieving impact.[2] A turn to nonacademic impact has the potential, therefore, of encouraging academics to engage more closely with the wider processes of social transformation. Attributing value to this type of impact is certainly intended to change research culture (Williams 2012). In the development field, the UK's DFID has made it clear that DFID-funded research programmes are expected to plan and implement a research uptake strategy and that research uptake strategies should encompass stakeholder engagement, capacity building, communication and monitoring and evaluation (DFID 2013a, b, c).

In addition to funding incentives, theoretical advancements in communication for development now favour a move from a top-down to a more inclusive communication style. Yet the former trickle down and transfer paradigms continue to guide and dominate the behaviours of academics (Shanley and López 2009). However, the use of social knowledge as a resource for policymaking has become a means to mobilise researchers and policymakers in new political alliances, over and above old ideological and partisan differences that have separated academia from engagement with practice (Fisher and Holland 2003). Nevertheless, within both worlds of academia and policy, there is still lack of clarity or consensus on the meanings of research impact or influence, and researchers have very different ideas about who they are trying to influence, to what end and using which methods (Harvey et al. 2012).

## 3 Thematic Overview of the Literature

In this section, we provide a brief analysis of the literature that highlights the major interlocking themes that have evolved in relation to the impact of research on development practice and policy formulation. In order to retain a contemporary

---

[2]http://www.ref.ac.uk/media/ref/content/researchusers/REF%20guide.pdf. Accessed 6 March 2013.

perspective, the referenced works are mostly from this century, with the exception of the seminal work of the late Professor Carol Weiss, whose observations underpin much of the rest of the literature. The compilation is dominated by grey literature from the practitioner and policy advisory domains.[3] We can only speculate as to why this is so, but it could reflect greater concern on the behalf of practitioners, policy-makers and their advisors regarding the role of research within their deliberations than exists within the research community. Such a concern would resonate with some of the observations that have emerged regarding the research-policy nexus.

A further aspect is that policy and practice are largely conflated into the same thing; arguably, as de Vibe et al. (2002) put it, the practical recommendations of NGOs mirror to a large extent the macro policy discourse – in areas such as building local institutions, supporting civil society and strengthening social capital. The new orthodoxy within development, they argue, that has as its mantras of participation and empowerment is shared not only among NGO practitioners but also among bilateral and multilateral donors and governments.

## 3.1 Intent

Among the conditions required for research to have an influence on policy and practice, several observers emphasise the need for researchers to have the intent for it do so. According to Sen (2005), the highest likely impact of research on development outcomes is when there is a clear demand from research users and there is an effective supply of high-quality policy-relevant research, backed by the intent to influence among researchers. Even if such intent exists, though, the lack of other conditions, such as leadership and capacity within the user community, and the impact of high-quality policy-relevant research will be limited. Accordingly, although, as Wheeler (2007) says, there are growing expectations within development that research should inform policy; intent to influence is a necessary but insufficient supply-side factor in determining the development effectiveness of research.

Carden (2009) highlights three principles behind the design of a research programme that may allow for the maximum impact, which include the intent to influence, along with the creation of networks and effective communications. Intent to influence must be expressly included among the research objectives. Other

---

[3]Grey literature is defined as "that which is produced on all levels of government, academics, business and industry in print and electronic formats, but which is not controlled by commercial publishers". It includes reports, theses, conference proceedings, bibliographies, technical and commercial documentation, official documents government reports and documents (Alberani et al. 1990).

essential elements of policy influence for development research are proposed by O'Neill (2005); they include intent, as the determination among researchers to do their work and report their results so as to inform policy decisions and improve policy outcomes.

## 3.2 Communication

Communication is by far the most cited factor in the literature on the impact of research on development policy and practice. The topic of communication for development is a subfield within international development studies, and it encompasses research communication, among other forms of communication.[4] The various forms of communication, who is involved in the communication process and when it occurs, are all themes that pervade the literature and intermingle with the other themes. For example, alongside the intent of a researcher to influence policy and practice, Ryan and Garrett (2003) and Sen (2005) stress the clear intent to communicate research as a supply-side factor for influence.

The need for communication between researchers and others is repeatedly stressed, sometimes implying an additional need for intermediaries to ensure it is done effectively. Newman et al. (2012) point to the recent interest in supporting evidence-informed policymaking in developing countries through building the capacity of researchers and research intermediaries to supply appropriately packaged research information (e.g. in the form of policy briefs) to policymakers. Court and Maxwell (2005) claim that successful evidence-based policymaking occurs when, among other things, the links are well made between researchers and policymakers, for example, through networks or by intermediaries.

The role of networks also emerges as a persistent theme throughout the literature, with Court and Young (2006) suggesting that there is often an underappreciation of the extent and ways that intermediary organisations and networks influence formal policy guidance documents. Research is more likely to contribute to policy, they say, if researchers and policymakers share common networks, trust each other and communicate effectively. Hovland's (2003) literature review of research communication for poverty reduction emphasises support for research networks, especially electronic and/or regional networks, while Masset et al. (2011) maintain that a networked policy research community is a precondition for increasing the likelihood of policy change. Stone (2009) suggests that the uptake of research is contingent on policy community networking, describing the long-term strategies of the Overseas Development Institute (ODI), a London-based think tank of policy

---

[4]See, for example, the C4D Network; "a non-profit organisation dedicated to supporting the communication for development sector" with more than 1,200 members. http://c4dnetwork.ning.com/

entrepreneurship that extends to longer-term influence through creating human capital, building networks and engaging policy communities.

The nature and quality of communication between and among researchers and policymakers is also important. A crucial capacity for researchers is the ability to communicate in a language that policymakers can understand (Greijn 2008) as well as being an effective communicator, with specifically, the ability to find common ground and to communicate well with various audiences. Modifying or creating policies based on evidence requires "translating" the technical language of research so that it is comprehensible for the relevant agents in the policymaking process. Good communication is also important when seeking partners, building alliances and working in networks (Langou 2008). Other measures for improving research communication include improving skills for achieving the right format and timing of communication, constructing appropriate platforms from which to communicate and promoting participative communication for empowerment (Hovland 2003).

For some, research communication is primarily a public relations or marketing exercise, the "communication" product that comes in the final stages of a linear research process. Increasingly, however, development practitioners and researchers have recognised the importance of iterative and participatory communication processes. This is according to Harvey et al. (2012), who reason that research communication has evolved away from solely linear and top-down models of influencing (e.g. getting research onto the desks of the most senior decision-makers) to more complex and multisited theories of change. They see a proliferation in roles and actors for communicating research in development which push the boundaries of conventional ideas of research and challenge how research agendas are set and how knowledge is generated and shared. For some researchers, this implies new and unfamiliar ways of working, as revealed in the survey of researchers by Shanley and López (2009) in which performance measurement systems revealed robust institutional preferences against communicating with the public. This is underscored by the finding of Jacobson, Butterill and Goering (2004) that plain language communication with the public is not widely accepted as a legitimate form of scholarship and also by Carden's (2009) claim that researchers are uncomfortable communicating with officials and politicians in the policy community.

Datta (2012) goes on to argue that researchers no longer have a monopoly over knowledge production and communication and that traditional approaches to communicating research to policymakers are inadequate. Researchers now share the field of knowledge production and communication with many others, and where appropriate, those who view their role in relation to policy should be prepared to engage with stakeholders affected by policy issues and to expose their findings to human interaction, review and scrutiny by others.

Good communication is vital for researchers (Saxena 2005), and policy is only influenced when the evidence is credible and well communicated (Court and Maxwell 2005). At its best, communication starts early in the research; it is designed into the research plan and is carried out as the project unfolds (Carden 2009). The UK's DFID has stipulated that many of the research programmes which it funds should spend at least 10 % of their budget on communication activities, and this appears to have had a positive impact on the uptake of research by both policy and practice (Shaxson 2010). Additionally important in the current context, aside from academic impact and impact on policy and practice, enhanced capacity is regarded as a legitimate research outcome. Among the enhanced capacities that are claimed to be particularly important for young scholars in developing countries are communication with policymakers (e.g. policy briefs), communication with the general public and communication with the media (OECD 2011).

## 3.3 Information and Communication Technologies

Another factor closely linked to communication is the rising use of information and communication technologies (ICTs), a development that some see as blurring the once-stark line dividing academia and professional and amateur writers; e.g. op-ed writers, bloggers, etc. (Lewin and Patterson 2012). Hovland (2003), in calling for improved communication between researchers and policymakers and other researchers and end users (i.e. the poor and organisations working with them), suggests incorporating communication activities into project design and using new ways of communicating through ICTs. Others see more of a transformative role of ICTs for those practitioners and researchers who are increasingly recognising the importance of iterative and participatory communication processes within development that use ICTs for the rapid, multisited, multimedia and participant-driven production and communication of research as it unfolds (Harvey et al. 2012).

DFID's report on social media engagement focuses on policy actors—people whose work is wholly or partially involved in developing or seeking to influence national and regional development policies—who use a range of ICTs to get information, including social media. The "echo-chamber" effect of social media, referring to the overlap between individuals and organisations working in allied or similar fields, works to amplify its content, giving rise to enormous reach. For instance, the 50 biggest followers of the Twitter account @DFID_Research have a combined reach of 2.4 million; @IDS_UK's 50 biggest followers number 3.6 million, and @odi_development's 50 biggest followers have a combined reach of 4.3 million.[5]

Despite such evidence of reach, it appears that many UK academics are reluctant to adopt Web 2.0 tools for their work. A major disincentive for the academic community to adopt them for research activities is the lack of institutional incentives for using them or for publishing online. One study found that UK researchers are

---

[5] @DFID_Research R4D is the open-access portal to DFID-funded research. It houses over 30,000 research documents on international development. http://www.dfid.gov.uk/r4d

@IDS_UK. The Institute of Development Studies is a leading global charity for research, teaching and communications on international development. Brighton, UK http://www.ids.ac.uk/

@odi_development. UK's leading independent think tank on international development and humanitarian issues London, UK http://www.odi.org.uk. All accessed 28 March 2013

discouraged from publishing online by the policy of having international peer-reviewed journal citations, rather than online citations, count towards academic promotion (Brown 2012).

DFID argues that online media accessed through digital devices—PCs, pads and mobile phones—play a central role in all areas of knowledge and research. It is therefore crucially important to understand the online behaviour of the target audiences for development research as well as the wide range of available platforms and tools which can be exploited by project teams. However, conventional wisdom holds that this kind of open sharing and joint activity is at odds with the nature of the research process, where the tradition is for solo teams of researchers to prepare their findings privately before putting them out to review and where, especially in an academic and commercial context, advancement and success is seen to depend on secrecy.

## 3.4 Intermediaries

Against a background of individual cultural differences, systemic inadequacies in professional and institutional incentive structures and the apparent weaknesses in academic communication skills and processes, it is not surprising to discover other people and organisations taking up the role of delivering research-based knowledge to practitioners and policymakers with the aim of strengthening their activities. The work of Court and Young (2006) emphasises the importance of links—of communities, networks and intermediaries (e.g. the media and campaigning groups)—in affecting policy change. Existing theory, they say, stresses the role of translators and communicators, a view echoed by Harvey, Lewin and Fisher (2012) who suggest that evolving notions of what constitutes expert or valid knowledge have affected development research institutes in the global north, bringing an increased focus on the roles of intermediaries and networks. In this context, researchers are being joined by other actors, such as research communication specialists, not necessarily involved in undertaking research but who seek to strengthen the use of research within change processes (Harvey et al. 2012; Court and Maxwell 2005). As researchers typically have little or no influence over the capacity of their audience to use their research findings, others maintain that further investment should be made in supporting the pull through and absorption of research through, for example, the use of intermediaries or knowledge brokers to mediate relationships or transmit knowledge between academics and research users (Stevens et al. 2013).

The role of knowledge intermediaries is examined by Shaxson (2010) in recognising the contribution that knowledge intermediary organisations make not only in synthesising, interpreting and communicating research results to individuals and organisations in policy and practice but also in understanding the demand for knowledge from them. The role of knowledge intermediaries in international development is discussed at length, encompassing: enabling access to and making information edible, creating demand for information, enabling marginalised voices, creating alternative framings, connecting spheres of action, enabling accountability, informing, linking, matchmaking, facilitating collaboration and building sustainable institutions. For each of these activities, there are measures of impact, and these do include citation analyses among many others. However it is noticeable that measures of impact are shifting from content analysis issues such as hit rates, downloads and citations and more to measures of inclusivity and stakeholder involvement in project and programme plans and institutional strategies (Shaxson 2010).

Of relevance in the current context, one observer points out that developing countries often lack the intermediary institutions that carry research to policy (Carden 2009). Rich countries have abundant research institutes, think tanks, university departments and independent media that perform as knowledge brokers—the transactors who connect research findings to policy issues—but which are often absent in developing countries. As a result, the mechanisms of policy influence are missing. As a means of overcoming this limitation, Carden (2009) suggests that in IDRC experience, there is often a South to South learning effect, with lessons from one developing country or region applied to another with IDRC's intermediary help. However, as Jones et al. (2013) point out, it is not necessary to be labelled as a knowledge intermediary in order to act as one; what matters is developing a clear understanding of the different intermediary functions that could be used and the resource implications of each.

## 3.5 Policy Entrepreneurs

In a refinement of the role of intermediaries, the concept of policy entrepreneurs has emerged as a role for researchers wishing to influence policy. A policy entrepreneur is an individual who invests time and resources to advance a position or policy. One of their most important functions is to change people's beliefs and attitudes about a particular issue (Stone 2009). Four critical skills have been identified: being able to understand politics and identify key players, being able to synthesise research by simple compelling stories, being a good networker and being able to build programmes that bring all these factors together (Masset et al. 2011). As the product of the researcher is not usually in a format that can be used by policymakers, an intermediary—research broker or policy entrepreneur—with a flair for interpreting and communicating the technical or theoretical work is needed. This is usually an individual but sometimes an organisation which plays such a role (Stone 2009). Additionally, as research-based evidence often plays a very minor role in policy processes, if researchers want to be good policy entrepreneurs, they also need to synthesise simple, compelling stories from the results of the research (Young 2008).

## 3.6 Networks

The theme of networks has already emerged in our discussions on communication and intermediaries, but there are additional aspects throughout the literature that heighten their relevance to the present discussion. Following Weiss (1977), it has been widely recognised that although research may not have direct influence on specific policies, the production of research may still exert a powerful indirect influence through introducing new terms and shaping the policy discourse. Weiss describes this as a process of percolation, in which research findings and concepts circulate and are gradually filtered through various policy networks. Some of the literature on the research-policy link therefore focuses explicitly on various types of networks, such as policy streams, policy communities, epistemic communities, think tank networks and advocacy coalitions.[6] Networks and inter-organisational linkages sit solidly among the determining influences as to why some ideas are picked up and acted on, while others are ignored and disappear (de Vibe et al. 2002).

Lewin and Patterson (2012) indicate that the diffusion of the Internet has transformed global news media and communication systems into interactive horizontal networks that connect local and global individuals and issues, and Stone (2009) argues that such networks facilitate the role of policy entrepreneur that is played by intermediary organisations such as the ODI. Given the importance of collaboration between researchers and policymakers within research programmes that are intended to influence policy and practice, it is no surprise to find an emphasis on the establishment and operation of networks that make such collaboration possible and more effective. As Carden comments, collaborations have proven the diverse and sometimes surprising rewards of organising research in networks of shared purpose (Carden 2009). National, regional and global networks are playing an increasing role in development policy, and two institutional models seem to be particularly effective, think tanks and national and regional networks, which are frequently cited as being influential (Young 2005).

## 3.7 Incentives

Among the interlocking factors that influence the impact of research on policy and practice in the international development literature, incentives stand out as a decisive determinant. Senior management and academics at research institutions need to provide strong leadership in supporting cultural changes around the impact agenda (Stevens et al. 2013). They should consider how best to accommodate impact within internal structures, job descriptions, annual appraisal and promotional criteria and pay awards and professional development opportunities. Other commentators have called on research institutions to provide researchers with the right incentives to engage effectively with users of research (Datta 2012), and a shift in incentive struc-

---

[6] An epistemic community consists of colleagues who share a similar approach or a similar position on an issue (Haas 1991). Advocacy coalitions consist of various different actors, including different government agencies, associations, civil society organisations, think tanks, academics, media institutions and prominent individuals (Sabatier and Jenkins-Smith 1999).

tures is called for that reward actual impact rather than only "high-impact" journals to ensure science is shared with those who need it. Incentives for researchers to produce outputs that reach a broader swath of society are so low that if engaged in at all, this occurs as an afterthought once results are published (Shanley and López 2009).

The incentives for officials also come under scrutiny insofar as research-policy links are dramatically shaped by the political context. The policy process and the production of research are in themselves political processes which are influenced by a range of factors including the attitudes and incentives among officials, their room for manoeuvre, local history and power relations. Understanding the degree of political contestation as well as the attitudes and incentives of officials is important in explaining some public policies (Young 2005).

## 3.8 Political Context

The influence that the political context has on the research and policy or practice nexus receives significant coverage in the literature, repeatedly identified as a determining factor for whether research-based and other forms of evidence are likely to be adopted by policymakers and practitioners. Research is more likely to contribute to policy if the evidence fits within the political and institutional limits and pressures of policymakers and if it resonates with their assumptions (Court and Young 2006). Accordingly, researchers must know and understand key stakeholders in the policymaking process and understand the way in which the door can be opened to politicians and public interest (Taylor 2005). They need to grasp and adapt to the dynamics of the political debate and bring to the fore relevant evidence at the right time (Greijn 2008). It becomes necessary therefore to create an enabling environment for improved communication of research as failure to use research is not always due to lack of communication but can instead be due to lack of a favourable political environment. In fact, the success (or failure) of communication at an individual, local or project level is largely determined by wider systems, including the political environment. It is noted that academics and think tanks have a far greater chance of being heard when there are like-minded influential politicians in the dominant advocacy coalition (Hovland 2003).

Understanding possible pathways of policy change, the role of formal and informal institutional checks and balances on power can help develop a clear road map for policy advocacy. This also means that knowledge producers need to be more self-aware of the political nature of their engagement in policy processes. Any act of producing knowledge is, by definition, a political one, and those producing knowledge need to engage with the policy process with their eyes wide open (Jones et al. 2013).

## 3.9 Demand

According to Mulgan and Puttick (2013), one of the most striking factors impeding the effective use of evidence is the absence of organisations tasked with linking the supply and demand of evidence. International development researchers are therefore encouraged to understand the demand for research among policymakers and practitioners, by, for example, mapping the existing information-demand and information-use environment. It has been said that if global public goods research is to be made applicable as well as accessible to national environments from the international system, it must be responsive to demand. This is one approach to engaging with users of research, by taking user realities and preferences into account in development research and communication and by gauging the extent of demand for new ideas by policymakers and society more generally. Some argue that to be effective, research must be located more securely within the context of wider knowledge or innovation systems, implying that the effectiveness and impact of research will be driven by continuous interactions between supply drivers and demand drivers (Hovland 2003).

Research on knowledge transfer, particularly in the field of policy development, has led to several models of the process. The science-push or knowledge-driven model conceptualises it as a unidirectional and logical flow of information from researchers to policymakers resulting in specific policy decisions, whereas the demand-pull or problem-solving model views the process as occurring through the commissioning of information from researchers by policymakers with the intent of addressing a well-defined policy problem. The interactive model construes knowledge transfer as a reciprocal and mutual activity, one that involves researchers and users in the development, conduct, interpretation and application of research and research-based knowledge (Jacobson et al. 2004). DFID acknowledges a preference to move from a linear, supply driven, transfer-of-technology model to a more interactive, demand-driven or collaborative model (Adolph et al. 2010).

Apart from understanding the demand for research, researchers are advised to participate in activities that would stimulate demand for their outputs, such as raising awareness and building capacity within policy circles. In this regard, Shaxson (2010) observes that we know more about how to improve the supply of evidence than we do about how to improve the demand for it, particularly in the policy sphere. Strategies that focus on improving awareness and absorption of research inside government and on expanding research management expertise and developing a culture of policy learning can ameliorate problems on the demand side (Stone 2009). Newman et al. (2012) address capacity to demand research evidence at three levels: individual, organisational and environmental. Capacity-strengthening interventions that stimulate research demand include diagnostic processes, training, mentoring, linking schemes, organisational policies and societal interventions. However, a better understanding is required of what type of mechanisms are most suitable to strengthen user demand for research and to encourage the development of new user participation models in research design and implementation (Adolph et al. 2010).

## *3.10 Engagement*

Several of the themes in the literature—such as effective communication, the role of intermediaries, participation in networks and stimulating demand—converge around the next emergent theme, that of engagement. Some observers use the term to denote the need for closer relationships between researchers and research users, especially policymakers. O'Neill (2005) suggests direct engagement by researchers with the policy community as one of three essential elements of policy influence for development research, saying that the research community must become participants in democratic governance, active at every level. Likewise, Hovland (2003) points towards platforms of broad engagement from which to communicate, such as a public campaign, for research to be more likely to be heard, a suggestion echoed by Datta (2012), who argues that public engagement processes that draw on a range of methods and approaches to elicit a diversity of views are likely to work better. A report by the DFID project on research to action regards engagement as individuals moving from simply accessing or consuming the content and services offered by an online platform to becoming more involved in the platform, recommending or promoting it and actively co-creating the content.

Despite these assertions, Datta (2012) notes that not all researchers see policy engagement as part of their role, suggesting that engagement processes may be more suited to those who see themselves as issue advocates who aim to influence policy in a particular direction and honest brokers who clarify and potentially expand the policy options available to decision-makers. Moreover, despite the new expectations that urge engagement in knowledge transfer, many researchers still accord it a low priority (Jacobson et al. 2004). As we have seen in Shanley and López's (2009) survey, fewer than 5 % of academics regard engagement with the media as an outlet for scientific findings as having any consequence for measuring scientific performance at their institutions. Also, according to performance measurement systems, scientists are intentionally discouraged from producing materials for civil society.

## 4 Summary

At the risk of oversimplifying a complex issue, we can summarise the major lessons to learn from the literature on the impact of research on development policy and practice as follows. It seems that development policy and practice can benefit from the knowledge that research generates, but several interlocking preconditions exist for it to do so:

- Researchers need to have the intent of influencing policy and practice.
- They need to produce high-quality research.

- Academic incentive and reward systems need to move away from a focus on publishing and citation counting and more towards the promotion of research that achieves social and economic impact.
- Research results need to be better communicated to wider audiences, including the public, civil society and policymakers.
- ICTs need to be used more effectively to improve research communication and to allow researchers to engage with other stakeholders in processes of knowledge sharing.
- Intermediaries between researchers and practitioners—individuals and/or organisations—effectively promote research findings to wider audiences where researchers themselves do not (for whatever reason).
- The role of policy entrepreneurs is fostered among suitable researchers and research institutions.
- Formal or informal networks of researchers, practitioners and policymakers exist to facilitate interchanges among stakeholders and promote the take-up of research results.
- Researchers engage with the political context of their work.
- Researchers engage with the users of their research in order to understand the demand-side dynamics of the use of their research in practice and policy circles.
- Policymakers, politicians and their advisers need to cultivate closer relationships with academic researchers in order to make full use of their capacity for producing evidence in support of policy decisions.
- There is effective engagement between researchers, practitioners and policymakers that serves to overcome the various barriers between them.

## 5 Implications for ICT4D Research

This chapter is premised on the claim that information society impact research in the global south has focused almost exclusively on the impact of ICTs, to the exclusion of the impact of the research outside academia. Even here, despite more than a decade of research, identifying the particular contribution of ICTs to specific development goals has proven to be extremely difficult (Kleine 2010). Furthermore, while the contribution in terms of technology diffusion and use—especially of mobile phones—is easy to detect, the focus has only recently shifted towards the question of development impact (Heeks 2010).

Heeks (2010) implies that the absence of ICT4D research impact on practice and policymaking is due at least in part to substandard research in the ICT4D field. He argues that the poor quality of ICT impact assessment to date derives from its lack of conceptual foundations. Furthermore, it seems that there are few researchers in ICT4D who are drawn from the development studies discipline, resulting in the use of an impoverished understanding of development within ICT4D research. Any subsequent discussion of ICTs' contribution to development in the absence of development studies' ideas to define and understand development may make little sense and could result in techno-centric project design as well as making it much harder to connect to development policymakers and practitioners (Heeks 2010). Such a condition contravenes one of the fundamental findings from the literature that ICT4D researchers need to produce high-quality research if they wish to influence policy and practice. But, as we have seen, this is only the starting point.

Beyond these findings, little evidence has been found of any impact of ICT4D research on development policy or practice. DFID and IDRC have been jointly engaged in the ICT4D Research and Capacity Development Programme (2007–2011) which claims a desired output of sustained policy dialogue, defined as "ongoing, evidence-based dialogue among regulators, policy makers, researchers, civil society and the private sector; leading to well informed decision-making on policy issues relevant to ICT4D".[7] According to the project documentation, there are numerous examples of national policies highlighting ICT in their delivery as a result of programmes funded through this ICT4D programme. However, it is not clear that these specific outputs were intended prior to the commencement of the programme. Neither did the SIRCA programme specify that any of the research projects it funded should declare a pre-existing intent to influence practice or policy or both, although a few actually did so. However, the SIRCA focus on building capacity for carrying out high-quality research clearly addresses the fundamental weakness of ICT4D research that Kleine (2010) and Heeks (2010) refer to. Nonetheless, with a better understanding of the conditions considered to be necessary for development research to influence policy and practice, it becomes an easier task to put forward some suggestions as to how ICT4D research could do the same.

In this regard, most of the lessons in the literature are as relevant to ICT4D as they are for international development research. They include (after Shanley and López 2009):

For research and academic institutions

- Restructure institutional incentives to take into account actual impact.
- Create incentives to invest in dissemination and an expanded range of research products.
- Raise awareness and encourage social change agents, knowledge brokers and linkage mechanisms.
- In hiring, balance consideration of publication record with capabilities such as originality, creativity, commitment, depth of field experience and impact orientation.

For researchers

- Interact with stakeholders at various levels to ensure relevance of research questions and outputs.
- Identify uptake pathways as part of project design.
- Design projects to meet end users' needs and aspirations.

---

[7] http://www.dfid.gov.uk/r4d/Project/60422/Default.aspx. Accessed 11 March 2013.

- Share and publish experiences of how research results have been "translated" or used for a non-scientific audience.

For journal editors and publishing organisations

- Challenge researchers to propose ways to evaluate the real impact of their work.
- Provide incentives to researchers to publish practitioner-oriented results of relevance to civil society.
- Break the language barrier by publishing "mirror" papers, translations of the complete paper into the language of where the research was undertaken.

For donors

- Recognise that sustainable change is a long-term process. Support longer-term project time frames (4–10 years) in which sufficient dialogue occurs at the initiation of projects.
- Expand proposal requirements to include the sharing of relevant research results in an accessible format to appropriate audiences.
- Verify that proposals designate sufficient funds for translation, printing, mailing costs and communication.
- Remember that originality often occurs at the fringes. Identify and support small but innovative, locally driven initiatives.

It seems overly optimistic to imagine any infusion of intent to generate nonacademic impacts into ICT4D research without sufficient incentives for researchers to take it up, associated with appropriate capacity building that would enable them to do so. The UK Government's 2014 Research Excellence Framework offers a model of how such an incentive scheme might work, although its effectiveness is yet to be proven, and difficulties can be foreseen in identifying and measuring the kind of impact that the scheme is seeking to induce. However, even with the financial incentives in place and with researchers formulating their strategies for closer engagement with practice and policy, with the backdrop of two communities and parallel universes described in the literature, it remains far from certain that the typical academic researcher will be either comfortable taking up the role of policy entrepreneur or even capable of implementing an effective communication strategy for presenting her research findings to a wider—nonacademic—audience. On top of this, there is the question of institutional incentives and the need to neutralise the obsession with academic *performativity*, citation counts and the tyrannical journal "impact factor", which of course, from the perspective of practice and policy, is nothing of the kind. Given, the entrenched nature of such phenomena, there seems little hope of any early moves away from them, but a start can be made by raising the issue and by further airing the debate that has already surfaced in our literature.

For any ICT4D academic researchers wishing to extend their influence into practice and policy, there seems to be merit in providing them with the guidance and capacity-building structures and processes that would make it possible and easier for them to do so. There are three examples from elsewhere that suggest a means of doing this. *Research to Action* is an initiative that caters for the strategic and practical needs of people trying to improve the uptake of development research, in particular those funded by DFID.[8] It is for development researchers in general who would like to be more strategic and effective in their communications. Two activities of relevance are a workshop on *Improving the Impact of Development Research Through Better Research Communication and Uptake* (Shaxson 2010) and *The Policy Influence Monitoring* project, which monitors and evaluates grantees' policy influence across Africa, South Asia, South East Asia and Latin America. It focuses on the factors and variables that inform how and when research influences policy.

Another interesting example of practical guidance for researchers intending to influence policy is the *Science into Policy* publication of the UK's National Environment Research Council,[9] which helps scientists to recognise the relevance of science to policymakers, identify available opportunities, routes and best practice to influence policymaking, and communicate science in an appropriate and accessible way to the right policymakers, showing how it fits their policy needs. It explains key aspects of the UK policymaking process and provides case studies from the impact of environmental research to illustrate good practice in science to policy. The final example is Canada's knowledge mobilisation network, *ResearchImpact*, that connects university research with research users across Canada to ensure that research helps to inform decision-making. Knowledge Mobilization Units work to match researchers with key policymakers in government, health and social service agencies to ensure that academic research is employed by policymakers and community groups to develop more effective, efficient and responsive public policies and social programmes.[10]

These examples illustrate how the use of relatively simple and low-cost, high-value knowledge-based mechanisms might stimulate and aid researchers towards practice and policy influence, especially those in the SIRCA programme as they become mature and experienced researchers. A particular advantage is that the researchers are already operating within a supportive and vibrant network that consists of other early career researchers from 18 developing countries in three continents as well as the seasoned collaborators and mentors who have been working with each of them, plus of course the combined technical, research and administrative expertise in the Singapore Internet Research Centre of the Wee Kim Wee School of Communication and Information at Nanyang Technological University and IDRC. An opportunity now exists to leverage the strength of the SIRCA network towards achieving the full potential of the research that it has conducted for instilling the capacity—both individual and institutional—for influencing practice and policy. In this regard, it seems that generating institutional capacity for practice and policy influence might be better organised within specialised research units as opposed to mainstream university faculties, where traditional processes are more entrenched.

---

[8] http://www.researchtoaction.org

[9] http://www.nerc.ac.uk/publications/corporate/documents/science-into-policy.pdf

[10] http://www.researchimpact.ca/home/. Accessed 12 March 2013.

ODI, for example, has established itself as an organisational policy entrepreneur by developing advisory ties to governments and international organisations and by institution building of policy communities via networking and partnerships.

## 6 Conclusions

The chapter has reviewed recent literature in the field of research on the impact of research on practice and policymaking in international development. The findings have considerable significance for ICT4D research, which has been assessed overall as lacking, firstly in that the general level of quality is questionable and secondly because there is little if any evidence of any impact on practice and policymaking in ICT4D. The first two phases of the SIRCA programme have successfully targeted the first problem. The second problem remains and is in need of major cultural and institutional shifts if a satisfactory solution is to emerge. However, some of the changes that are necessary, those relating to intent, communication, engagement and networking, can be initiated relatively easily by promoting the transitions that ICT4D researchers will have to make in order to increase the relevance of their work to wider audiences.

**Open Access** This chapter is distributed under the terms of the Creative Commons Attribution Noncommercial License, which permits any noncommercial use, distribution, and reproduction in any medium, provided the original author(s) and source are credited.

## References

Adolph, B., Jones, S.H., & Proctor, F. (2010). *Learning lessons on research communication and uptake*. London: Triple Line Consulting Ltd for DFID.

Alberani, V., Pietrangeli, P. D. C., & Mazza, A. M. R. (1990). The use of grey literature in health sciences: a preliminary survey. *Bulletin of the Medical Library Association, 78*(4), 358–363.

Brown, C. (2012, August). Are southern academics virtually connected? *GDNet*. http://depot.gdnet.org/cms/files//GDNet_study_of_adoption_of_web_2_tools_v2.pdf. Accessed 28 Mar 2013.

Carden, F. (2009). *Knowledge to policy: Making the most of development research*. Los Angeles/Ottawa: SAGE Publications Inc/IDRC.

Court, J., & Maxwell, S. (2005). Policy entrepreneurship for poverty reduction: Bridging research and policy in international development. *Journal of International Development, 17*, 713–725.

Court, J., & Young, J. (2006). Bridging research and policy in international development: An analytical and practical framework. *Development in Practice, 16*(1), 85–90.

Datta, A. (2012). Deliberation, dialogue and debate: Why researchers need to engage with others to address complex issues. *IDS Bulletin, 43*(5), 9–16.

de Vibe, M., Hovland I., & Young, J. (2002, September). *Bridging research and policy: An annotated bibliography* (ODI Working Paper 174). London: ODI.

DFID. (2013a). *Assessing the strength of evidence* (DFID Practice Paper). https://www.gov.uk/government/uploads/system/uploads/attachment_data/file/158000/HtN_-_Strength_of_Evidence.pdf

DFID. (2013b). *Social media engagement*. A report of the activities on the R4D project. https://www.gov.uk/government/uploads/system/uploads/attachment_data/file/200088/Research_uptake_guidance.pdf. Accessed 28 Nov 2013.

DFID. (2013c). *Research uptake: A guide for DFID-funded research programmes*. http://r4d.dfid.gov.uk/pdf/outputs/Communication/R4D%20Social%20Media%20Engagement%20Report_HR.pdf. Accessed 28 Mar 2013.

Fisher, E., & Holland, J. D. (2003). Social development as knowledge building: Research as a sphere of policy influence. *Journal of International Development, 15*, 911–924.

Gendron, Y. (2008). Constituting the academic performer: The spectre of superficiality and stagnation in academia. *European Accounting Review, 17*(2), 97–127.

Grein, H. (2008). Linking research-based evidence to policy and practice. *Research, Policy and Practice, Capacity*, (35), 3.

Haas, E. B. (1991). *When knowledge is power: Three models of change in international organisations*. US: University of California Press.

Harvey, B., Lewin, T., & Fisher, C. (2012, September). Is development research communication coming of age? *IDS Bulletin, 43*(5), 1–8.

Heeks, R. (2010). Do information and communication technologies (ICTs) contribute to development? *Journal of International Development, 22*, 625–640.

Hovland, I. (2003). *Communication of research for poverty reduction: A literature review* (Working Paper 227). London: Overseas Development Institute.

Hovland, I. (2007). *Making a difference: M&E of policy research*. London: Overseas Development Institute.

Jacobson, N., Butterill, D., & Goering, P. (2004). Organizational factors that influence university-based researchers' engagement in knowledge transfer activities. *Science Communication, 25*(3), 246–259.

Jones, H., Jones, N., Shaxson, L., & Walker, D. (2013, January). *Knowledge, policy and power in international development: A practical framework for improving policy*. London: ODI Background Note.

Kleine, D. (2010). ICT4WHAT? Using the choice framework to operationalise the capability approach to development. *Journal of International Development, 22*(5), 674–692.

Langou, G. D. (2008). Developing capacities for policy influence. *Research, Policy and Practice, Capacity*, (35), 14.

Lewin, T., & Patterson, Z. (2012). Approaches to development research communication. *IDS Bulletin, 43*(5), 38–44.

Masset, E., Mulmi, R., & Sumner, A. (2011). *Does research reduce poverty? Assessing the welfare impacts of policy-oriented research in agriculture*. Brighton: Institute of Development Studies, the University of Sussex.

Mulgan, G., & Puttick, R. (2013). *Making evidence useful. The case for new institutions*. The Economic and Social Research Council and NESTA Foundation. http://www.nesta.org.uk/library/documents/MakingEvidenceUseful.pdf.

Newman, K., Fisher, C., & Shaxson, L. (2012). Stimulating demand for research evidence: What role for capacity-building? *IDS Bulletin, 43*(5), 17–24.

Nutley, S., Walter, I., & Davies, H. (2007). *Using evidence: How research can inform public services*. Bristol: Policy Press.

O'Neil, M. (2005). What determines the influence that research has on policy-making? *Journal of International Development, 17*, 761–764.

OECD. (2011, April). *Opportunities, challenges and good practices in international research cooperation between developed and developing countries*. Paris: OECD.

Ryan, J. G., & Garrett, J. L. (2003). *The impact of economic policy research: Lessons on attribution and evaluation for IFPRI*. Washington, DC: International Food Policy Research Institute.

Sabatier, P., & Jenkins-Smith, H.C. (1999). The advocacy coalition framework: An assessment. In P. Sabatier (ed.), *Theories of the policy process. Boulder*: Westview Press.

Saxena, N. C. (2005). Bridging research and policy in India. *Journal of International Development, 17*, 737–746.

Sen, K. (2005). *Rates of return to research: A literature review and critique.* Manchester: DFID.

Shanley, P., & López, C. (2009). Out of the loop: Why research rarely reaches policy makers and the public and what can be done. *Biotropica, 41*(5), 535–544.

Shaxson, L. (2010, November 29–30). *Improving the impact of development research through better research communications and uptake.* Report of the AusAID, DFID and UKCDS funded workshop, London.

Stevens, H., Dean, A., & Wykes, M. (2013). *DESCRIBE project final project report.* Exeter: University of Exeter. http://www.exeter.ac.uk/media/universityofexeter/research/inspiringresearch/describeproject/pdfs/2013_06_04_DESCRIBE_Final_Report_FINAL.pdf

Stone, D. (2009). RAPID knowledge: Bridging research and policy at the overseas development institute. *Public Administration and Development, 29*, 303–315.

Sumner, A., Ishmael-Perkins, N., & Lindstrom, J. (2009). *Making science of influencing: Assessing the impact of development research* (IDS Working Paper 335). Brighton: IDS.

Taylor, M. (2005). Bridging research and policy: A UK perspective. *Journal of International Development, 17*, 747–757.

Weiss, C. (1977). Research for policy's sake: The enlightenment function of social research. *Research, Policy Analysis, 3*(4 Fall), 531.

Wheeler, J. (2007). *Creating spaces for engagement: Understanding research and social change.* Brighton: Development Research Centre on Citizenship, Participation and Accountability.

Williams, G. (2012). The disciplining effects of impact evaluation practices: Negotiating the pressures of impact within an ESRC–DFID project. *Transactions of the Institute of British Geographers, 37*, 489–495.

Young, J. (2005). Research, policy and practice: Why developing countries are different. *Journal of International Development, 17*, 727–734.

Young, J. (2008). Impact of research on policy and practice. *Research, Policy and Practice, Capacity*, (35), 4–7.

# Constructing Theories of Change for Information Society Impact Research

Alexander Flor

## 1 Introduction

It was a cold February morning in 2010 when a select group of development workers assembled at Raamweg 5, The Hague. The motley group represented major sectors, themes, stakeholders, and continents in the Global South. Hosted by the International Institute for Communication and Development (IICD) and sponsored by the Food and Agriculture Organization of the United Nations (FAO), they convened with a singular purpose in mind: to put their heads together and attempt to bring coherence to the information and communication technology for development (ICT4D) discourse.

### 1.1 A Question of Impact

It has been 10 years since the G8 nations announced their intentions in Okinawa to help bridge the digital divide and, in doing so, alleviate poverty in the Global South. However, the past decade had been met with uncertainty and mixed feelings about the promise of information society. An oft-cited indictment is the preponderance of anecdotal (and the lack of hard) evidence that directly correlated information and communication technologies to development and poverty alleviation. *Do digital access and opportunities really contribute to achieving the Millennium Development Goals?* Five years before the end of the MDG timeframe, this question remained largely unanswered.

---

A. Flor (✉)
Faculty of Information and Communication Studies, University of the Philippines
(Open University), Los Baños, Laguna, Philippines
e-mail: aflor@upou.edu.ph

It was not the intention of The Hague gathering, however, to answer this question. Although unarticulated, the group's collective experience and gut feeling pointed towards the affirmative. They would instead attempt to craft the logic behind this affirmative answer and do so with the strength of their convictions.

## 1.2 Contradictions

In the past, ICT4D programmes have tried to address the lack of evidence-based impact studies by strengthening capacities for research. However, increased capacity in the conduct of disciplined inquiry responds to methodological challenges only, not to the substantive. The substantive challenge involves addressing the innate contradictions encountered when embarking on information society impact research.

The first of these contradictions deals with the dual nature of information society. By 2005, the international development assistance community had seen it fit to classify development programmes under two major categories: sectors and themes. Sectors include agriculture, health, education, environment, natural resources, etc. Themes involve crosscutting concerns such as governance, gender, poverty, sustainability, and climate change. When situated within the development arena, the information society is *both* a sector and a theme. Its sectoral dimension covers hardware design, software development, infrastructure expansion, universal access, information and communication policy, and knowledge products and services. Its thematic nature, on the other hand, is manifested in its crosscutting applications.

ICT4D may be applied in any development sector. When used in agriculture, it becomes eAgriculture; in health, it may take the form of telemedicine; in education, it is referred to as ICT4E; and in the environment, its major application is geospatial information systems (GIS). Generally, when assessing development performance, programme evaluators would consider ICTs as belonging to a different sector altogether (information technology or telecommunications) and, hence, beyond their purview and concern. Although ICT impacts significantly on these programmes, sectoral evaluations tend to class it along with other control (i.e. ceteris paribus) variables and, thus, not sufficiently looked into.

A second contradiction deals with the fact that information society impact cannot be disaggregated from outcomes generated by purely sectoral interventions. To be fair, information society impact research has been conducted early on (e.g. Batchelor 2003, 2006, Batchelor and Norrish 2005) and widely enough (see Heeks and Molla 2009). Proponents submit that the difficulty in establishing a direct link between information society and the MDGs is not due to the lack of impact evidence but to difficulties in disaggregating and attributing such impacts to ICTs alone. For instance, it is difficult to sift through macro-level economic data and directly link ICT interventions to significant change in income, equity, or environmental quality indicators. Information society impact may be somewhat likened to a coefficient whose exact value cannot be factored separately from a given product. A postmodernist would refer to this contradiction as the *invisibility* of information society within the development arena.

Thirdly, information society impact is often not immediate and hardly tangible. ICT4D may facilitate development processes. Openness may empower development actors. But neither technology nor openness generates instant economic returns unless these are directly linked to a service within the value chain. This lack of immediacy or concreteness of information society impact further contributes to the difficulty in its documentation and measurement as hard evidence.

These contradictions are exacerbated by the fact that, as Matías Dodel points out elsewhere in this book, information society discourse lacked a comprehensive, universally accepted framework that establishes clear causal links between ICTD interventions and MDG outcomes. It was with the development of such a framework that The Hague summit invested its efforts into.

## 1.3 Purpose

The purpose of this chapter is to take The Hague discussion a step further by constructing *theories of change* based on the framework developed. The practical implication of building these so-called theories of change will be the availability of alternative results chain models for the conduct of information society impact research relative to the Millennium Development Goals. The chapter has three major sections. The first one discusses the framework which resulted from The Hague ICT4D meeting in 2010. The second elaborates on the theories of change. The third presents an essay on evaluating impact and unintended outcomes.

## 2 The Hague Framework

The Hague discussion was chaired by Stephen Rudgard of FAO Rome and facilitated by his associate, Michael Riggs, backstopped by IICD's Denise Senmartin. Rudgard was among the first advocates of evidence-based research (EBR) for evaluating impact within the international development assistance community, while Riggs was the driving force behind the global eAgriculture community. Senmartin, on the other hand, was involved in impact studies on ICTs for rural livelihoods (IICD 2006). Attendees included: Africa-based information science expert Peter Ballantyne, representing the international agricultural research community; evaluation specialist Kay Leresche; gender and communication advocate Anriette Esterhuysen; and myself as the sole representative from Third World academia.

## 2.1 Antecedents and Assumptions

Rudgard, Riggs, and Senmartin acknowledged that ICT4D frameworks have been proposed in the past. These models, although useful, have not been generally adopted since some were too complicated while others were context specific. Some were donor driven or biased towards an ideological point of view. Many were focused on technology neglecting the sociological, cultural, and developmental dimensions. At the very onset, the group was in agreement on the need to revisit these models and incorporate them into a unified framework that is universally acceptable and applicable for all development sectors and themes.

*Senmartin* says the framework should:

- Be versatile and be applicable for planning, implementation, monitoring and evaluation, impact analysis, or scientific research
- Cater to development practitioners, planners, policymakers, academics, and donors
- Have sound theoretical bases
- Possess the qualities traditionally associated with good frameworks, i.e. coherence, comprehensiveness, parsimony, and elegance

## 2.2 ICT4D Model

In spite of these agreements, our disparate backgrounds came into play and eventually led to contentious debates that, upon the prompting of Ballantyne, we conceded to be attendant to the process of convergence and synthesis. The opposing discussions, in fact, generated a model that many will consider coherent yet comprehensive, elegant yet parsimonious. Characteristic of the development discourse that we represented, there was an initial tendency to use the terms model, theory, framework, construct, and concept interchangeably. Indeed, the interfaces between some of these terms were quite significant. As we progressed, however, the differences became more distinct. We began to refer to a framework as a structured set of conceptual boxes wherein we can situate a narrative or approach the study of ICT4D. By theory, we meant an explanation of the causal relationships among the elements that make up the ICT4D phenomenon. By ICT4D model, we meant a visual representation of how elements making up the ICT4D phenomenon interacted. At The Hague meeting, the ICT4D model became the visual representation of the ICT4D framework. The elements are, in effect, ICT4D concepts, and a statement of relationship between two or more of these concepts was considered a construct.

**Constructs** Theoretically, the model takes off from a generally acknowledged construct: *Communication can effect developmental changes in societies*. This proposition had been tried and tested since the 1950s through the 1970s in the works of Beal et al. (1957), Rogers (1962), and Quebral (1973). In fact, this idea is being actively revived in the current C4D (communication for development) initiative within UN agencies.

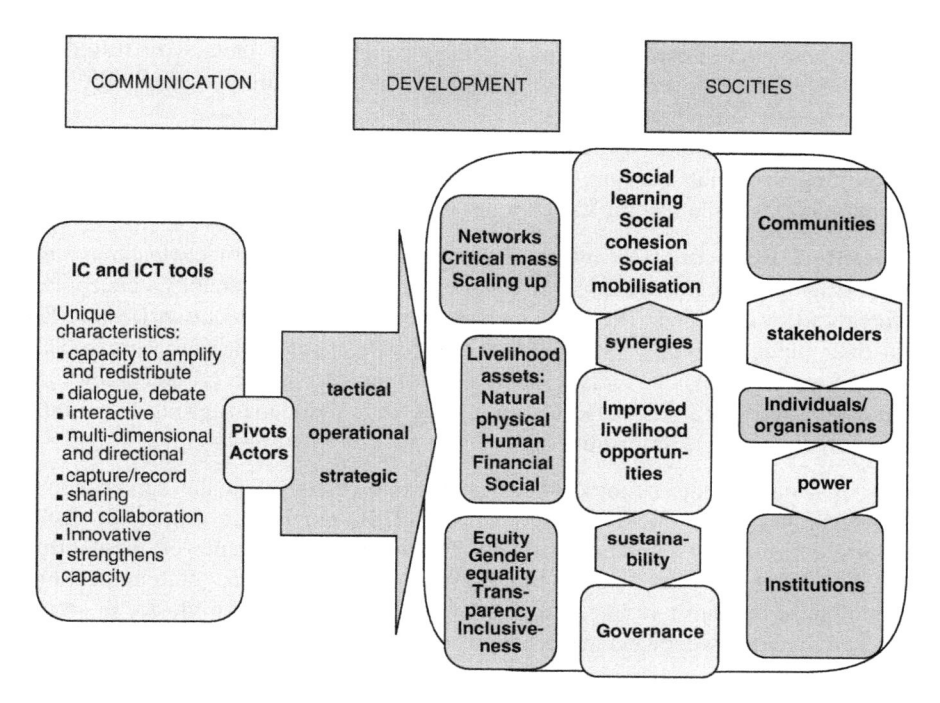

Constructing Theories of Change for Information Society Impact Research 49

Fig. 1 The Hague framework

We supplemented this basic construct further by the following commonly shared assumptions:

- Communication is a social process.
- Communication is the exchange of information or the sharing of knowledge.
- The essence of ICT4D is not technological but social; the emphasis should be on the "C" and the "D" instead of the "I" and the "T." In eAgriculture, for instance, bandwidth may be important but food security far outweighs it.

Additional constructs that made up the framework submit that ICT, as a social phenomenon, has the following attributes: knowledge capture, storage, distribution, amplification, interactivity, multidimensionality, multidirectionality, sharing, collaboration, and technological innovation (Fig. 1).

Furthermore, individual, organizational, or institutional pivotal actors (otherwise referred to as intermediaries, catalysts, change agents, focal points, or champions) link ICTs to the beneficiary. There are primary, secondary, and higher-order impacts of ICTs through these pivotal actors:

- ICTs contribute to natural, physical, financial, human, and social capital resulting in improved livelihood outcomes for individuals and organizations (Flor 2008).

- ICTs lead to transparency, equity, gender equality, and inclusivity resulting in strengthened governance for institutions while providing sustainability to improved livelihood outcomes.
- ICTs create networks, critical mass, and enable upscaling of interventions resulting in social learning, social cohesion, and social mobilization among communities while feeding its synergies to improved livelihood outcomes.

**Conceptual Referents** The major elements of this model are *communication*, *developmental changes*, and *society*. For each of these major elements, conceptual referents were identified. The conceptual referents for *communication* were ICTs and their unique features. The referents for *developmental changes* were social learning, social cohesion, social mobilization, improved livelihood outcomes and improved governance. The referents for *society* were individuals, groups, communities, organizations, and institutions.

After 2 days of discussions, The Hague group decided that the agreed upon constructs and visual model constituted an ICT4D framework that held much promise and that it was time to disband. Theoretically, the framework explained how development sectors could benefit from information and communication technologies. Furthermore, it enumerated the elements required to do so. In other words, it made us understand the dynamic behind the impact.

## 2.3 Gaps

However, as the group wound up the discussion and prepared to adjourn, there was a sentiment that something had been left undone. In particular, Leresche, a veteran of multilevel impact evaluations in Africa, raised concerns that the framework did not provide for the specifics of evidence. It visually and narratively laid out the phenomenon in conceptual terms, but it did not do so in empirical terms. The framework needed much more than conceptual referents but empirical referents as well. It required an articulation of the impacts, an elaboration of the outcomes, and an identification of interventions, but most of all, a determination of the causal links between these three. In other words, The Hague framework needed *theories of change*.

# 3 Theories of Change

## 3.1 Definitions

What are theories of change?

The questioning of impacts articulated at the beginning of this chapter was not unique to information society discourse. It was true to the entire development assistance community as well. Even before the Millennium Development Goals were proposed at the turn of the century, stakeholders in the development process—donors, beneficiaries, governments, and civil society organizations—have become more critical, more circumspect, and more vigilant with the results of development aid. A variety of approaches and tools were introduced by the same stakeholders to ensure that impacts were appropriately monitored and evaluated. Apart from the *evidence-based approach* championed by DFID came *results-based management* (RBM) of the World Bank, *management for development results* (MfDR) of the Asian Development Bank, *performance-based assessments* (PBA), and *theories of change* (ToC). The latter, as defined, is a set of beliefs that guides the thinking about how and why a complex change process will unfold (Clarke 2004). A ToC is not a theory in the academic sense of the word. It is not a general theory on how change occurs but a "theory" specific to an intervention (Clouse 2011), which may take the form of a programme or a project. Thus, although adequate within the general context of ICT4D, The Hague framework required ToCs when it came to specific impacts of particular information society interventions, projects, or programmes, which stakeholders are on the lookout for.

The New York-based Center for Theory of Change explains that:

... a Theory of Change defines all building blocks required to bring about a given long-term goal. This set of connected building blocks–interchangeably referred to as outcomes, results, accomplishments or preconditions is depicted on a map known as a pathway of change/change framework, which is a graphic representation of the change process. (http://www.theoryofchange.org. Accessed 3 May 2013)

The main tool of the ToC is a *map* of causes and effects, which, the Center describes, is constructed in six stages (http://www.theoryofchange.org/what-is-theory-of-change/how-does-theory-of-change-work/. Accessed 3 May 2013):

1. Identifying long-term goals
2. Backwards mapping and connecting the preconditions or requirements necessary to achieve that goal and explaining why these preconditions are necessary and sufficient
3. Identifying basic assumptions about the context
4. Identifying the interventions that the initiative will perform to create desired change
5. Developing indicators to measure outcomes to assess the performance of the initiative
6. Writing a narrative to explain the logic of the initiative

For the sake of clarity, definitions of impact, outcomes, and outputs specific to information society context are provided here. In the simplest terms, *impact* would refer to the positive or negative contribution of information society intervention on a Millennium Development Goal. *Outcome* is an immediate or intermediate condition caused by the intervention that may lead to a goal and thus linked to an impact. *Output* is a product or a service delivered by the intervention that may lead to an outcome.

How does information society impact on the Global South? This section will attempt to answer this question using the theories of change approach initially in general terms, then in specific ones.

## 3.2 Changing the Global South

**Long-Term Goals** As discussed by Andrea Ordonez in a separate chapter, the term *development*, as used within the context of ICT4D, is often associated with project-based goals. Nowadays, development interventions for the Global South are designed and implemented to contribute to the long-term targets of any one of the eight Millennium Development Goals. Identified during the Millennium Summit of 2000, these goals are targeted to be achieved by 2015. With 193 UN member states and 23 development organizations endorsing the MDGs, the international development assistance community has patterned its plans and devoted its resources to the fulfilment of these goals, rightly or wrongly. If the information society is to be recognized as having a positive impact on the developing world, then it will have to be situated within the MDG framework.

**Change Narrative** What are the requirements of these long-term goals that are provided by information society? Based on The Hague group's collective experience, there are four attributes of the information society that may be seen as preconditions or requirements necessary to achieve any one of the MDGs: openness, equity, quality, and scale. Openness refers to access to natural, physical, economic, intellectual, and knowledge resources. Equity pertains to the distribution of these resources. Quality relates to the value of these resources. Scale indicates the levels, range, and degree of availability of these resources.

The information society runs on an information-based economy. It functions with and operates on networks. Information or knowledge becomes the primary wealth generating, enabling, and empowering resource. Networks become the dominant platform for any wealth generating, enabling, and empowering activity. The continued development, utilization, and application of information and communication technologies perpetuate these conditions and become both the cause and effect of an information-based economy.

The Hague framework emphasizes the liberating nature of information and communication technologies with their innate characteristics: dialogue, debate, and interactivity; documentation, capture, and recording; multidimensionality and omnidirectionality; amplification and redistribution; and sharing and collaboration. To this list of qualities, the following may be added: augmentation, automation, mobility, speed, integration, and synergy.

Dialogue, debate, and interactivity; documentation, capture, and recording; and multidimensionality and omnidirectionality contribute to transparency, which leads to *openness*. Amplification and redistribution result in inclusivity. Sharing and collaboration lead to enablement. Integration and synergy bring empowerment.

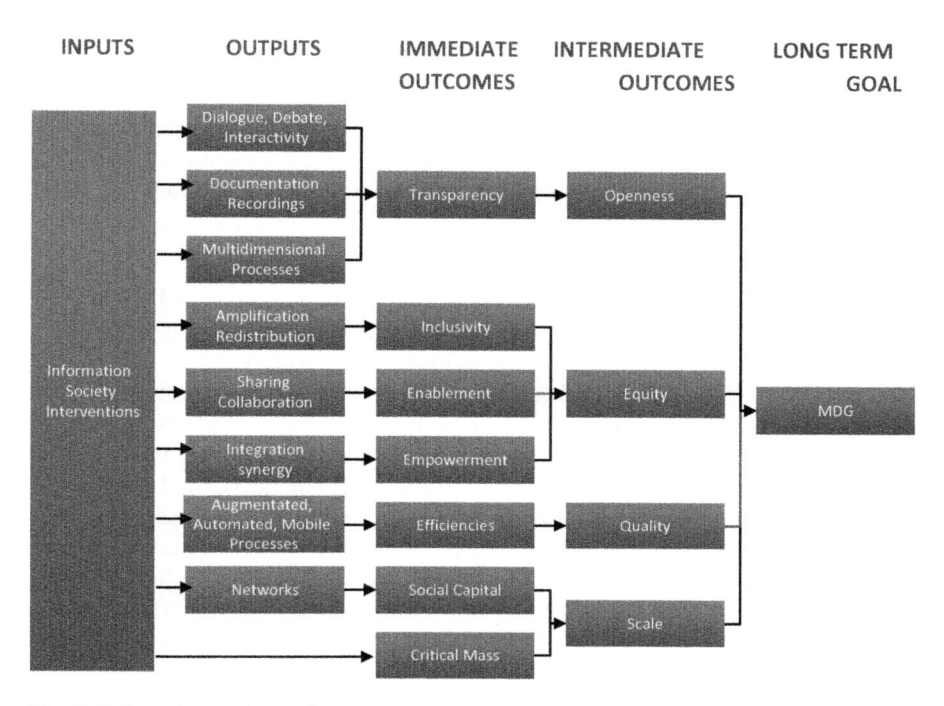

Constructing Theories of Change for Information Society Impact Research 53

Fig. 2 Information society pathway to change

Inclusivity, enablement, and empowerment generate *equity*. Augmentation, automation, mobility, and speed improve effectiveness and generate efficiencies that lead to *quality*. On the other hand, networking develops social capital and a critical mass that contributes to *scale* (Flor 2005). A not-so-specific pathway to change resulting from this narrative is found in Fig. 2.

Pathways to change maps are usually drawn vertically. With the long-term goal on top, one works down to the intermediate outcomes, immediate outcomes, outputs, and interventions. Figure 2 was drawn horizontally to show the logical sequence from interventions to long-term goal not suggestive of hierarchy. Furthermore, pathways generally portray causal links between specific interventions that lead to particular outputs that result in definite outcomes linked to explicit goals. The above general results chain was constructed for illustrative purposes only. Found below are three examples that focus on specific goals and are patterned after conventional ToCs. They do not represent any specific project on the ground. The elements are a composite based on my own previous field experiences. Nevertheless, these are indicative of authentic information society ToCs as applied to the Millennium Development Goals. Note that the sample ToCs are structured with an upward vertical logic from the intervention to the output, through the immediate and intermediate outcomes ending with the long-term goal on the top.

## 3.3 Indicative Information Society ToCs

**Pathway to Change 1: Agriculture Sector** MDG 1 declares that extreme poverty and hunger should be eradicated by 2015. This places the agriculture sector as a top priority in national and international development programmes. How can ICTs contribute to the achievement of MDG 1? An indicative pathway is found below (Fig. 3).

To contribute to the long-term goal of ending extreme hunger, a hypothetical national eAgriculture programme would have two explicit outputs: a crop decision support system (CDSS) and a marketing information system (MIS) for agricultural products. The CDSS, if designed appropriately and used efficiently, will result in maximized production of staple crops such as rice, corn, and tubers as well as minimized wastage of agricultural inputs such as water and fertilizer. On the other hand, the MIS will assist policymakers in formulating an inclusive food distribution policy and will result in transportation and distribution efficiencies. Maximized production and minimized wastage will lead to the intermediate outcome of food security operationalized through the self-sufficiency ratio indicator. Similarly, an inclusive food policy and efficient transportation and distribution system will lead to the intermediate outcome of equitable food distribution. Food security and equitable distribution will contribute to Millennium Development Goal 1, which is to end extreme poverty and hunger by 2015. Such is merely an indicative pathway. For a deeper understanding, Kabran Aristide describes in Part II of this book *on the ground* experiences on eAgriculture in the Ivory Coast.

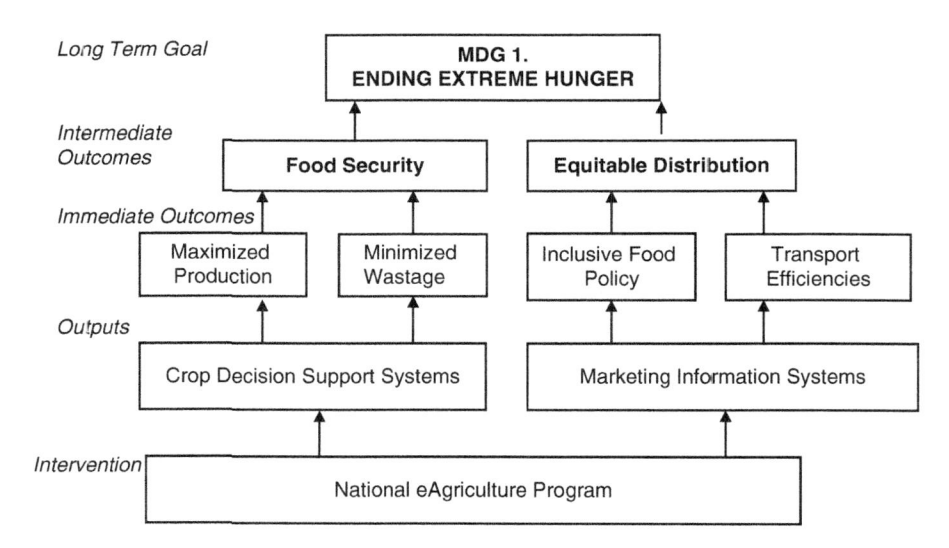

**Fig. 3** eAgriculture indicative pathway to change

## Figure 4

**Long-Term Goal:** MDG2. UNIVERSALIZATION OF PRIMARY EDUCATION

**Outcomes:** Improved Access | Assured Quality | Increased Equity

**Outputs:** eLearning Platforms/Materials | Online Teacher Training Courses | EMIS

**Components:** ICT for Pedagogy | ICT for Teacher Training | ICT for Education Governance

**Intervention:** National ICT4E Program

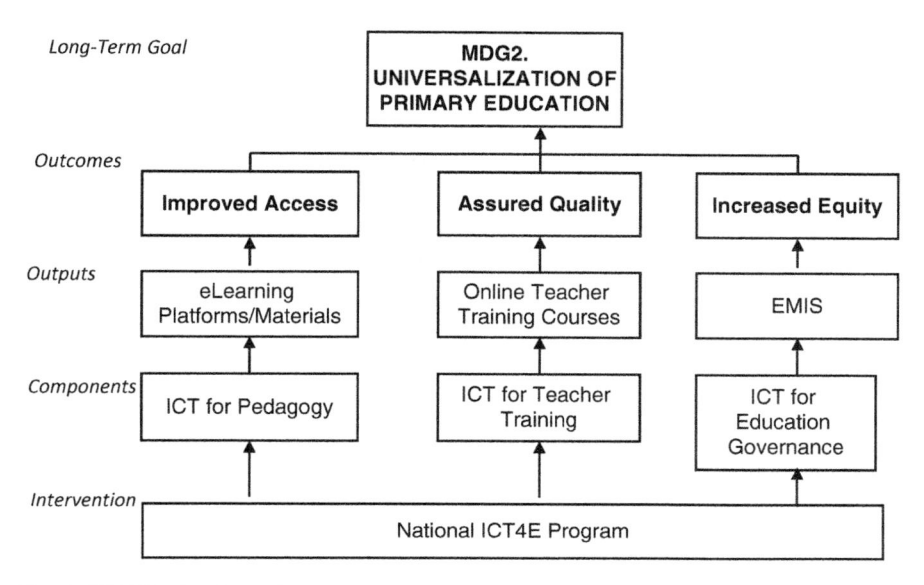

**Fig. 4** ICT4E indicative pathway to change

**Pathway to Change 2: Education Sector** MDG 2 states that universal primary education should be achieved by 2015. Elementary school education should be made available for all in every country in the world (Fig. 4).

The information society can contribute to the goal of universalization of primary education via a National ICT4E Programme with three components: ICT for pedagogy, ICT for teacher training, and ICT for education governance. The ICT for pedagogy component will produce eLearning platforms and educational resources. The ICT for teacher training component will generate online teacher training courses that may be taken by actively serving teachers during their free time. The ICT for education governance component will design and develop an inclusive educational management information system (EMIS) that would cover depressed, disadvantaged, and underserved communities. The eLearning platforms and educational resources will result in improved educational access operationalized by enrolment indicators. Online teacher training courses will bring about quality education measured as completion rates. An inclusive EMIS will result in increased educational equity. All three outcomes—improved access, assured quality, and increased equity—will contribute to MDG 2, the universalization of primary education by 2015. This hypothetical pathway, specifically its teacher training thread, is validated by Olivera, Ale, and Chib in their Peru case study found in the second part of this book.

**Pathway to Change 3: Climate Change Theme** The previous examples—eAgriculture and ICT4E—focused on interventions that contribute to long-term *sectoral* goals. ICT4D theories of change may also contribute to long-term *thematic*

```
Long-term Goal                    CLIMATE CHANGE RESILIENCY
                                              ▲
Intermediate              ┌───────────────────┼───────────────────┐
Outcomes      Improved Access to      Increased Climate       Decreased Risk
                 Resources          Change Responsiveness       Exposure
                      ▲                      ▲                      ▲
Immediate                              Knowledge Gain
Outcomes   Increased Social Capital    Attitude Change       Enhanced Readiness
                                       Practices Adopted
                      ▲                      ▲                      ▲
Outputs                                Climate Change
              Social Networks         Knowledge Products    Early Warning Systems
                      ▲                      ▲                      ▲
Intervention          └──────────────────────┼──────────────────────┘
                    Knowledge Management System for Climate Change Adaptation
```

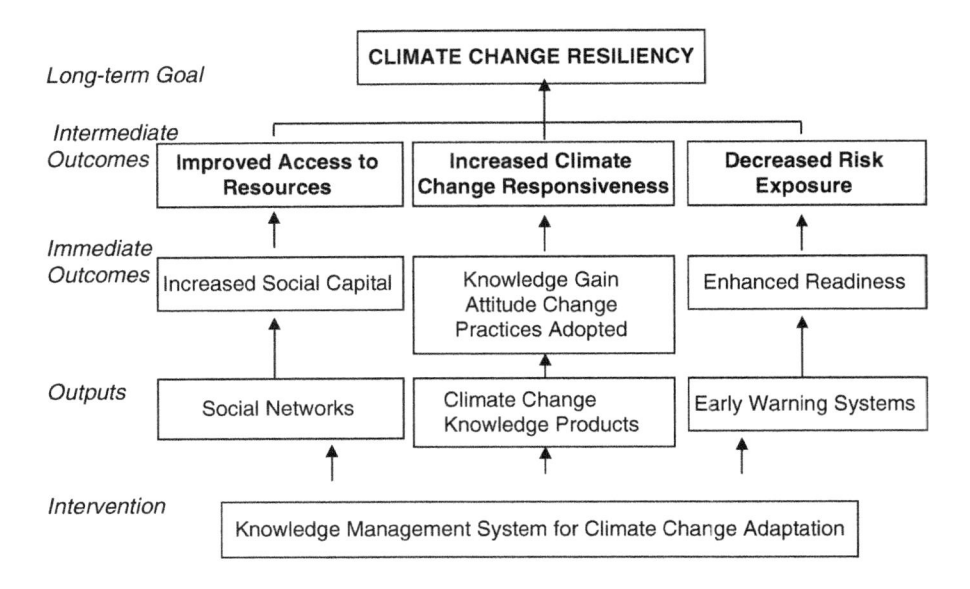

**Fig. 5** Climate change KM indicative pathway to change

goals, such as gender and governance, which cut across sectors. Under MDG 7, ensuring environmental sustainability, we can situate a subgoal that qualifies as a thematic one and that may be addressed by information society interventions.

Climate change resiliency is a thematic goal since it cuts across the environment, agriculture, health, education, and national security. Given the fact that the Global South is the most adversely affected by this phenomenon, it is included here as a long-term goal for information society ToC (Fig. 5).

The information society can contribute to the long-term goal of climate change resiliency, at the community level, with the design, development, and testing of a knowledge management system for climate change adaptation. This system will generate three outputs: social networks, climate change knowledge products, and early warning systems for natural disasters such as tsunamis, typhoons, floods, droughts, and forest fires. An immediate outcome of social networks would be an increase in social capital among community associations, which would eventually result in improved access to climate change adaptation resources. Immediate outcomes of climate change knowledge products are knowledge gain, attitude change, and practices adopted, which result in increased climate change responsiveness. An immediate outcome of early warning systems is increased community readiness leading to decreased risk exposure. Improved access to climate change adaptation resources, increased community responsiveness, and decreased community exposure to risks will collectively result in long-term climate change resiliency (Gonzalez-Flor et al. 2013).

## 4 Evaluating Impact

The three pathways illustrated above are visual representations of information society theories of change. They provide indicative elements (i.e. interventions, outputs, and outcomes) that impact on the Global South. Moreover, the pathways establish clear causality between the elements. The pathways or ToC maps may guide the planning and design of information society interventions. More importantly, they serve as bases for monitoring and evaluating results, impacts, and contributions to the MDGs. Identifying the elements of change and tracing their causal links establish a logical relationship between information society attributes and Global South development. However, ToC maps alone do not address the need for evidence.

### 4.1 Evidence of Change

Indicators provide evidence of change. Indicators are empirical measurements of outputs and outcomes. Outputs, in the form of goods or services produced by the intervention, are generally quantified or qualified output indicators or both. In the eAgriculture example given above, the outputs are a Crop Decision Support System and a Market Information System. The output indicators are quantified as the "*number* of knowledge/information systems" or qualified as "*operational* knowledge/information systems." Similarly, in the ICT4E example, the outputs are eLearning materials, online teacher training courses, and an EMIS. The output indicators then become the number of eLearning materials, the number of online teacher training courses, and an inclusive EMIS.

Outcomes, being immediate or intermediate conditions or circumstances brought about by the outputs of an intervention, are measured in relative terms instead of absolute numbers. Outcome indicators come in the form of rates, ratios and proportions or products, coefficients, and quotients. In the KM for climate change example, immediate outcome indicators involve the differences between baseline, midterm, and final measurements of social capital within community organizations; adaptation of knowledge, attitudes, and practice (KAP) among families; and disaster preparedness of neighbourhood associations.

Each output and outcome identified in a ToC should have at least one corresponding observable and measurable indicator to fulfil the evidence-based standard.

### 4.2 Unintended Outcomes

And now for the caveat: while well-constructed theories of change on information society impact may guide us in establishing causal links between ICTs and the MDGs, we can be blindsided by them. The parameters set in our ToCs provide us with the focus and vision required for objective disciplined inquiry. Furthermore, it will provide us with solid evidence of significant change. However, these parameters, derived mostly from results-based indicators exemplified above, may act as horse blinders, robbing us of our peripheral vision and preventing us from appreciating the totality of the information society phenomenon under study. As such, impact studies on information society and openness themes, be these ex post regulatory or final project evaluations, require us, at a certain point in the process, to remove these blinders and to explore the so-called unintended outcomes or impacts, negative or positive.

Presented herein are unintended ICT4D outcomes discovered in three impact evaluations conducted in 2011 in three separate countries (Indonesia, Philippines, and Nepal), at three evaluation points (ex ante, midterm, and final evaluation), and on three information society domains (eLearning, openness, and information systems). The subjects of impact evaluation—the eLearning component of JICA's Maritime Education and Training Improvement (METI) Project in Indonesia, the indigenous knowledge system (IKS) component of EU's Focused Food Production Assistance to Vulnerable Sectors (FPAVAS) in the Philippines, and the MIS component of UNESCO's Strengthening National Capacity on Planning, Monitoring and Evaluation of Literacy and NFE Programmes (CAPEFA) in Nepal—provide us with surprising outcomes that have not been incorporated in the project design but may eventually be considered as a project's saving grace.

**Case 1. The Economics of eLearning** This case study may not be directly related to an MDG, but because of its implications on GDP, it may be indirectly linked to MDG 1: the eradication of poverty. Indonesia is the largest Muslim country in the world. It is also home to some of the poorest communities particularly in its coastal areas (Badan Pusat Statistik 2010), which may be a contradiction of sorts. It is the world's biggest archipelago covering three time zones. Compared to any other country, Indonesia has the longest coastline. The country is strategically located in one of the world's busiest waterways: east of the Indian Ocean, west of the Pacific Ocean, south of the China Sea, and north of the Corral Sea. It is within the planet's largest center of marine biodiversity, the Sulu Celebes Sea. Thus, Indonesia has been described as the "Sleeping Giant" in the global maritime sector. Most certainly, Indonesia has a legitimate claim of becoming a global maritime power. The country can potentially become one such power by establishing the strategic presence, in terms of quantity as well as quality of its merchant marines and officers within the global fleet. In other words, it can achieve its goal of becoming a leading maritime nation by producing a well-trained highly educated maritime workforce manning flagships from different countries. Strategically, it can make its mark through quality maritime education and training. From 2009 to 2011, the Indonesian Ministry of Transportation implemented a JICA-funded undertaking, Maritime Education and Training Improvement Project, to develop policies, curricula, and delivery systems to improve the quantity and quality of the country's maritime workforce.

Compliance to the International Convention on Standards of Training, Certification and Watchkeeping for Seafarers (STCW Convention) and the Seafarer's Training, Certification and Watchkeeping Code (STCW Code) has been the global framework, rationale, and driving force for maritime education and training improvement. STCW 2010, the latest version of this international agreement, provides for the standardization of maritime distance education programmes. Section A-I/6 on Training and Assessment states that "Each Party shall ensure that all training and assessment of seafarers for certification under the Convention is structured in accordance with written programs, including such methods and media of delivery, procedures, and course material as are necessary to achieve the prescribed standard of competence." This implies a tacit acceptance of distance learning, in general, and eLearning, in particular, as an MET delivery system provided that certain standards are set, adhered to, monitored, and enforced. Hence, eLearning as a strategy and delivery system was incorporated into the project. In the ex ante evaluation of the strategies introduced by the project, the potential impact of eLearning on education and training improvement was assessed. Having been assigned this task, the evaluation employed a pathway of change model that aimed at quality education and training as an intermediate outcome. While conducting the evaluation, it became clear that the potential impact of eLearning extended beyond the confines of education and training improvement and infringed upon the economic aspect.

The economic argument hinges upon opportunity costs incurred when an Officer Third Class leaves his or her current onboard assignment to undergo advanced training for his or her upgrade to Officer Second Class. When Indonesian maritime academies enrol Officers Third Class for upgrading for eight months, Indonesian families collectively incur an estimated opportunity cost of USD 3.96 million per year based on the onboard salaries of their breadwinners. When Officers Second Class go onshore for upgrading for another six months, an opportunity cost of USD 8.4 million per year is further incurred due to their higher pay scales. These opportunity costs can be mitigated with onboard instead of onshore training via eLearning. Opportunity costs pertain to the families of Indonesian officers only and hence are incurred at the micro level. At the macro level, Indonesian GDP is deprived of as much as USD 30 million annually due to the onshore training for upgrading. Additionally, maritime eLearning programmes are ten times cheaper than residential or on campus programmes (Azuma et al. 2011). A classroom to student ratio of 1:30 is no longer necessary. Maritime academies need not be constrained to accept applicants on the basis of limited facilities. The educational expenditure of the trainee is likewise reduced. Transportation costs, educational materials costs, and tuition fees would be significantly reduced. These are positive unintended economic impacts of introducing eLearning into the Indonesian MET sector (Flor 2013b).

**Case 2. The Downside of Open Knowledge Resources** From 2010 to 2011, the EU implemented the Focused Food Production Assistance to Vulnerable Sectors (FPAVAS) Project in the Philippines. Incorporated in the project design was the development of an indigenous knowledge system that would capture climate change adaptation best practice among the indigenous tribes in the six project areas for sharing and reuse. This paved the way for a piggyback study titled "Design, Development and Testing of an Indigenous Knowledge Management System Using

Mobile Device Video Capture and Web 2.0 Protocols" that was implemented during the FPAVAS midterm. The theoretical basis for proposing that mobile devices may lead to the active participation of IPs as ICT4D Web content providers is founded on the relationships among three concepts: social capital, the network effect, and critical mass theory. The study's philosophical basis rested upon open access and open learning resources assumptions (Flor 2013a).

The primary ICT4D intervention was the mobile device—GPRS-enabled mobile phones, with audio-video capture and Internet browsing functionalities. In a nutshell, the vertical logic employed in the study predicted that the technological intervention would enable and empower IP communities to capture their indigenous practices on climate change adaptation and share these via the Web among other IP communities for their reuse, resulting in increased climate change resiliency. During the conduct of the study, however, marked reluctance from organized indigenous people's groups to participate in the initiative was observed. It soon became apparent that interfacing indigenous knowledge with open access concepts held complicated issues.

Firstly, indigenous belief systems covering knowledge transfer, sharing, and reuse are guided by a tradition of hierarchy. Indigenous communities, as a rule, have tribal elders, chieftains, and healers who regard themselves as custodians of knowledge, which may only be shared with prudence, responsibility, and, on occasion, sanctity. An unintended consequence of the intervention was to contradict the prevailing belief system that indigenous knowledge on feeding (agriculture) and healing (medicine) cannot be made openly available to any person who may misuse it.

Secondly, interventions in the past have failed to respect the privacy of IP communities. This brings to mind a field experience while developing and testing the ethnovideographic methodology 20 years ago. Fieldwork was being conducted among the indigenous tribes of Central Mindanao under an International Potato Center grant, video documenting the indigenous agricultural practices of the Talaandig-Higaonon tribe in the slopes of Mt. Kitanglad, Bukidnon. One practice in particular was the planting of sweet potato which is one of their staples. Like many of their counterparts from all over the world, the members of the tribe plant the crop during the full moon, naked. For purposes of academic research, the video capture of such an event may be acceptable and may even be repackaged into a rich media knowledge product. However, uploading this knowledge product to YouTube showing the tribe members unclothed would be ethically indefensible. Based on this, another unintended result of the intervention may be the violation of privacy of IP communities.

Thirdly, non-IP users of indigenous knowledge are prone to prejudice and value judgments. Mainstream cultures have often prejudged indigenous peoples as uncivilized, lazy, unlearned, superstitious, primitive, and dirty (Buasen 2010). A possible outcome of making indigenous practices openly available on the Web that has not been considered was the reinforcement of prejudice among non-IP users.

**Case 3. Championing Non-formal Education Through MIS** Strengthening National Capacity on Planning, Monitoring and Evaluation of Literacy and Non-formal Education Programmes was a 2-year undertaking of the UNESCO Office in Kathmandu in cooperation with the Non-formal Education Centre (NFEC) of the Nepal Ministry of Education and Sports (MOES). It was designed to directly contribute to the implementation of the National Education for All Action Plan 2003–2015, which explicitly identified Non-formal Education (NFE) as one of the four priority areas requiring strategic interventions. The main intervention of this project was the introduction of an NFE management information system and the training of its users from the national to the provincial and district levels. The project commenced in December 2009 and ended in June 2011. The author was tasked to conduct the final evaluation of the project.

The evaluation found that the intended outcomes of the project were not achieved. However, the project resulted in four positive unintended outcomes not necessarily associated with capacity development. Firstly, there was an increased awareness and interest on NFE programmes among communities that participated in the project. Adjunct to this observation was the increased commitment and sense of fulfilment among NFE stakeholders. Thirdly, the project resulted in an increased number of NFE knowledge products and content. Fourthly, the project validated the importance and need for an NFE-MIS among policymakers, programme planners, and implementers and renewed their commitment to establish one. None of these four outcomes were identified as immediate or intermediate outcomes in the project document.

## 5 Conclusion

The debate on the impact of the information society on the Global South is ongoing and the jury is still out. In this chapter, we attempted to provide context and perspective to this debate.

The misgivings of development planners, governments, and donors regarding the outcomes of ICT4D projects were reviewed. The contradictions attendant to evaluating ICT4D impacts were articulated. We discussed The Hague framework, described how it came about, and proposed a way forward, which was to construct theories of change for specific projects or programmes.

The chapter presented a generic theory of change that established causal links between attributes of information society and the Millennium Development Goals. It developed indicative ToC maps or pathways to change applicable to eAgriculture, ICT for education, and KM for climate change adaptation.

We capped the chapter by describing a possible downside to the use of theories of change. We said that ToCs may guide us in establishing causal links between ICTs and development in the Global South, but we can get blindsided by them. The parameters that we set in our impact evaluations sometimes act as horse blinders, denying us of our peripheral vision and preventing the full appreciation of the information society phenomenon. At a certain point in the evaluation process, we will have to remove these blinders and explore unintended outcomes or impacts.

**Open Access** This chapter is distributed under the terms of the Creative Commons Attribution Noncommercial License, which permits any noncommercial use, distribution, and reproduction in any medium, provided the original author(s) and source are credited.

## References

Anon. (2010). *Keadaan ketenagakerjaan Agustus 2010* (Official Report No. 77/12/Th. XIII). Jakarta: Badan Pusat Statistik.
Anon. (2013a). *The millennium development goals*. http://ec.europa.eu/europeaid/what/millennium-development-goals. Accessed 9 Jan 2013.
Anon. (2013b). *What is theory of change*? http://www.theoryofchange.org/what-is-theory-of-change/how-does-theory-of-change-work/. Accessed 3 May 2013.
Batchelor, S. (2003). *ICT for development contributing to the millennium development goals*. infoDev. Washington, DC: World Bank.
Batchelor, S. (2006). *Independent assessment of catia (Catalyzing Access to ICT in Africa)*. Unpublished Evaluation Report.
Batchelor, S., & Norrish P. (2005). *Framework for the assessment of ICT pilot projects*. infoDev. Washington, DC: World Bank.
Beal, G. M., Rogers, E. M., & Bohlen, J. M. (1957). Validity of the concept of stages in the adoption process. *Rural Sociology, 22*(2), 166–168.
Buasen, C. P. (2010). *IKSPs in the midst of engaging IPs*. PowerPoint presentation.
Clarke, H. (2004). *Getting from here to there: Creating roadmaps for community change*. ActKnowledge. PowerPoint presentation.
Clouse, M. (2011). *Theory of change basics*. ActKnowledge. PowerPoint presentation.
Flor, A. G. (2005). Social capital and the network effect: Implications of China's eLearning and rural ICT initiatives. In *Building eCommunity centers for rural development* (pp. 67–80). Tokyo: Asian Development Bank Institute and UNESCAP.
Flor, A. G. (2008). *Scoping report: ICT and rural livelihoods* (Southeast Asia and the Pacific). IDRC.
Flor, A. G. (2013a). Exploring the downside of open knowledge resources: The case of indigenous knowledge systems in the Philippines. *Open Praxis. Journal of the International Council for Open and Distance Education, 5*(1), 75–80.
Flor, A. G. (2013b). Economics of open, distance and eLearning for the Indonesian maritime sector. *ASEAN Journal of Open and Distance Learning, 4*(2), 51–57.
Gonzalez-Flor, B. P., Baguinon, N. T., Concepcion, R. N., Flor, A. G., Reaño, C. E., & Sison, E. C. (2013). *Climate change adaptation among farm families and stakeholders: A toolkit for assessment and analysis, Ver 2.0* (World Bank Philippine Climate Change Adaptation Project, pp. 10–15). Asian Institute for Development Studies Inc.
Heeks, R., & Molla, A. (2009). *Impact assessment of ICT for-development projects: A compendium of approaches* (Paper No. 36. Development Informatics Working Paper Series, pp. 10–105). Institute for Development Policy and Management, University of Manchester.
IICD. (2006). *ICTs for agricultural livelihoods. Impact and lessons learned from IICD supported activities*. The Hague: IICD.
Makoto, A., Flor, A., & Isogai, K. (2011). Strategic options for course delivery and curriculum improvement for deck and engine departments of Indonesian maritime academies. In E. Asaka, M. Tesoro, & W. Padang (Eds.), *Maritime education and training improvement*. Jakarta: Ministry of Transportation.
Quebral, N. C. (1973). Development communication. In J. F. Jamias (Ed.), *Readings on development communication* (pp. 36–60). Los Baños: University of the Philippines Los Baños Press.
Rogers, E. M. (1962). *Diffusion of innovations*. Glencoe: Free Press.

# A New Set of Questions: ICT4D Research and Policy

Andrea Ordóñez

There is a growing interest within researchers to find ways for their work to be relevant to society. The possibility of influencing policy is one option to catalyse change in the use of information and communication technologies for development (ICT4D). This paper proposes a set of questions to aid the ICT4D community in exploring the complexity of the policy processes where their research could be of use. The final aim is to better inform all stakeholders by understanding the context where they participate in policymaking. The main argument is that influencing policy requires intent from the onset of a research project and not only ex post communication strategies. After all, not all research can or should influence policy. In the case of ICT4D, the review of the existing literature shows that policy has not been an explicit area of interest in the domain due to the notions of "policy" and "development" that prevail. The framework developed in this chapter is aimed at allowing the research community interested in policy impact to take into consideration aspects of the policymaking process and to not only communicate results wisely but also identify meaningful and timely research questions and their connection with policies and pinpoint appropriate methods.

## 1 From the Ivory Tower to the Wild Policy Arena

There is a growing interest within the research community on its link with the broader society and its actual relevance in the search for answer to complex issues. It thus has become important for researchers to demonstrate their impact on various arenas and through a myriad of means (i.e., quantitative or qualitative evaluations

A. Ordóñez (✉)
Independent Scholar, Fatima OE2-102 y Guapan., Quito 170312, Ecuador
e-mail: andrea@andreaordonez.org

or rankings). The perspective of universities and other research institutions as ivory towers is challenged, and new visions of more active and engaged scholarship are being developed.

These concerns on the purpose and relevance of research motivate researchers to reconsider their role in a broader context of society and how they carry out their work. Although these reflections brought by this broader approach to research might be positive, a possible downturn is the appearances of "silver-bullet" solutions that arguably allow researchers to become more relevant with simple and concrete steps without significantly changing their work or their relation to society. These solutions to increasing the impact of research are common in the form of guidelines, tips, and step-by-steps that, arguably, correct the problem. This approach to increasing the relevance of research may oversimplify influence to the marketing of ideas, communication strategies, and other chores to make "my" research accepted. Without disregarding the relevance of work carried out to help researchers become better communicators and discussants, these solutions are likely to be only partially successful if a more complete analysis of the contexts and the research itself is not carried out along the way.

The concept of ICT4D in itself includes impact when it states "for development" in its name. This is why it is not surprising that there is a relevant discussion about the connection between the research carried out and its own impact on development. After all, as Díaz Andrade and Urquhart (2012) state, there is a belief in both ICT4D practitioners and researchers that ICTs can change lives for the better. Within a wider perspective of research's impact on development, there is a subset of questions pertaining to the impact of the literature on policy. The logic behind this line of inquiry is that policies are one way in which the expected catalysing power of ICTs can be realised. As a result, there is a growing interest within the ICT4D community to inform policy, to find ways to measure, and to evaluate such impact.

International agencies clearly stated such objectives in their calls for research and their financing priorities. The "ICT4D Grants Programme" carried out by the Nairobi University clearly stated that the research dissemination will be aimed at reaching decision makers. DFID's and IDRC's joint "ICT for Development (ICT4D) Research and Capacity Development Programme" stated policy dialogue as one of its objectives. They summarised it by stating that they aimed at an "ongoing, evidence-based dialogue among regulators, policy makers, researchers, civil society and the private sector; leads to well informed decision making on policy issues relevant to ICT4D". Sinha et al. (2012), on a reflection on the SIRCA programme, also concurred that its objective was rigorous and relevant from the onset, with a strong focus of a transition from research to practice or policy.

This showcase of initiatives portrays a general sense of urgency to reach out to the policymaking communities to allow research to achieve its potential impact. Although this focus on promoting relevant knowledge is powerful, it can motivate two notions that may negatively affect the goal of linking research and policy: first, that all research can influence policy and second, that the dilemma lies solely on the communication and diffusion strategies carried out for research. This chapter and proposed framework challenge these two premises.

First, it challenges the premise that all of the ICT4D research can and should inform policy. As it will be discussed later on, ICT4D has traditionally been framed through questions that do not necessarily inform policymakers in a direct instrumental manner. This is not a negative trait of this research per se, since research might have a broader impact on society than just on policymakers such as informing practitioners and even on each individual's actions directly. Research is a complex task, and freedom should be given for researchers to take on questions that are relevant to a variety of audiences.

Furthermore, the aim of impacting policies that may affect the whole populations should not be taken lightly. There are dangers from expecting all research to inform policy and for that to be a rule for measuring success through short-term gains: being instrumental in bringing about immediate changes in policy. This measurement of success could push researchers to present their results as being much more conclusive than they are in attempt to attract policymakers. In research processes, there is a need for space for inconclusive results, and for further inquiry if necessary, and such characteristic must be acknowledged. Researchers might also lose interest on research that seems to be relevant in the long term and in the possibility of "enlightling" policy processes in the long term with new concepts and frameworks (Weiss 1977).

The second premise this paper challenges is that impact is not a matter of the research process but mainly a matter of the researchers' communication abilities and skills. Aligned with the previous premise, if all research can influence policy, the issue is not the research itself but the communication processes carried out afterwards. Until now, the issue on whether research reaches policymakers has been generally analysed ex post. This means that research is expected to be carried out within the usual academic parameters and, later, be communicated and packaged in ways in which it can ease "uptake" by policymakers. As a result of this ex post perspective, most of the debate on the impact of research on policy refers to the aspects of dissemination and policy engagement once the research has been carried out (Lewin and Patterson 2012). This perspective has led to a growing marketing-style communication model based on a linear model where researchers produce and policymakers consume knowledge (Correa and Mendizabal 2011).

This chapter is based on the conception that not all research should inform policy, but rather, a subset of research should inform policy, while others inform other researchers, practitioners, and technology users. Chib and Harris (2012) have developed a typology of research impact. This typology includes impact on the research community and impact on the wider society. Within this latter category, three possible impacts are identified: capacity development, socioeconomic benefits, and policy impact. This chapter acknowledges the existence of all these impacts but focuses on the last one.

Within the subset of research that is prepared to inform policy, the strategies to accomplish this goal should not start once the project is finished, but before it begins. The framework developed in this chapter is aimed at allowing the research community interested in policy impact to take into consideration aspects of the policymaking process and to not only communicate results but identify meaningful and timely research questions and their connection with policies and pinpoint relevant methods.

The objective of such framework is twofold. First, it seeks to provide a new set of lenses to analyse the body of knowledge formulated in the ICT4D realm. The second objective of this framework is to assist researchers in framing research questions and projects that deliberately link empirical and theoretical research with policy dilemmas from the onset. Ultimately, the objective is not to bridge a "gap" between research and policy after the research has been carried out but to invite researchers to take into consideration policy and political dimensions before carrying out a given project. As O'Neil (2005) proposes, the first requirement for policy influence to occur is intent. Researchers should be interested in working on policy issues. Then again, once this intent is identified, how should researchers approach the challenge? The following critique and framework seek to shed light on that path.

The chapter first explores the existing frameworks to understand ICT4D research in order to have a wide perspective of the research available in the domain. Subsequently, the paper explores the notions of two concepts in the domain: development and policy. This analysis sheds light on why policy has not been a central aspect of the ICT4D domain. Finally, a framework is introduced which sets the scene to explore the policy context and its link with research carried out in the policy domain.

## 2 Existing Frameworks to Analyse ICT4D Research

ICT4D research has been under self-scrutiny since its inception. As a result, a variety of authors have focused on finding ways to conceptualise, find categories, and identify gaps in research. An overview of these existing categorisations and reviews reveals the underlying assumptions in ICT4D research.

Walsham and Sahay (2006) want to make sense of the landscape of literature on ICTs and development categories that could also guide a future research agenda. Their study concludes that this area of research has matured since 2000 when their survey began in terms of theories, methodologies, and results. In the survey, they are able to classify research into four major foci of inquiry. The first line of work centres on the contribution of ICTs to development. Within this category is the work related to the link between technologies and economic and social development in specific countries or domains. The second line seeks to understand cross-cultural working through the use of ICTs. Articles in this category pinpoint the challenges of collaborating internationally and transferring technologies. A third category of work focuses on local adaptation of technologies. How this adaptation takes place, the role of globalisation, and the challenges faced by those who act as brokers in these processes are some of the questions which articles in this category respond to. Finally, the fourth category of research focuses on particular groups which they describe as those "outside the margin of the digital divide" or those that have the least contact with technology.

Although some specific topics in these four categories might be of interest for specific policy processes, the categorisation and examples provided suggest that researchers do not include policy as a significant concern in their inquiry process. As the authors conclude, "topics and issues in developing countries are normally deeply intertwined with issues of power, politics, donor dependencies, institutional arrangements, and inequities of all sorts. These are precisely the type of issues where critical work can open up the 'blackbox' as an aid to deepen understanding, and a stimulus to appropriate action" (Walsham and Sahay 2006: p. 13).

Brown and Grant (2010) simplify Walsham and Sahay's (2006) model by summarising it in two broad categories: ICT *for* development and ICT *in* developing countries. This categorisation and the survey of 184 articles in peer-reviewed journals conclude that there is an over-representation of research on the "ICT in developing countries" rather than the "ICT for development" category. The authors identify that there is a mismatch between the goals of research questions and the expected goal of impacting development. Is it correct to assess research that is focused on understating technology within developing countries contexts by whether they create or promote more development even if that is not the way it is framed, the authors ask (Brown and Grant 2010). They expose the mismatch between the research questions and the public perception of what ICT4D should achieve. Brown and Grant (2010) thus support Heeks' (2002, 2007) perspective of a disconnection between ICT4D and development studies from a more theoretical perspective. These classifications, however, do not include a perspective of policy as a vehicle or development or a clear category of work linked to political research questions.

Another categorisation is Avgerou's (2008) proposal that focuses on information systems' innovation in developing countries. She analyses the discourses behind ICT innovations in developing countries. She identifies three discourses. The first one assumes that the issue is "catching up", which acknowledges a country divide must be bridged by the adoption of existing technologies from the developed world. A second discourse assumes that the issue is constructing new technologies for the different contexts. This suggests a view that technologies must be embedded in societies. The third discourse is concerned with creating the possibilities for technologies to become significant catalysts for change in the lives of people. One could argue that the first two discourses are related to what Brown and Grant (2010) called "ICT in developing countries" and the third one is related to what they called "ICT for development research". One conclusion that Avgerou (2008) arrives at is that, in this field of research, there is rarely any engagement with macro-political analysis, a required aspect of inquiry especially when discussing the transformative power of ICTs.

From the perspective of assessing ICT in development, Heeks (2009) constructs a model that links technologies with development through a chronological categorisation of issues: readiness, availability, uptake, and impact. He calls the first two foci—readiness and availability—ICT4D 1.0, the early agenda of infrastructure, digital divide, and supply of services. ICT4D 2.0 includes the other two categories: uptake and impact. He argues that this progression is necessary to reframe the poor and, instead of situating them on the margin of technology, to put them in the centre. His vision implies that the progression of the ICT4D domain must move towards what has previously been called "ICT for development", to be able to track and have strong evidence on the impact of ICT on development. In this vision of the work carried out in ICT4D, little is said on the policy and political aspects of the impact of technology on development. Policymakers are portrayed as receivers and implementers of externally created knowledge and options, where the focus of interventions is strengthening their capacities rather than approaching them as decision makers within a political context.

As a synthesis, the current reviews of research on the ICT4D domain show that its concern has shifted towards understanding, distilling, and interpreting the D in its name: development. This change has not included a systematic line of inquiry on policy or politics which are absent from the reflections of what ICT4D is and how it impacts development. Although there are some exceptions, they are rare and have not become a solid category in any of the reviews analysed. An exception, for example, comes from the field: Hilbert (2012) shares a conceptual though practical framework that intertwines the policy and technological and social aspects of what he calls the transition towards the knowledge society. As a conceptual model, it is a tool to understand changes, plan interventions, and research priorities and has been used by the United Nations Regional Commission for Latin America and the Caribbean (UN-ECLAC) on planning and studying policies at different levels of government in the region.

## 3 Notions of Development in ICT4D

As the ICT4D domain shifts towards understanding the impact of ICT on development, revising how development is understood will shed light on the apparent disconnection between current research and policy. It seems as though the glue that holds ICT4D together is the premise of a catalysing effect of ICT in development (Avgerou 2008), although within the field there are a variety of conceptions of what "ICTs" and "development" mean. This is a view that has been constructed over time. As Avgerou (2008) recalls, in a panel in 1997, the notion of information systems in developing countries was analysed with positions that ranged from the untapped market conception to the ethical imperative of such research. Since then, others such as Heeks (2009), Walsham (2013), and Avgerou (2010) have supported the view to focus more on development. Despite this assessment of a detachment from development outcomes, or the lack of an explicit definition of development in research projects, authors have notions of development in their work. These notions of what and how development is achieved may have affected the possibility of its applicability on policy contexts.

Heeks (2009) depicts three concepts through which development is understood in ICT4D literature: economic growth, sustainable livelihoods, and freedom. Within the freedom perspective, Sen's capability approach has gain traction in the ICT4D community. In this approach, development is primarily achieved through the direct interaction of the individual with technology. Consequently, the relationship between the individual and the ICTs is prioritised over the broader context, which explains the lack of explicit reference to policies or politics in this framework.

Beyond the concept that is used to define development, there are underlying conceptions of development in ICT4D research. I would argue that the overarching characteristic of the research carried out so far is that it is developed from an external or foreign perspective. As Coward (2007) analyses, in the case of Asia, there is an over-representation of external researchers in the field. In addition to the number of researchers involved, the frameworks used are many a time also external. Furthermore, as Traxler (2012) reflects, ICT4D is described in terms of north and south. By analysing the challenges of a research project in Cambodia, he concludes that this dichotomy makes it difficult for researchers in the south to conceptualise their own experience. This external perspective has three main characteristics. Firstly, development is viewed as project-based interventions, and thus, research reflects on the concrete experiences of those specific cases. Secondly, development is understood from the modernity of western countries. Lastly, development is considered an apolitical endeavour.

The project-based view is observed through the research carried out as well as the domain's reviews. Such is the relevance of specific projects that Heeks and Molla (2009) carried out a compendium that covered an extensive variety of evaluations of ICT4D projects, with a variety of methodologies employed. The one characteristic of this review is the sense that they evaluate specific projects. Scholars have been particularly concerned with the failure of ICT4D interventions (Avgerou and Walsham 2000), and questions have been raised on whether academicians may be failing to provide adequate and relevant research to the practitioners. As a result, there has been a need for reflection on frameworks for successful ICT4D projects (Heeks 2009) that include aspects of governance, design, and sustainability. The inquiry on ICT projects has been strongly focused on identifying "what works". Without disregarding the validity of this question, a wider perspective on the contexts and causalities for failures and success could be identified. For instance, Chib et al. (2012) have proposed adding stakeholder perspective to the analysis of the project including practitioners, researchers, policymakers, and donors. This is indeed a more holistic perspective on project implementation.

In the case of ICT4D research in Africa, Thompson and Walsham (2010) find the same trend: "point" implementation of projects instead of strategic engagement with broader issues at the societal level. As a result of this project-based approach, the need for evidence of success from projects is specially aimed at the international development community and, within it, international aid agencies that financed many interventions (Heeks 2009). Although some countries in the developing world are still dependent on foreign aid, there is a growing understanding that the priorities of donors are not the same as policymakers and thus should not be treated as the same audience.

The second underlying characteristic is that development is usually aligned with a view of modernity and is the link to the global world (Díaz Andrade and Urquhart 2012). This means that, for the most part, ICT4D researchers and practitioners bring a view of development as the one of the western societies. In an analysis of the development discourse adopted by Internet scholars in India and China, Zhang and Chib (2014) identify that in India, the modernisation discourse is dominant and in China, it is steadily growing. In the case of India, the authors also note the relevance of a technocratic perspective and the focus on achieving goals such as economic growth, industry development, and governance. This external view of development does not acknowledge that development is a process both at the social and political spheres where interests, positions, and view must be confronted, discussed, and agreed upon (or not). In other words, development is understood as a goal or, in research jargon, as a dependent variable and not as a destination involving process, negotiation, and trade-offs. Nonetheless, this bias has led to the emergence of different approaches. As Flor (2012) has identified, a more critical theory tradition is also present in ICT4D studies, probably as a response to this existing modernity bias. Accordingly, participatory and action research methods have gained relevance.

Lastly, development is seen as apolitical, and in this context, ICTs are tools that bypass politics reaching the community or individual directly through project interventions. Others have arguably conceived ICTs impact for development mainly through a market system (Avgerou 2003). At the end of the day, this view, aligned with the applications of Sen's capability approach, yields an understanding of development primarily as a personal or grass-roots process that can be achieved in spite of the broader political context. Circumventing the discussion of politics, however, hides the power struggles and the unequal distribution of benefits of the introduction of ICTs in developing countries. Furthermore, a lack of understanding on the politics of ICTs gives the ICT4D community little knowledge of the incentives behind the success, failures, and scalability of projects being implemented. Politics is also a high component of what the context is. Although various authors suggest that context should be taken into consideration both in the implementation and the research of ICT4D projects, these are seen superficially at the most.

The underlying notions of development in the ICT4D research domain might be a reason why there is an apparent disconnection between the work carried out in the research domain and policy processes.

## 4 Policy in ICT4D Research

While, as debated in the previous two sections, the discussion of what development is within the research field has gained momentum, the discussion on policies has lost traction. As portrayed by Heeks (2009) in the evolution of the research domain from ICT4D 1.0 to ICT4D 2.0, the broad issues of policy were considered in the former.

Topics such as infrastructure or service supply were studied when the research field was starting. Heeks (2009) further criticises the research carried out in these topics as one of a "menu" that established rules and regulations that policymakers could choose from, with little regard for appropriateness or implementation.

Another set of possible questions are those related to the political economy of ICT promotion and adoption. The evolution of the research in the ICT4D toward a search for impact on development and the self-inquiry on development has overshadowed the relevance of ICT policies which are a strong way to actually realise the catalysing effect of ICTs. The predominant view of policy from the ICT4D research perspective has been narrowed to those specific aspects of availability, supply, and other basic requirements for further progress to be made in the specific projects that ICT4D practitioners and researchers implement. Cecchini and Scott (2003), for instance, after examining different cases of ICT initiatives, reflect on the necessary prerequisites for such initiatives to work, including macro-policies to achieve low-cost connectivity. Few of such studies have been identified. Furthermore, the vision of policy in developed countries only as the requirement for successful interventions to be successful further strengthens the perspective that the research domain has been biased toward the implementation of projects rather than policies.

This narrow view, however, contradicts what developed countries carried out and that now other countries are implementing. ICT policies go beyond the availability of the technology and link technology to a bigger picture of changing society to the ideal of the information society or knowledge economy (Hall and Löfgren 2004). Policies are not only statements of what will be done but a narrative of values, perceptions, principles, and aims. In the case of ICT policy, it is not only about availability of technologies, but mostly aspirational statements of how these technologies will allow societies to transform. This means that while most research see ICTs as progressive, policies state disruptive visions of ICTs (Avgerou 2008).

Although policy is mostly absent in the ICT4D research domain, there are individual researchers that have focused on understanding ICT policy, with a focus on developing countries. Kendall, Kendall, and Kan (Kendall et al. 2006), for instance, have analysed discourses in ICT policy debate within online communities. Duncan-Howell and Lee (2008) take a particular case of ICT for education policy and pinpoint the urgency that policymakers in the developing country have, due to a sense of catching up. As a result, the authors conclude that policy processes many times entail finding models from other countries that have succeeded and transferring them to the country in dispute. This, however, may lead to inefficient or even negative policies.

Through the use of different approaches and theoretical frameworks, researchers have examined particular country cases such as Egypt (Stahl 2008), Pakistan (Baqir et al. 2009), India (Dabla 2004), and Bangladesh (Hasan 2012). These cases explore the difficulties of the actual implementation of policies and the gaps between policy objectives and outcomes and the constraining factors or the positive spillovers encountered.

What this area of inquiry has in common is that the relationship between ICT and policy is seen to be sectoral. This means that in the cases described above, the focus area of study is the particular ICT policies. Hafkin (2002) identifies 21 policy issues ranging from networking architecture, technological choices, tariffs, regulations, services, and e-government. As a result, these policies are considered a specialised area of interest. Developing countries might be repeating the issue that Hall and Löfgren (2004) observe in Sweden: the dominance of experts in the field has made it hard for other nonexperts to get interested and to participate in such policies.

Understanding the policy process and the discourses behind policies is relevant for researchers and practitioners who want to influence policy. Without identifying the context in which decisions are being made, the actors involved, and the interests at stake, there is very little opportunity for these topics to change. These questions of how policymaking is actually carried out, however, are not as relevant for policymakers themselves who are actively engaged in the process and know of their workings tacitly. Furthermore, research on how ICTs are incorporated in other thematic policies has not been encountered. Although some insight might be available from those particular disciplines, efforts from the ICT4D perspective to understand the role of ICT4D in other sectoral policies are not explicit. Malapile and Keengwe's (2014) research is an example of such analysis for the education policy debate.

## 5 Challenges for Researchers

Why is it so challenging to change these aspects of the ICT4D realm? It is likely that the researchers face challenges to fit their research in the context of policy debates. As De' (2012) has analysed, the types of research questions carried out in the ICT4D field often face complex scenarios where both theories and methods might require adaptation. He urges researchers to acknowledge the difficulty to work in messy environments. Things get even more complex when a research is trying to frame research questions and projects within a wider political scenario to inform policy changes. Young and Mendizabal (2009) recall some of the main challenges to becoming, what they call, policy entrepreneurs—those that can navigate and alter their policy context. These challenges include: understanding policy changes and its different dimensions, identifying the decisive aspects of the context that require attention, and recognising the factors that cause policy to change or new ones to be adopted. These challenges are likely to be tackled both by practice and involvement in policy processes as well as addressing policy and political questions that can shed light on the process of influencing policy.

## 6 Outlining ICT4D Research for Policy

In this section, I will present a framework for researchers to reflect on how to better approach the challenge of informing policies. I argue for more research which considers politics and policies that is strategic in supporting policymakers. The need for this has already been clearly stated (Avgerou 2008; Walsham 2013; Thompson 2008).

This objective is not something that can be solely achieved through communication strategies but through a change in how research projects are planned and implemented. This does not mean that research should lose its independence from the political powers but that it understands the policy scenario, challenges, and possibilities. As Vialle (1981: p. 315) reminded us, the "purpose of a research project depends on the very real game of interests, on the needs and desires of individuals and groups who play a part in research or gain some benefit from it. From this perspective, the assumed 'neutrality of scientific research' is a myth or, to say the least, an ideal that is difficult to reach".

For researchers who aim at changing or creating new policies, this framework suggests a set of questions that could assist them navigate these policy contexts. As suggested by Chib and Harris (2012), policy influence requires researchers to focus not only on academically interesting questions but policy-relevant questions. Along these lines, I propose for research aimed at influencing policy be framed in an integral way. Firstly, it is essential to understand the relationship between ICTs and policies. Researchers in ICT4D should understand the political context in which they are planning to interact. Secondly, researchers are encouraged to explore beyond the ICT policy realm and also consider sectoral policies in which ICTs can be catalysts of change. Thirdly, researchers should consider not only the academic rational of their research project but how it can link with policy requirements. This entails consideration of research questions that could better fit the requirement to inform policy.

These considerations will place an additional burden on researchers, but it would allow a better link between their work and the policy debates they wish to participate in. Taking into consideration these aspects of policy will enable researchers to produce knowledge that is better suited to enter the policy debates. This, however, is not a silver-bullet solution. Policymaking is a complex endeavour, where ideas are not only validated through peer reviews but by public debates and consensus of a variety of stakeholders with different interests and positions. Research will become only one source of innovation and policymaking in a more complex scenario.

### *6.1 Politics and Policies*

Politics and policy are intertwined, and it is unlikely policy choices can be understood without the politics surrounding them. Therefore, it is relevant to carry out research that can help us better understand the arenas where the decisions regarding ICTs and policies are being carried out, the power struggles behind them, and the opportunities available for policy change. The goal of exploring this topic is to examine policy processes critically to identify and maximise the spaces where research can be influential. A research approach that tackles the politics and the policy aspects is the combination of two dimensions: policy cycle and political context (Ordóñez et al. 2012). While the policy cycle takes on a more rational approach, the political context considers the emotions, interest, and values of the variety of actors involved. Combining these two dimensions of change tackles what Hall (1989) has described as the three factors for policy adoption: policy viability, administrative viability, and political viability.

The policy cycle is a model that depicts policymaking as an ongoing process of stages that keep evolving. It has been criticised for poorly depicting the complex nature of policymaking as overly rational. However, it can be employed as a framework that spells out different stages of policy and, if considered as a flexible framework, portray basic concepts of policymaking: various decision makers and high degree of competition among policy advocates or advisers (Howard 2005). The concept of the policy cycle allows researchers to reflect on how research can be used to set the agenda, define a problem, or facilitate implementation. It is likely that the different stages of the policy cycle require different approaches to research and communication.

It is equally important to understand the locus of the debate since not all policies are decided upon in the same scenario. Grindle (2007), for instance, has made a distinction between reforms that occur in the "political arena" and those that occur in the "bureaucratic arena". In the first, political interests are primordial and changes can be slow but more long lasting; in the second one, "bureaucratic arena" changes are carried out de facto with a focus on implementation and technical viability but with no lasting impact due to changes in staff or reversal due to lack of political support. The extent to which research can be used differs according to the locus of the policy debate as well, and while research might have less relevance in highly political debates, it could be better received by the implementers of policies.

In this sense, understanding where and how ICT innovations and adoptions are being carried out within governments and the champions and coalitions that are enabling these changes would be interesting lines of work. It would also shed light on the distinctions between the actors involved on their take on technology and what they see their role to be. Furthermore, the locus of the policy debate for ICT-related policies can be taken for granted, or can be strategically determined by the actors involved, considering the strengths and limitations of each.

Finally, it is relevant to understand the policy space (Radin 2013) for ICT-related policies. Policies, by definition, are carried out in a world of constraints where implementing one could leave other options out. As discussed by Heeks (2009), a variety of ICT initiatives have proven to be unsuccessful. Combining this unfavourable fact with constraints due to budget allocation issues results in probably little political space for ICT policies. Policy space, however, is not static and can be created when actors work together, frame issues creatively, and are able to pose the subject at hand not only as another issue in a world of constrained budgets but as part of the solution to such issues.

While the discussion on the policy cycle and locus might be more clearly stated, the discussion of the context for policies tends to seem elusive. As Avgerou (2010) has mentioned in the case of ICTs, oversimplifying context as a different "local culture" adds little value on understanding the interactions of people and technology in the developing world. With regard to policy context, it may involve understanding the interaction between policy and research, political systems, electoral processes, structure of governments, and so on.

For the purpose of simplifying our understanding regarding the politics of implementing ICT4D initiatives and the role research can play in the process, I would argue for focusing on understanding the rational and value-driven aspect of policy problems. Hoppe (2010), for instance, looks into two dimensions of a policy problem: on the one hand, the level of certainty regarding relevant knowledge for the policy process and, on the other hand, the level of consensus on relevant norms and values. The first dimension is rational and relates to what is known about the problem at hand and how stakeholders react to such knowledge. Is knowledge valid, trustworthy, and relevant? The second dimension refers to values surrounding the problem and whether stakeholders agree or disagree on how the problem is defined and the values that should guide its solution. This way of thinking about the political context focuses on the relationship among the stakeholders involved in relation to their rational and value-based interpretation of the problem at hand. Some of the questions that could be seen from this perspective involve both the evidence and the value surrounding decisions on the role of the state and the provision and support of ICTs.

Furthermore, it would be interesting to explore the perceptions of stakeholders. For instance, it could be interesting to examine how policy actors see technology, either as something that should be imported from the developed world or constructed locally or the result of the articulation of imported and local knowledge. It could also be interesting to explore if ICTs are seen as a disruptive or progressive force of development (Avgerou 2010), the expected uses of ICTs (Harindranath and Sein 2007), and how they would gauge the success of an ICT policy.

Research in the realm of ICT4D-related policies and its politics is the basis to be able to plan research programmes that can respond to the challenges of public policy. This research allows understanding of the setting where ICT4D research would interact with policy and politics and sheds light on the complexities of policymaking and the adoption of ICTs in public programmes, projects, and regulations.

## 6.2 Beyond ICT Policy

As discussed in previous sections, ICT4D has moved away from thinking about ICT policy partially because of its narrow conception. For this reason, in a new outlook on the relationship between ICT4D research and policy, I suggest considering two types of policies: ICT policies and sectoral policies where ICT4D research can have a catalyst effect.

By ICT policies, I refer here to those policies of infrastructure, access, and availability of ICT that the government puts in place. It refers to what has been previously analysed and what is traditionally considered the arena of proposed changes for those working on ICT and development. This arena is still important since countries constantly delineate and adjust their ICT policies in accordance to the context and the appearance of new technologies.

After the revision of the literature in the ICT4D domain, however, it is salient that its outcomes can inform other policies as well. For example, the work carried out on the impact of access to price information through mobiles (Islam and Grönlund 2007) could inform agricultural systems' policies and other agricultural policies. These sectoral perspectives go beyond the traditional ICT sectoral policy perspective and view ICTs as possible disruptive forces in other fields. Informing these policies might be a way in which the catalysing effect of ICT policy can be realised. The challenge of approaching other policy debates, however, is that it might require sectoral experts who understand the specificity of that given debate.

## 6.3 Knowledge for Policy

The two previous subsections have been focused on setting some guidelines of how to approach the broad questions of politics and policy within the ICT4D and other development studies that face similar concerns. These questions can guide researchers on how to approach the policy process, but it is not research alone that will change policy. This is why this last section examines the types of research that could be carried out.

As stated in the first section of this chapter, this document is based on the conception that not all research can or should influence policy. If such is the case, what are the types of research that become useful in the policy process and how? Vialle (1981) produces a typology of educational research based on its primary objective. As the author summarises, the problem with typologies is that they are not clear-cut categories, but they can help researchers determine their objective and approach their work with more clarity. Based on Vialle's (1981) work, I present five types of research according to their objectives in the policy process: conceptual, planning, implementing, action research, and monitoring. Researches in ICT4D that aim at reaching policymakers could benefit from reflecting on how their work can be used in the policy process before it is launched.

Conceptual research refers to the academic knowledge that explains phenomena, uncaps relationships between different variables, and creates categories and concepts to simplify complex trends. This type of research, often referred to as "blue sky" or "pure" research, is valued in the academic community, but it might be the most distant to policymakers that face day-to-day decisions. Despite this tendency, conceptual research can be extremely important for the development of policy narratives that convey the reasons for decisions being put forward (Bellettini and Ordóñez 2011). In the case of ICT4D, the broad conceptual frameworks in relation to the development process have been discussed before. However, there is space for developing the theory about ICT policies that respond better to the needs of developing countries.

Conceptual research identifies trends and phenomena but does little to identify why or how they happen. Planning research is the category that seeks to explain the factors that cause or hinder a given outcome. This type of research sheds light on what the policy priorities could be and how these could become crucial for the expected outcome. An interesting case of this type of research is Cecchini and Scott's (2003) which prioritises public policies for ICT4D initiatives to be successful. This type of research frames the possibilities of action for policymakers to consider.

Research in the planning category leaves a blueprint of what should be done and in what sequence, at best. The next challenge which policymakers face is actually making a decision and implementing a policy, programme, or such to address the issue at hand. The research for planning refers to the one that identifies the key factors that affect a development outcome. In the case of ICT4D, research that finds key aspects to make a technology useful or an intervention successful would be relevant knowledge for planning. To plan a policy, the debate should focus not only "on what works" but also on the factors for success. Is it the capacity to use a technology, its availability, or its price? These questions allow policymakers to focus, from a myriad of options, on those that could have the most impact. Ty et al. (2012), for example, have discussed the use of ICTs for environmental planning. In their analysis, they conclude that it is not only necessary to integrate more data with the use of technologies but that, for it to be meaningful, changes in the planning process must occur. This research could inform policymakers of the need to change internal processes and not only introduce new technology.

Instrumental research refers to identifying new actions or reforming current existing programmes. It is probably the most innovative aspect of research for policy. This type of research aims at creating solutions. Considering local context, specific needs, and constraints, it creates options. In this sense it is inventive and creative. Many ICT4D projects could be framed as instrumental research. These projects, however, are usually carried out outside the governmental arena. Research that takes into consideration the limitations, possibilities, and requirements for scaling up could support policymakers to view some of these ideas as valuable policy options.

Action research is the fourth proposed category. Its primary focus is changing behaviours or actions through direct interventions. These types of research seek to connect researchers and practitioners directly in solving issues encountered during the implementation of an idea and tweaking issues in the process. These are usually endeavours best carried out in smaller settings with direct and constant interaction between researchers and those involved in the policy (i.e., teachers, bureaucrats, extensionists). For such research to be meaningful, strong links are necessary and researchers become not only observers but participants. In this sense,

**Table 1** Examples of types of questions in different types of policies and research objectives

| Research objective/type of policy | ICT policy | ICT for sector policies |
|---|---|---|
| Conceptual | Models for the regulation of ICT in developing countries, including the costs, issues, and incentives for the expected consequences | Analysis of how ICTs increase capacities in different aspects of development: education, health, and productivity |
| | | Broad conception and categories of the role of ICTs in sectoral policies |
| Planning | Analysis of the social, economic, or political factors that incentivise or hinder the use of ICTs | Implications of the introduction of ICTs in sectoral policies |
| Instrumental | Concrete research that can yield information regarding good mechanisms for governance, pricing, regulation, promotion of services, competition, etc. | Prototypes, technology options for specific policy needs |
| Action research | Pilot programmes of new regulations, prices, or governance structures | Joint implementation of programmes with constant research to shift courses or change policies to achieve a successful implementation |
| Monitoring and evaluating | Evaluation of compliance with the law, quality of services, who are the beneficiaries, and how and whether policies are achieving their expected outcomes or not | Evaluation of the role of ICTs in the sectoral policies and if they are having the expected outcomes |

Gitau et al. (2010)s have pointed to the relevance of the action research approach in ICT4D and the role NGOs can play in them.

The last type of research is for monitoring and evaluating or for impact assessment and seeks to answer the question of whether a policy is delivering on its expected goals. The primarily goal of these initiatives is accountability, focusing on the accomplishments of goals. This type of research might become influential in trying to strengthen a successful policy or eliminate useless ones. This research, however, tends to lack solution or alternatives since its primary focus is defining and measuring success. Many project evaluations would fall in this category. The following table summarises how research questions could be framed in terms of the policies it will inform and the objective of the research process (Table 1).

## *6.4 Actors Involved in ICT4D and Policy Research*

The proposed framework for ICT4D research and policy above encompasses many dimensions of the inquiry needed for sound policymaking. Understanding the politics and policy processes of ICT4D adoption as well as setting out an agenda with an explicit focus on policy is necessary. To cover this wide range of issues and disciplines, various actors could and should be involved. ICT4D already has a tradition of contribution from the perspective of practitioners; similarly, the new set of questions with a policy focus will require a wide participation. Furthermore, ICT4D is also a market, where telecommunication companies compete for their share of customers and their right to operate in given countries thereby adding yet another layer of complexity.

ICT4D and policy research is not an arena that will be successfully covered by academia alone. The interface between policy and research is complex, with more actors participating in the knowledge production process including NGOs, think tanks, government research department, consultants, and others (Young 2005). A wider analysis of the knowledge that is being generated in this arena would not only require a review of the work present in journals but also in grey literature that involves research that does not appear in the usual venues. This further analysis can depict the existing knowledge and the gaps of ICT4D and policy research. Furthermore, as Chib et al. (2012) have suggested, it is important to understand the interactions among the various actors in ICT4D. This work should not only be carried out as an ex post analysis but an ex ante evaluation to determine power structures and struggles that may allow or prevent research from being used.

## 7 Conclusion

This chapter has critically analysed the knowledge production within ICT4D where the concepts of policies and politics have not been specifically considered. The variety of existing literature reviews point to the diversity of research that has emerged in the field but also acknowledge a lack of focus on the power struggles and the decision-making processes surrounding policies related to ICTs. An overemphasis on proving a link between ICT and development has overshadowed other research that focuses on finding policy options and understanding the factors that may affect them and successfully implement them.

The proposed framework seeks to challenge the external vision of development currently mainstreamed in ICT4D and proposes to embed research not only broadly in the local context but most importantly in the political context. This implies taking a critical view of both the politics and policy aspects of ICT not only in the ICT policy realm but also in other sectoral policies' debates. Furthermore, it argues for framing projects not only in the context of expanding the field's knowledge but from the perspective of policy choices and political constraints. Widening the space of research on ICT4D and policy questions also implies the inclusion of other actors whose research is not always published in international journals or, for that matter, on the specific ICT4D-related journals. An exercise of a wider sample of sources is suggested to better understand all research involved in the process of informing policymaking.

Research, as has been described throughout this chapter, has political implications; it can set new agendas, change the way problems are depicted, and shed light on its solutions. The proposed set of questions is an attempt to acknowledge this in order to help researchers navigate the political contexts they participate in. By spelling out the motivations of policymakers, the drive of researchers, the complexity of the context, and the types of policies being changed, researchers are better equipped for entering a political debate. Researchers, however, are well advised to recognise the variety of reasons why policies are being carried out, including political and economic benefits for certain groups. In this context, research is one aspect where many others are intertwined.

**Open Access** This chapter is distributed under the terms of the Creative Commons Attribution Noncommercial License, which permits any noncommercial use, distribution, and reproduction in any medium, provided the original author(s) and source are credited.

## References

Avgerou, C. (2003). The link between ICT and economic growth in the discourse of development. In *Organizational information systems in the context of globalization* (pp. 373–386). New York: Springer.
Avgerou, C. (2008). Information systems in developing countries: A critical research review. *Journal of Information Technology, 23*(3), 133–146.
Avgerou, C. (2010). Discourses on ICT and development. *Information Technologies and International Development, 6*(3), 1.
Avgerou, C., & Walsham, G. (2000). *Information technology in context: Implementing systems in the developing world*. Brookfield: Ashgate Publishing.
Baqir, M. N., Palvia, P., & Nemati, H. (2009). *Evaluating government ICT policies: An extended design-actuality gaps framework* (Vol. 14). In Second annual SIG GlobDev workshop, Phoenix.
Bellettini, O., & Ordóñez, A. (2011). *Translating evidence into policies: Two cases in Ecuador* (Working Paper, 1) (pp. 1–19). http://www.grupofaro.org/sites/default/files/archivos/publicaciones/2012/2012-05-11/wp1_evidence_and_policy.pdf
Brown, A. E., & Grant, G. G. (2010). Highlighting the duality of the ICT and development research agenda. *Information Technology for Development, 16*(2), 96–111.
Cecchini, S., & Scott, C. (2003). Can information and communications technology applications contribute to poverty reduction? Lessons from rural India. *Information Technology for Development, 10*(2), 73–84.
Chib, A., & Harris, R. W. (Eds.). (2012). *Linking research to practice: Strengthening ICT for development research capacity in Asia*. Singapore: Institute of Southeast Asian Studies.
Chib, A., Ale, K., & Lim, M. A. (2012). Multi-stakeholder perspectives influencing policy-research-practice. In A. Chib & R. Harris (Eds.), *Linking research to practice: Strengthening ICT for development research capacity in Asia*. Singapore: ISEAS Publishing.
Correa, N., & Mendizabal, E. (Eds.). (2011). *Vínculos entre conocimiento y política: el rol de la investigación en el debate público en América Latina*. Lima: Universidad del Pacifico/CIES.
Coward, C. (2007). *Research capacity in Asia: A literature review in the Information Society*. Paper presented at Workshop on Research in the Information Society, 25–27 April 2007. Philippines.
Dabla, A. (2004). The role of information technology policies in promoting social and economic development: The case of the State of Andhra Pradesh, India. *The Electronic Journal of Information Systems in Developing Countries, 19*, 1–21.

De', R. (2012). Messy methods for ICT4D research. In A. Chib & R. Harris (Eds.), *Linking research to practice: Strengthening ICT for development research capacity in Asia.* Singapore: ISEAS Publishing.

Díaz Andrade, A., & Urquhart, C. (2012). Unveiling the modernity bias: A critical examination of the politics of ICT4D. *Information Technology for Development, 18*(4), 281–292.

Duncan-Howell, J. A., & Lee, K-T. (2008). *Policy catch up: Developing nations and developing ICT policy documents.* In: Proceedings British Educational Research Association (BERA) annual conference 2008, Herriot-Watt University, Edinburgh.

Flor, A. (2012). ICTD PRAXIS: Bridging theory and practice. In A. Chib & R. Harris (Eds.), *Linking research to practice: Strengthening ICT for development research capacity in Asia.* Singapore: ISEAS Publishing.

Gitau, S., Diga, K., Bidwell, N. J., & Marsden, G. (2010). *Beyond being a proxy user: A look at NGOs' potential role in ICT4D deployment.* Proceedings of the Information and Communication Technology and Development Conference (ICTD'10), 13–16 December 2010. London.

Grindle, M. S. (2007). Good enough governance revisited. *Development Policy Review, 25*(5), 533–574.

Hafkin, N. (2002). *Gender issues in ICT policy in developing countries: An overview.* Expert Group Meeting on "Information and communication technologies and their impact on and use as an instrument for the advancement and empowerment of women" 11–14 November 2002. Seoul, Republic of Korea.

Hall, P. A. (1989). *The political power of economic ideas: Keynesianism across nations.* Princeton: Princeton University Press.

Hall, P., & Löfgren, K. (2004). The rise and decline of a visionary policy: Swedish ICT-policy in retrospect. *Information Polity, 9*(3), 149–165.

Harindranath, G., & Sein, M. K. (2007, May). *Revisiting the role of ICT in development.* In Proceedings of the 9th international conference on Social Implications of Computers in Developing Countries, São Paulo.

Hasan, M. (2012). *Social equity and integrity through ICT: A critical discourse analysis of ICT policies in Bangladesh.* Doctoral dissertation, Uppsala University.

Heeks, R. (2002). Information systems and developing countries: Failure, success, and local improvisations. *The Information Society, 18*(2), 101–112.

Heeks, R. (2007). Theorizing ICT4D research. *Information Technologies and International Development, 3*(3), 1–4. Cambridge: MIT Press.

Heeks, R. (2009). *The ICT4D 2.0 manifesto: Where next for ICTs and international development?* Manchester: University of Manchester Institute for Development Policy and Management (IDPM), Development Informatics Group.

Heeks, R., & Molla, A. (2009). *Impact assessment of ICT-for-development projects: A compendium of approaches.* University of Manchester Institute for Development Policy and Management (IDPM), Manchester.

Hilbert, M. (2012). Toward a conceptual framework for ICT for development: Lessons learned from the cube framework used in Latin America. *Information Technologies and International Development, 8*(4), 243.

Hoppe, R. (2010). From 'knowledge use' towards 'boundary work': Sketch of an emerging new agenda for inquiry into science-policy interaction. In R. J. in't Veld (Ed.), *Knowledge democracy: Consequences for science, politics, and media.* Heidelberg: Springer.

Howard, C. (2005). The policy cycle: A model of post-Machiavellian policy making? *Australian Journal of Public Administration, 64*(3), 3–13.

Islam, M. S., & Grönlund, Å. (2007). Agriculture market information e-service in Bangladesh: A stakeholder-oriented case analysis. In *Electronic government* (pp. 167–178). Berlin/Heidelberg: Springer.

Kendall, K. E., Kendall, J. E., & Kah, M. M. (2006). Formulating information and communication technology (ICT) policy through discourse: How internet discussions shape policies on ICTs for developing countries. *Information Technology for Development, 12*(1), 25–43.

Lewin, T., & Patterson, Z. (2012). Approaches to development research communication. *IDS Bulletin, 43*, 38–44. doi:10.1111/j.1759-5436.2012.00361.x.

Malapile, S., & Keengwe, J. (2014). Information communication technology planning in developing countries. *Education and Information Technologies, 19*(4), 691–701.

O'Neil, M. (2005). What determines the influence that research has on policy-making? *Journal of International Development, 17*(6), 761–764.

Ordóñez, A., Bellettini, O., Mendizabal, E., Broadbent, E., & Muller, J. (2012). *Influencing as a learning process*. In Think Tank Exchange Conference. Working paper for the Think Tank Initiative, South Africa, June 18–20, 2012.

Radin, B. A. (2013). *Beyond Machiavelli: Policy analysis reaches midlife*. Washington, DC: Georgetown University Press.

Sinha, C., Elder, L., & Smith, M. (2012). SIRCA an opportunity to build and improve the field of ICT4D. In A. Chib & R. Harris (Eds.), *Linking research to practice: Strengthening ICT for development research capacity in Asia*. Singapore: ISEAS Publishing.

Stahl, B. C. (2008). Empowerment through ICT: A critical discourse analysis of the Egyptian ICT policy. In *Social dimensions of information and communication technology policy* (pp. 161–177). New York: Springer.

Thompson, M. (2008). ICT and development studies: Towards development 2.0. *Journal of International Development, 20*(6), 821–835.

Thompson, M., & Walsham, G. (2010). ICT research in Africa: Need for a strategic developmental focus. *Information Technology for Development, 16*(2), 112–127.

Traxler, J. (2012). The challenge of working across contexts and domains: Mobile health education in Rural Cambodia. In A. Chib & R. Harris (Eds.), *Linking research to practice: Strengthening ICT for development research capacity in Asia*. Singapore: ISEAS Publishing.

Ty, P. H., Heeks, R., & Choung, H. V. (2012). Integrating digital and human data sources for environmental planning and climate change adaptation. In A. Chib & R. Harris (Eds.), *Linking research to practice: Strengthening ICT for development research capacity in Asia*. Singapore: ISEAS Publishing.

Vielle, J. P. (1981). The impact of research on educational change. *Prospects: Quarterly Review of Education, 11*(3), 313–325. Springer.

Walsham, G., & Sahay, S. (2006). Research on information systems in developing countries: Current landscape and future prospects. *Information Technology for Development, 12*(1), 7–24.

Walsham, G. (2013). Development informatics in a changing world: Reflections from ICTD2010/2012. *Information Technologies & International Development, 9*(1), 49.

Weiss, C. H. (Ed.). (1977). *Using social research in public policy making* (Vol. 11). Lexington: Lexington Books.

Young, J. (2005). Research, policy and practice: Why developing countries are different. *Journal of International Development, 17*(6), 727–734.

Young, J., & Mendizabal, E. (2009). *Helping researchers become policy entrepreneurs*. London: Overseas Development Institute.

Zhang, W., & Chib, A. (2014). Internet studies and development discourses: The cases of China and India. *Information Technology for Development, 20*(4), 324–338.

# Progress Towards Resolving the Measurement Link Between ICT and Poverty Reduction

**Julian May and Kathleen Diga**

This chapter provides a review on the debate and latest literature around Information and Communication Technologies (ICTs) and its connection to poverty. The review first acknowledges the trend of global poverty, which today can be measured in a multitude of dimensions. This multidimensional poverty measurement approach has emerged within ICTs and Development (ICTD) research alongside a new contribution called "digital poverty". When looking at the empirical linkages between the concepts of poverty and ICTs, the literature reveals heterogeneity in the measurement choices as to who are the poor and whether the poor have ICTs across developing countries. Yet in various cases where the poor have ICTs, some are found to be sensitive to changes of price and see variability within equity of affordability. Furthermore, only few studies have been able to show causal inference to make the micro-level impact linkage between ICTs and poverty. In reviewing this literature, we provide some of the major themes, gaps, and recommendations towards improving the understanding of ICTD and poverty.

## 1 Introduction

In January 1961, the United Nations (UN) declared its first "decade of development", focusing on the increasing growth rate of aggregate national income in developing countries while recognising the need to provide some benefit to the

---

J. May, Ph.D. (✉)
Institute for Social Development, University of the Western Cape, Bellville, Western Cape, South Africa
e-mail: jmay@uwc.ac.za

K. Diga
School of Built Environment & Development Studies, University of KwaZulu-Natal, Durban, South Africa
e-mail: digak@ukzn.ac.za

© The Author(s) 2015
A. Chib et al. (eds.), *Impact of Information Society Research in the Global South*,
DOI 10.1007/978-981-287-381-1_5

poorer sections of the population. Commenting on the poor record of this first decade of development in 1970, Robert McNamara, then president of the World Bank Group, argued for a "... whole generation of development that will carry us to the end of the century" (cited in Meier 1970: p. 4). In the decade to follow, another World Bank President, Alden Clausen, stated that "... a key and central aim of the World Bank is the alleviation of poverty" (World Bank n.d.), while in 1980, 1990 and again in 2000 and 2001, "Poverty" was within the title of the World Development Reports (World Bank 1980, 1990, 2001). At the start of the fifth decade after President Kennedy's inaugural address, yet another World Bank President, James Wolfensohn, emphasised the need to "... create an environment in which you can ... give opportunity and empowerment and recognition to people in poverty" (Wolfensohn 2000). Lending support to these statements, numerous international declarations have been made since the General Assembly's resolution 1710 (XVI) of 1996 committing most countries in the world to a range of laudable goals, all of which are appropriate if poverty is to be eliminated. Of these, the United Nations Millennium Declaration in 2000 and the commitment by 189 countries to the eight Millennium Development Goals (MDGs) were especially noteworthy. The expiry of the MDGs occurs in 2015, and current reflection is being made as to whether much has changed in the reduction of poverty since the first declaration over 50 years ago.

After over five decades of sentiments, there has been evidence of dramatic shifts in global poverty. The USD 1.25-a-day absolute poverty rate in 2012 was 19 % (or 1.1 billion people) compared to the previously high rates of 43 % in 1990 (Ravallion 2013). This lower rate of global poverty has been the result of dramatic decline in China's poverty levels as well as steeper poverty declines among other developing countries (Ravallion 2013). For some, however, lower global poverty rates are not sufficient. The ongoing and persistent levels of poverty must be addressed, yet it remains active within many regions of the world, especially in sub-Saharan Africa. The UN has committed to driving the rate of extreme poverty to 0 % by 2030 (United Nations 2013). This ambitious goal of poverty eradication has been supported by ongoing global changes. This includes the mix of improved economic growth policies in developing countries, dramatic gains in human capital both in terms of health and education and the roll out of government social policies such as cash transfers.

Contestation however is widespread as to which policy strategy mix would effectively tackle poverty eradication. In a world shrouded with the global financial crisis and a wide variety of economic and social programmes, one can be left uncertain as to the most effective way forward for the end of the poverty. In India, researchers Drèze and Sen (2013) seek continual improvements of social welfare programmes to uplift the poor, while Bhagwati and Panagariya (2013) concentrate on market deregulation as the growth solution to end poverty. In such contexts, countries are left with decisions to develop their most appropriate policy combinations for future long-term growth (Rodrik 2013), using their evidence-building tools of measurement.

During this period, developing countries are also experiencing dramatically improved access and use of Information and Communication Technologies (ICTs). Among these policy decisions, one may ask to what degree should ICTs be considered within the development policy mix. While some argue uncertainty around the next industrial revolution after this latest growth of ICT innovation and services (Gordon 2012), others are more optimistic of the continuation of IT revolution (Byrne et al. 2013) and that the growth of ICTs within developing countries can continue unbounded in its potential economic prospects. The measure of ICT growth may well be a necessity as one may underestimate the ICT opportunities and policies which support inclusive growth for a national economy. In other words, the inclusion of ICTs as part of a country's inclusive growth policies may well provide another answer as to what factors can contribute towards the reduction of poverty.

The acknowledgement of ICTs as a contributing element in poverty reduction has not been instant. Much of the earlier 1990s, literature on ICTs focused on economic growth, acknowledging mainly descriptive results around gross domestic product (GDP) changes in relation to ICT growth (Röller and Waverman 2001; Teltscher and Korka 2005; Waverman et al. 2005). Furthermore, the efficiencies in industrial production via technological advancement leads to increased labour productivity or business-driven solutions and the way towards national economic growth (Oliner and Sichel 2000). While these studies have shown some evidence linking ICT to economic growth, such growth results may not necessarily be linked to poverty reduction. With that said, less emphasis was placed on the social analysis between people, structures and the ICTs within developing country communities and ICT's disruption to people's everyday life (Adeya 2002). Within this literature, we now have a poverty and ICTs literature baseline within ICTD research at a time when mobile phones were mainly held by the wealthy population due to high cost (Adeya 2002).

Since this initial ICT and poverty literature review, communication technology has rapidly become available across the globe. Citizens of various income levels and geographical regions have shown numerous cases of having some ICT access. What is less clear is the consistent choice of ICT measurement and poverty measures used by researchers when exploring the nexus between ICTs and poverty. Understanding the choices available and what has been used round measurements for ICTs and poverty are important in order to steer a common language particularly when working in a multidisciplinary area like ICTD. While there is a good base of literature which now covers ICTs and poverty (Spence and Smith 2009), this review particularly covers what research ground has been covered around ICTs and poverty measurement. Firstly, the literature brings readers up to date on the accepted multitude of approaches and indicators for measuring poverty. The section which follows further explores what ICTD researchers have used to measure poverty. Finally, the last section addresses the various indicators around ICTs which are being used in the poverty and ICT literature.

## 2 Poverty Measure

Before this chapter delves into the recent work around poverty and ICTs, we briefly look at the current trends around poverty and inequality research. There is a paucity of ICTD researchers who are experts in poverty research; it is thereby important to unpack the relevant tools and concepts around poverty measurement. In better understanding poverty measures, one can then choose the appropriate tool and thus understand its relationship to ICTs. Appropriate poverty measure would in future help government and institutions make evidence-based decisions around strategies of poverty reduction. Relevant poverty measurement highlighted here embraces three trends: the multiple dimensions of poverty, the ways in which one can build a composite index of poverty and the dynamic nature of poverty measured over time.

Hulme (2013) raises the point that positioning and conceptualisation are important ideas to consider if one wishes to understand poverty better. In this chapter, we take on Lipton's (1997) definition where poverty is "the inability to attain an objective and absolute minimum standard of living and that this can be reflected by a quantifiable indicator applied to a constant threshold that separates the poor from the non-poor" (found in May 2012a: p. 64). This measure of poverty is also inspired by Sen's (1999) human development approach whereby one can be deprived based on his or her capabilities to meet some set basic human needs. With this approach in mind, the conceptualisation of poverty measurement has been evolving, and preference is paid to measuring poverty beyond the sole indicator of income among individuals and households. As mentioned in the introduction, the absolute poverty rate has been on a decline. This global poverty line allows one to compare across countries.[1] For example, in cross-country comparison, one interesting development to date is that the absolute poor (those living on USD 1.25 a day) are in majority located in middle-income countries (Sumner 2012). Besides this global poverty line, some countries measure their own relative income poverty line which assists to better serve the needs of their citizens. This relative poverty line usually consists of a cost for a basket of basic needs. A subjective poverty line where one determined deprivation by self-perception has been inspired by Bhutan's 'Happiness Index' and has gained global attention in poverty research.

Moving away from singular measures, there is much consensus among poverty researchers that poverty must be reviewed within a multidimensional lens (Alkire and Santos 2013; Moser 1998). Poverty is not only about one's level of income. Some of the recognised and important poverty dimensions besides financial include: human capital (including health and education levels), physical capital, welfare services (i.e. living standards levels) and social capital. The Human Development Index (HDI) has attempted to report on multiple social indicators (not necessarily poverty measures) and combine the indicators together to develop one index indicator which can compare low to very high human development across countries. At the micro household level, only recently have there been attempts to look at HDI among subgroups (Harttgen and Klasen 2012) and the further step to aggregate a country poverty index which brings multiple poverty elements together into one index measure (Alkire and Santos 2013).

Another evolution in poverty research is poverty dynamics. While yearly poverty and inequality statistics provide important cross-sectional baseline of populations, authors like Carter and Barrett (2006) have challenged these static models of poverty

---

[1] Taking into account that the quality of national statistics varies around the world.

and moved towards monitoring poverty over time (chronic vs. transitory). Rather one can follow the same households over time (i.e. panel survey) and see whether these households transition in or out of poverty. The multiple dimensions of poverty measure and looking at poverty over time are relatively new phenomena in the empirical work of poverty researchers. These recent developments around poverty measurement are clearly improving the world's understanding of human deprivation through a more holistic manner. ICTD researchers who wish to examine populations in low-income communities and with poorer households would gain immensely in reflecting on these evolutions of poverty measurement.

There are various applied poverty reduction interventions such as improved education, health and cash transfers initiatives mixed with an ICT component, but this chapter is limited to describing each of these studies. Rather, the premise of this chapter instead is to concentrate on poverty measurement choices taken by these ICTD studies. As mentioned earlier, the measurement choices should help governments and institutions appropriately evaluate socio-economic improvements and thereby best inform evidence-based policy development.

## 3 ICTs, Growth and Development

The acceptance of these poverty measurement trends come at a time of ICT proliferation, more specifically one sees the abundant resources of affordable mobile phones and the ever accessible Internet which are effectively changing the way one communicates. These ICT tools have generated much interest in their ability to reach the hands of even the most poor, and this evidence has opened up heated debate on understanding whether poverty change and human development can be brought about with such tools (follow the expert discussion by Spence and Smith 2010; Toyama 2012). The optimistic see its availability to the poor as transformational in social relations and business functions, while others are less hopeful as they see little direct wellbeing changes of say reduction of hunger or better welfare facilities. ICT tools have generated much interest given their ability to demonstrate usage among the poor, and some studies do touch upon some of the new poverty dimensions such as concepts of empowerment, inclusion and connectedness in poor communities (see a rich list of literature in Baron and Gomez 2013). The next section unravels some of the ways deprivation is measured in ICTD research, along with how ICTs are measured.

The lower costs to mobile phone access in the early 2000s were seen in many developing countries. Mobile Internet and broadband infrastructure continues to reach across regions alongside a variety of Internet-enabled devices. The potential of using ICTs in creative ways to generate or access income and other assets by the least privileged has become more and more realistic. Resource-poor smart- or feature-phone owners could also participate and navigate through Internet social networking applications such as Facebook or Twitter. Moreover, the myriad of prepaid and micropayment service packages continue to expand usage at relatively lower costs than early 2000s. While the expansion of access and usage are becoming reality for developing country citizens, one is limited in understanding the actual levels of participation by the poor. Due to this limited knowledge, one is further uncertain of how close we are to the universal reach of ICTs. First, one must understand how ICTs and poverty are measured today in order to better determine a way forward to reach access for all, even to rural and poor members who can benefit from the improvements of communications infrastructure.

## 4 Measuring ICTs in Developing Countries

As following the guidance from poverty research, the theoretical use of Sen (1999) capability approach has dominated recent ICTD landscape. Utilising the human development approach, ICTs are explored theoretically as a broader and more holistic way to understand wellbeing as helping to expand the choices and freedoms of the actors themselves and their "functionings" or actions which in turn can lead to changes of wellbeing (Attwood et al. 2013). Further acceptance of multiple dimensions of poverty is seen in ICTD literature.

Data before 2007 was found to be sparse in providing accurate ICT usage information in datasets available such as in Africa (James and Versteeg 2007) and more so among the poor. The ICT statistics administered by International Telecommunications Union (ITU) have helped to portray global supply of ICT. Developing countries today are outweighing the growth of ICT uptake over developed countries (ITU 2012). The ITU also hosts the ICT Development Index (IDI) which ranks countries based on a composite number of ICT readiness, intensity and impact. The IDI includes countries classified on the United Nations' HDI ranking as "low" or "medium". Table 1 presents some of these aggregate ICT supply side indicators for some "low" to "medium" human development countries in Africa.

From this African region set, South Africa is leading in the ICT provision and HDI rank across indicators. The HDI ranks seem to also align with the sequence of the country's IDI rank. Furthermore, the other three African countries may have low HDI and IDI but now have over 50 % of inhabitants with mobile phone subscriptions. Nevertheless one sees regional disparity of ICT access. In review of this global data, some "low" human development countries are experiencing high uptake in mobile phone subscriptions but low uptake of Internet usage (ITU 2011; Stork et al. 2013):

> Technology is the tool, NOT the outcome. Judith Rodin (Rockefeller Foundation at the Social Good Summit, New York, September 2013)

While the ITU statistics may distinguish between HDI and IDI levels, they do not distinguish between rich and poor households or individuals within countries. Since 2007, much work has been done to rectify the paucity of available data and research around ICT usage by the poor and its role in poverty reduction. Descriptive micro-level ICT statistical research has been conducted in ICT access and usage by

**Table 1** Access indicators (2011)

| Country (HDI rank—2013) | (IDI rank—2012) | Fixed-telephone subscription per 100 inhabitants | Mobile-cellular subscriptions per 100 inhabitants | International Internet bandwidth bits/s per Internet user | Percentage of households with computers | Percentage of households with Internet access |
|---|---|---|---|---|---|---|
| South Africa (121) | 91 | 8.2 | 126.8 | 18.874 | 19.5 | 9.8 |
| Kenya (145) | 114 | 0.7 | 64.8 | 4,544 | 7.8 | 6.9 |
| Nigeria (153) | 122 | 0.4 | 58.6 | 368 | 9.3 | 4.6 |
| Zambia (163) | 135 | 0.6 | 60.6 | 452 | 2.7 | 2.4 |

Source: ITU (2011)

the poor (including this non-exhaustive list: Agüero et al. 2011; Barrantes 2007; de Silva and Zainudeen 2008; Galperin and Mariscal 2007; Gillwald and Stork 2008; May 2012c). Today we have some knowledge of actual demand for ICTs by the poor. This micro-level work starts with first finding how the poor are classified in ICTD studies and how many of these "poor" individuals and households have ICTs now available to them.

## 5 Classification of Poor Within ICTD Studies

When designing research involving poverty measures and ICTs, one must be upfront of the way in which these indicators will be done. The classification of "poor" without measurement is found in ICTD research, and reasons to not measure may be due to inconvenience and avoiding the need to ask uncomfortable questions about poverty to their respondents. This incomplete information does no justice for decision-makers or research in using findings towards social welfare improvements or resource allocations. The field of ICTD now has a wide range of measurement unit(s) of analysis choices when examining the poor, and therefore, there is no reason to not complete an appropriate measurement design for poverty.

At a country and community level, the "poor" enumerating areas or regions can be targeted, and households can fall in an area where the average household income is below some determined threshold. In some cases, a group of countries are assigned poor as a result of their cross-comparison rank definition of "low" based on GDP, GNP or their HDI (e.g. James n.d.). The result of James' (n.d.) study of 11 African countries shows that the relatively low GNP countries find households having stated more intensive usage (i.e. Ethiopia, Kenya and Uganda). Furthermore, intensive usage is also found among higher GNP countries for the reasons of communication for safety reasons (i.e. Botswana, Namibia and South Africa). In measuring among poor areas, shared ICT facilities can assist to fulfil underserved areas as was done in Yu'an, China for telecentres (Soriano 2007).

For individual and households, there is a variety of available ICTD statistical micro-level studies which attempt to classify the poor. ICTD studies have used income as a mechanism to measure the absolute poverty line (USD 1.25 or USD 2.00 per capita per day, such as May (2012c)) and relative poverty lines (expenditure per capita, Barrantes 2007, or national poverty data in South Africa, infodev 2012b). Other studies have chosen to measure the proportion of a subpopulation such as the lower 25 % income bracket of the population (Gillwald and Stork 2008) or in other words, the bottom or base of the pyramid (de Silva and Zainudeen 2008; infodev 2012a, b). In following the multidimensional poverty research trend, a team of researchers in the project titled "poverty and ICTs in urban and rural eastern Africa" (PICTURE Africa) reviewed multiple dimensions of poverty through the lens of financial, human, physical, social and digital assets. What is found among these studies is that the poor are unlikely be a homogeneous group across regions given the variation of contexts. In ICTD research specifically, there is heterogeneity in the trend of choosing the measurement of the poor, and therefore, the choice of poverty threshold is just as broad as the multiple approaches to understand poverty itself.

## 6 Defining Reach of ICT to the Poor: Access, Ownership and Usage Among the Poor

While the choice of how to measure the "poor" among ICTD studies is heterogeneous, the classification of ICTs used by the poor has been just as wide ranging. The measurement of ICTs has included the count of physical products (i.e. radio, television, video recorders, computer, mobile phone, SIM cards, landline), those with connectivity (Internet connection—both mobile and fixed) and access to applications, services (e.g. email, Internet usage) and systems. In particular, there has been much progression in the thinking around three elements of ICTs and the poor: ICT access, ownership and usage at the community, household and individual level.

## 7 ICT Access and Ownership of the Poor

Predominantly, ICTD research has spent much time deriving ICT access indicators by asking poor households and individuals their level of access. ICT access from a micro-level demand survey has allowed for broad acceptance of access to include both private and shared access among household members (Rashid 2011). Public access computing (including telecentres and cybercafés) is deemed out of range for the poor (and also those with little to no education and the elderly) (Gomez 2013). Some of these access indicators have been gathered and provided by the government in order to support ICT infrastructure among poor communities, as well as to support their universal access policies.

Recently, ICT ownership or appropriation has been defined at the household or individual with low-income levels as part of a household's asset portfolio. For example, approximately three-quarters of those earning under South Africa's relative poverty line (USD 1.80 per person per day) have a mobile phone (infodev 2012b). Ownership across poor households in Latin America varied from high 90 % ownership in Jamaica and Colombia to 30 % access in Mexico in 2007 (Galperin and Mariscal 2007). Selected bottom of the pyramid households in Asia were monitored in 2006 for mobile phone ownership with countries like Pakistan, India and Sri Lanka having less than one-quarter ownership and relying on shared access (de Silva and Zainudeen 2008) but growing significantly by 2008 (Sivapragasam and Kang 2011). Finally, while one may have determined access or ownership ICT indicators or both, further understanding of the depth of usage has been the least understood, and today, it is asked in studies at varying degrees.

# 8 Usage

There are many cases and researches around the usage of ICTs especially mobile phones in development but few attempt to measure ICTs among a measured low-income population. Measuring ICTs can be understood for its usage to directly or indirectly improve the lives of the poor. Directly, we understand that direct cash transfers facilitated by ICTs could be an immediate approach to lift one out of income poverty. GiveDirectly is one institution providing direct mobile money transfers to a household phone, and the group's preliminary work finds the mobile money spent on basic food (GiveWell 2012). Indirectly, improving food security, financial inclusion and employment opportunities are three of the major research contributions available in describing ways in which the poor individuals or households use ICTs in attempt to improve their activities and livelihood. Citizens can help report on irregularities and therefore improve accountability on basic food distribution systems via SMS as is the case in India (Nagavarapu and Sekhri 2013). ICTs can also be used to improve rural livelihoods (which most likely occur in poor communities) through improved market access for produce as well as the lessening of food wastage (Grimshaw and Kala 2011). ICTs are observed as being used to help bring about changes to one's everyday life. These changes are then being attributed to the possible changes in one's level of poverty. In a review of mobile money or mFinance, new forms of banking facilities are now available which were not previously available to the poor, and in some cases, insufficient inputs (i.e. financial and literacy skills) are seen among the poor as well as some of their mixed perceptions around costs and risks (Leon et al. 2015 in Part II of this book). As for output, few studies have tried to understand cost savings and changes in business outputs among the poor (Leon et al. 2015). As for digital employment to the unemployed and the poor, we are also in the early days of this understanding. In terms of tackling poverty through improving income generation and work opportunities, groups such as Jana (or formerly txtEagle) and Samasource utilise microwork or the opportunity to offer small piecemeal work over the mobile phone to unemployed low-income personnel. These small earnings give even those most poor an opportunity to earn some meagre mobile phone credit which can help to diversify one's income earning portfolio. Despite these ICT studies in seeking its usage in improving the livelihoods and poverty levels of the poor, social ties and security or safety are seen as reasons for strongest usage (Galperin and Mariscal 2007). Awareness and usage of the Internet among the poor in selected countries in Asia were very low in 2008 (Zainudeen and Ratnadiwakara 2011). As seen above, many of the ICT usage demonstrations are found on small scale, without the use of rigorous methods of measuring changes particularly in indicators around poverty reduction (Kenny and Sandefur 2013).

## 9 Factors Affecting ICTs and Poverty: Affordability

As part of usage, individuals or households commit to ICT usage through the purchase of ICT goods and services. Affordability of ICTs is another element which has come through within studies around poverty and ICTs. The concept is important particularly in ensuring a fair cost for communication which allows all citizens the opportunity to communicate. The three elements of access, ownership and usage all depend on whether ICTs are considered affordable among the poor. Barrantes and Galperin (2008) explore how far the poor were willing to spend by looking at an affordability threshold for the mobile phone (i.e. 5 % of personal income of a basket of monthly mobile costs). Their multiple Latin America study found the poor had high basket monthly costs (e.g. 30–45 % in Brazil and Peru). These disturbingly high costs for mobile phone usage also showed lower mobile penetration in comparison to Latin American countries with lower monthly mobile costs (Barrantes and Galperin 2008). In one study of individuals in Africa, those individuals at the bottom 75 % had a share of 10.9 % of their monthly mobile expenditure in relation to income and those at the top 25 % were spending 4.8 % of their mobile expenditure (Gillwald and Stork 2008). In a later study in selected Asian countries, household data was compared and found that the poorest quintile exceeded 24 % of their proportion spent towards mobile services over total monthly expenditures (Agüero et al. 2011) (see Table 2). As we look further down the quintiles, we also see that the spending proportion reduces; we see the richest quintile (top 20 %) spend far less than 10 % on mobile services over total monthly expenditures.

From the demand of mobile phones, the researchers strongly suggest that communication functions as a necessity despite high costs. Further costs such as taxes on mobiles which increase mobile service expenditures may truly burden the most poor (Agüero et al. 2011). Even the most basic or everyday needs like food are in some cases being held back in order to afford the costs of mobile phone expenses (Diga 2007; Duncan 2013; infodev 2012a). In an economically depressed community in South Africa, the household respondents who earned a monthly income of between R300 to R5,000 (USD 37–USD 625) state that they on average use 26 % of their income on cell phones (handsets and airtime) (Duncan 2013).

Table 2 Percentage of expenditure in mobile services in selected Asian countries by income quintiles (%)

| Quintile | Bangladesh | Pakistan | India | Sri Lanka | Philippines | Thailand |
|---|---|---|---|---|---|---|
| 1 (Bottom 20 %) | **29.7** | **45.8** | **24.3** | **27.0** | **57.0** | **24.4** |
| 2 | 11.5 | 17.2 | 11.3 | 11.7 | 28.8 | 11.4 |
| 3 | 7.8 | 9.9 | 8.4 | 6.5 | 18.4 | 7.3 |
| 4 | 6.5 | 6.8 | 5.7 | 4.7 | 11.7 | 5.2 |
| 5 (Top 20 %) | **3.8** | **5.1** | **4.4** | **3.1** | **6.3** | **3.7** |

Source: Aguero et al. (2011)

Majority of the study's respondents from this same township perceived both the mobile and the airtime to be expensive (Duncan 2013). One unanswered question is whether the high costs of ICTs are trapping people in poverty as suggested by Duncan (2013). This affordability question needs further exploration as regards to poverty and ICT.

As mentioned earlier, the IDI has been helpful in comparing ICT uptake through an index across countries; however, limitations are raised in trying to measure a subpopulation such as poor households. Barrantes (2007) attempts to further the study at a micro- or household level in calculating how many of the income poor were also failing to have ICTs in what she called "digital poverty". Digital poverty is defined as "the minimum ICT use and consumption levels, as well as income levels of the population necessary to demand ICT products" (Barrantes 2007: p. 33). In conceptualising digital poverty, the extreme digitally poor are households who are deficient of all forms of ICT connectivity and have little capability or mean to accept or deliver electronic messages or to participate actively (two-way interaction) with information. On the other hand, the digitally wealthy participate fully through electronic media both in receiving or sending information usually through the Internet (Barrantes 2007). All the various ICT access, ownership and usage indicators are combined together and are composed into the ICT household index. In an example of over 17,000 Peruvian households (in 2003), she identified 68 % of the sample to be extremely digitally poor households. She then identifies the poor as those without sufficient income to cover the basic food basket of Peru, and this subpopulation was made up 17.59 % of the selected sample. Those who were extremely poor were nearly all extremely digitally poor in 2003. While this is an older study, it is one of the few trying to determine a composite indexed definition of ICT deprivation in relation to income poverty.

This early study was limited in household data around ICTs as each of the household members can have their own individual range of digital wealth or impoverishment. The main changes from 2003 were to remove household telecentre and computer usage and add more specific individual Internet usage such as whether someone was either ICT active or passive. Active Internet users are those defined as having the ability to have two-way interaction through the use of ICT transactions (Barrantes 2010). Thus, in this later study, the topology of digital poverty is updated to the following individual indicators in Table 3 (Barrantes 2010).

Table 3 Revised classification criteria according to their digital poverty level

| Digital poverty level | Indicators in survey |
| --- | --- |
| Digitally wealthy | Telephone user, active Internet user |
| Connected | Telephone user, passive Internet user |
| Digitally poor | Telephone user, no Internet |
| Extremely digitally poor | No telephone, no Internet |

Source: Barrantes (2010), prepared by authors

**Table 4** Digital poverty status of households by financial poverty status and geolocation (%)

| ICT | Not poor | Poor | Urban | Rural |
|---|---|---|---|---|
| No ICT | 7.0 | 23.4 | 9.1 | 21.6 |
| Digitally poor | 14.7 | 27.4 | 14.6 | 27.1 |
| Connected | 50.8 | 38.0 | 48.8 | 40.1 |
| Digitally wealthy | 27.5 | 11.3 | 27.5 | 11.3 |
| $n =$ |  | 1,473 |  | 1,508 |

Source: May (2012c)

When comparing the sample of 1,500 individual Peruvians of the digitally poor to the digitally wealthy, the demographic findings showed that the digitally poor were with lower annual incomes and lower levels of education and lived outside of Lima (urban capital). This study is also limited as a result of a small sample size, and it does not go further to identify the economically poor or nonpoor in this study and point out the subgroup's digital assets.

A digital poverty or ICT index has also been applied in East Africa (May 2012c). The features of May's (2012c) digital poverty are also different from both Barrantes' (2007, 2010) studies as May utilises a count in the number of ICT access or usage observations per capita. Taking a look at digital poverty from East Africa in 2007 and 2008, the economically poor (i.e. those below the absolute poverty line of USD 2.00 per capita per day) had certainly a larger proportion of the households without ICTs or being extremely digitally poor than those identified as not poor. Interestingly, there is nearly 15 % of not poor who are also identified as digitally poor and in reverse around 11 % of the poor who are digitally wealthy. One can also note similarities of the not poor percentages to that of urban geolocation and for the poor and rural (Table 4).

From the Barrantes (2007, 2010) and May (2012a, b, c) findings, those with few educated members in the household, lower-income levels and with few young people in the household may need further consideration in ways to increase their ICT participation. As a final note from the evolution of thinking around digital poverty is the distinguishing possibility of developing a digital poverty threshold. Barrantes takes a relativity stance by arguing that such a set target is impractical to monitor and review given the ever-changing ICT environment. The point is made that the sole monitoring of statistics on the insufficient ICT supply in poor areas will not be effective in moving people out of digital poverty.

All in all, there is no consistent rate within ICTD studies as to the poor's access, ownership and usage of ICTs. Furthermore, access and ownership concepts are further being solidified by the growing yet uneven rates of adoption by the poor in the various global subpopulations. ICT usage (including depth and quality of usage) among the poor, on the other hand, still appears up for debate and not well understood.

The possibilities of short- to long-term socio-economic changes or technological changes are vast within heterogeneous contexts and situations, and this particular review tries to delineate today's usage and ownership around ICT and controlling for certain factors, whether one can show ICTs' relationship to poverty reduction. Through this work, researchers reveal the heterogeneity of ICT demand by low-income individuals, households and communities. Nevertheless, across developing countries, we see growth of ownership and access to ICTs especially among poor countries and among the poorer population of those countries.

## 10 Impact That Poverty Has on ICTs

Some background papers have now covered the literature around ICT and poverty (Adeya 2002; Diga 2013; Spence and Smith 2009). One important feature to distinguish is the understanding of the causal inference of ICTs impact on poverty. One step is to first understand the direction in which we are examining impact. In one case, we can ask whether one's socio-economic status has an impact on one's ICT access, ownership or usage. In following up with the Barrantes and May research above, one can examine whether an individual's income level has a causal effect on ICT access. One theory can be that one with greater income can now afford say a mobile phone and thereby have a strong motivation to own and use various ICTs for their everyday use. Through her analysis, Barrantes (2010) confirms that the lower poverty level of the household, the improved likelihood for the household to be connected either via Internet access or mobile phone. As the data was only collected in 1 year, one is limited in the ability to measure impact or changes over time. Another attempt to test whether one's deprivation level had an effect on ICT access was done within poor communities in East Africa. While looking at all selected dimensions of poverty and ICT access, there was a positive and significant association (May 2012c). Upon closer observations, the findings showed better odds of ICT access when the household had at least one member with secondary education and living in urban areas (May 2012b). From the same 2007–2008 cross-section of this study, financial capital (through the per capita household expenditure indicator) relative to the absolute poverty line (USD 2.50 per person per day) appears to also be an important predictor of ICT access (May 2012c). There are still very few studies which have looked at this causal relationship among a larger aggregate population.

## 11 Impact That ICTs Have on Poverty Reduction

Improved reach of communication technology to the poor cannot be the only outcome in the debate around development. In looking at the other direction, research is observing the linkages of ICTs leading to poverty reduction. Toyama (2011) asks "how might mobile phones be exacerbating, rather than alleviating, poverty?" in his editorial. One theorises that through participating in ICT inclusive tasks, behaviour change occurs (whether it be obtaining income effectively or improving work processes efficiency), and through that, one hopes to find some level of longer-term impact on the individual's socio-economic status. An investment of time and effort in changing one's approach to work and generate a livelihood through the usage of ICTs could lead to an improvement of income and assets for individuals and their respective households. Ultimately, these improvements would see an individual or household move out of poverty. In looking at the previous pre-2002 literature (Adeya 2002), the evidence was inconclusive and remained with uncertainty whether mobile phones and ICTs were causing more harm than good and vice versa.

Impact studies on poverty and ICTs while not abundant have sprouted in the last 10 years. One literature review (Duncombe 2011) examines studies on mobile phones, development and impact. Of the 18 studies he reviewed, four highlighted long-term impacts through mobile phones, while the others measured more short-term indicators. The impact studies reviewed by Duncombe had low-income respondents or took place in low resourced communities, yet none of the reviewed studies in fact measured the changing levels of poverty among individuals or households.

One study which looks at multiple poverty dimensions, Aminuzzaman et al. (2003) mention that the ICTs measured in the study had less economic empowerment effects on users than compared to say transportation effects. Souter et al. (2005) highlight negative economic value of mobile phones by lower-income groups, while positive economic value was found with higher-income groups. Again this study did not necessarily address measurements of poverty level. Studies have become more sophisticated in trying to observe the changes over time of the same people or communities. Muto (2012) uses panel data to find Ugandan households more likely to leave rural areas for job-seeking migration when there is mobile phone ownership within the household.

One Tanzanian study conducts qualitative quasi-experimental work on small businesses using ICTs over time. The researchers monitored the changes of poverty over time among a randomly selected group of small business owners in two similar Tanzanian towns. One town's group of microbusinesses received a free mobile handset, mobile airtime (approximately USD 20 a month) and paid Internet email access of 1 h per week at an Internet café for 5 months (Mascarenhas 2014). The other town received none of these items. Both towns started with a similar poverty level of around 55 % taken based on the income of the selected sample of businesses. After implementing the intervention, the one town with the ICT provisions saw poverty level drop to 16.1 %, while the other similar town without ICT provisions saw poverty level drop to 38.9 % (Mascarenhas 2014). The study also examined multiple dimensions of poverty, with the treatment group (or the group with ICT provision) improving in five dimensions, and the control group without ICT provision only saw improvement in two dimensions. Within a short term, ICT usage had a clear effect on the small businesses compared to those who continued status quo.

The Tanzanian study above was part of the Poverty and Information and Communication Technology in Urban and Rural Eastern Africa (PICTURE Africa) project.

Further applied statistical analysis by PICTURE Africa was completed to improve the impact understanding between poverty and ICTs. At the micro-level, the panel study measures the same household's multiple dimensions of poverty and ICTs over time. Households were randomly selected within a nationally representative sample of the poorest enumeration areas in four East African countries. The survey findings showed that the ICT index statistically causes change in per capita expenditure. Furthermore, that with every one unit increase of ICT access, one sees a 3.7 % improvement in one's poverty status from 2007 to 2008 and 2010 in the four Eastern African countries (May et al. 2014). During the same period, the proportional expenditure change per capita in a household with ICTs was felt more strongly by the poorest than the nonpoor surveyed (May et al. 2014). The study thereby sees a slight movement of convergence between the poor and nonpoor based on the gains resulting from ICT access. In other words, the poverty level change is moving in a pro-poor direction. One must however be cautious of the results in that the gains made through the availability of ICTs to the very poor would only be seen in the medium term (6–10 years). This panel study represents a first in incorporating the poverty trends of looking at multiple dimensions of poverty and ICTs which can impact on the poor over time.

In the Duncombe (2011) review of mobile phone and impact, the one methodological gap was with the lack of participatory research methods. His concern was addressed through another applied participatory research case on ICT and changes in wellbeing among resource-poor communities, the community-based learning, ICTs and quality-of-life (CLIQ) project (Attwood 2013). The CLIQ project reviewed changes of self-perceived wellbeing of the same individuals in four poorer South African communities over time. This participatory research asked participants how their usage of computer training, free Internet and computer hours and goal setting affected their quality of life. The findings showed that in those participants who had high participation in the various intervention activities throughout the period and within telecentres with good functionality and process, one saw a greater response to quality-of-life change (Attwood et al. 2014). This unique study shows an innovative way of measuring ICT usage and wellbeing changes in a human-centred way. Furthermore, one takes this subjective status of participants, and it is the participants themselves who decide whether or not they have used the ICT tools to expand their choices and freedoms and thereby change their quality of life.

These quality-of-life impact findings as well as the PICTURE Africa findings are the first of its kind in exploring panel survey data and applied research analysis on the relationship between ICT and poverty. Both studies have given us a micro-level depiction of the nuanced mixed results of income, expenditure and self-perceived life impact changes over time. One can highlight that the findings are part of integrated Sen-inspired human development frameworks and assists in providing a more holistic understanding of the complexities around poverty reduction. This includes exploring the integration of ICT policy which supports human development where literature is limited (Diga et al. 2013). Furthermore, while these micro-level studies have certainly helped bring about ways to test an ICT composite index against poverty levels, these studies still need further refinement to include crucial indicators in the index such as less reported ICT skills. The ICT Development Index identifies ICT skills but only uses school enrolment and adult literacy as proxies to this ICT skills indicator. These findings as well as those which have been provided through descriptive findings above are part of the growing contributions of applied research and theory towards ICTs and poverty.

## 12 Conclusion and Way Forward

This study reveals the current progress within empirical description and analysis around measuring the nexus between ICTs and poverty. In looking at the literature, earlier reviews around the theory of ICT and development showed fairly simplified constructions of ICTs (either via access, ownership or usage) without well-measured indicators on the poor. Furthermore, earlier literature before 2002 concentrated on the macro-level of economic change, and less emphasis was placed on understanding the micro-level impact changes on poverty reduction. Research today acknowledges that ICTs could dually serve as tools for both economic development and poverty reduction.

Through this current review, one acknowledges that research in measuring poverty reduction at the micro-level has further developed in ICTD literature. Today, the variety of poverty measures being utilised by ICTD researchers appear to be aligned with the current concepts used by poverty experts. For example, both discipline streams are approaching poverty and ICT measurement in multiple dimensions and are attempting to analyse its transitions and impact over time. Despite this congruent nature of contemporary poverty theory and ICT research, there are still few studies within social welfare and poverty research trying to build on the measurement link between ICTs and poverty. With the importance of statistical analysis, less research has been done on ICTs and poverty (or wellbeing) through a participatory perspective. Participatory approaches and subjective wellbeing measures in ICT and poverty studies would add to the knowledge contribution in this field. The incorporation of the participatory methods which substantially involve the participants and where their aspirations and wellbeing are being asked is recommended. Applied techniques and refinement of indices for ICTs and the various poverty composite measures are also necessary to provide realistic recommendations to stakeholders on the future of ICT infrastructure or social policy development.

As a way forward, we are far from reaching the end on the war against poverty. Various approaches, interventions and participation would need to coordinate together to reach an end goal of improving the lives of the poor. Developing countries today have learned that national policy requires commitment to finding the ideal balance of inclusive growth—economic growth alongside social welfare policies—within one's limited national budget, and it is important to build on the strength of measurement in order to see the true nature of poverty. In reflecting on ICTs and poverty, Toyama (2011) however argues that communication technology will help to amplify the success or failure of existing institutional capacity towards development. Institutional competence, alongside varying costs and levels of training and capability will be elements which will make ICTs access and usage possible even for those of the underserved population.

While the ICTD community has produced evolving contributions towards understanding the connection between ICT and poverty, unfortunately less can be said for the social development community. The Millennium Development Goals (MDGs) reach their end in 2015, yet current suggestions around a next round of MDGs have little mention of ICTD in playing a contributory role or being in any way measured. Nevertheless, emerging research is observing some of the ICTD interventions in developing countries targeted at the goals and indicates some contributions towards poverty alleviation (Kaino 2013). More research work and advocacy for understanding the nexus between ICT and poverty will need to be raised in moving forward into the future. Furthermore, one must make note of global trends of economic instability and changing industrial development within a holistic development approach which may fundamentally change the way research is done around ICT and poverty.

The pronounced voice which comes from the south on ICT and poverty may suggest that there is great value in understanding the lived experience of using mobile phones in the everyday lives of people especially within resource-poor communities. Yet with all the various measures of ICTs, behaviour change and impact, the gap remains in further work in the south to understand this evolution of ICT and poverty over time.

While this study concentrated around poverty measurement, poverty cannot be viewed without looking at inequality. Massive global income disparities are still clearly found between countries and within countries. The improvement of work which distinguished whether there is a convergence of income and less inequality as a result of ICTs or vice versa would also be a move forward in ICT and poverty measurement research.

Today's soon to expire Millennium Development Goals are being re-evaluated within a time of global instability, pushing countries to make dramatic policy choices to that of the past. In other words, countries are taking recessionary initiatives which prioritise growth through economic policy. Finally, for the poor to truly benefit in wellbeing change, a country's economic growth strategy would likely need the support of complementary ICTs and other poverty reduction strategies through redistributed resources such as social welfare grants, health care, improved educational facilities etc. The fight to ensure that ICTs find their place within a balanced frame of inclusive growth will be the challenge moving forward during uncertain times.

**Acknowledgements** We would like to express our gratitude to those who reviewed and provided comments on this paper including: Roxana Barrantes, Arul Chib, Roger Harris, Mary Luz Feranil, Andrea Ordonez and Matias Dodel Schubert.

**Open Access** This chapter is distributed under the terms of the Creative Commons Attribution Noncommercial License, which permits any noncommercial use, distribution, and reproduction in any medium, provided the original author(s) and source are credited.

# References

Adeya, C. N. (2002). *ICTs and poverty: A literature review*. Ottawa: IDRC.
Agüero, A., de Silva, H., & Kang, J. (2011). Bottom of the Pyramid expenditure patterns on mobile phone services in selected emerging Asian countries. *Information Technologies and International Development, 7*(3), 19–32.
Alkire, S., & Santos, M. E. (2013). A multidimensional approach: Poverty measurement & beyond. *Social Indicators Research, 112*(2), 239–257.
Aminuzzaman, S., Baldersheim, H., & Jamil, I. (2003). Talking back! Empowerment and mobile phones in rural Bangladesh: A study of the village phone scheme of Grameen Bank. *Contemporary South Asia, 12*(3), 327–348. doi:10.1080/0958493032000175879.
Attwood, H. E. (2013). The influence of quality-of-life research on quality-of-life: CLIQ case studies from KwaZulu-Natal, south Africa. In M. J. Sirgy, R. Phillips, & D. Rahtz (Eds.), *Community quality-of-life indicators: Best cases VI* (Vol. 4, pp. 1–18). Dordrecht: Springer.
Attwood, H., Diga, K., Braathen, E., & May, J. (2013). Telecentre functionality in South Africa: Re-enabling the community ICT access environment. *Journal of Community Informatics, 9*(4).
Attwood, H., May, J., & Diga, K. (2014). Chapter 8: The complexities of establishing causality between an ICT intervention and changes in quality-of-life: The case of CLIQ in South Africa. In E. O. Adera, T. M. Waema, J. May, O. Mascarenhas, & K. Diga (Eds.), *ICT pathways to poverty reduction: Empirical evidence from East and Southern Africa*. Rugby: Practical Action Publishing.
Baron, L. F., & Gomez, R. (2013). Relationships and connectedness: Weak ties that help social inclusion through public access computing. *Information Technology for Development, 19*(4), 271–295.
Barrantes, R. (2007). Analysis of ICT demand: What is digital poverty and how to measure It? In H. Galperin & J. Mariscal (Eds.), *Digital poverty: Latin American and Caribbean perspectives* (The Regional Dialogue on the Information Society (REDIS-DIRSI), Eds., pp. 29–53). Ottawa: International Development Research Centre.
Barrantes, R. (2010, September). *Digital poverty: An analytical framework*. Paper presented at the Chronic Poverty Research Centre conference, University of Manchester, Manchester.
Barrantes, R., & Galperin, H. (2008). Can the poor afford mobile telephony? Evidence from Latin America. *Telecommunications Policy, 2008*(32), 521–530.
Bhagwati, J., & Panagariya, A. (2013). *Why growth matters: How economic growth in India reduced poverty*. New York: Public Affairs.
Byrne, D. M., Oliner, S. D., & Sichel, D. E. (2013, March 27). *Is the information technology revolution over?* Available at SSRN: http://ssrn.com/abstract=2240961
Carter, M. R., & Barrett, C. B. (2006). The economics of poverty traps and persistent poverty: An asset-based approach. *The Journal of Development Studies, 42*(2), 178–199.
de Silva, H., & Zainudeen, A. (2008). Teleuse at the bottom of the pyramid: Beyond universal access. *Telektronikk, 2*(2008), 25–38.
Diga, K. (2007). *Mobile cell phones and poverty reduction: Technology spending patterns and poverty level change among households in Uganda*. Masters in Development Studies, University of KwaZulu-Natal, Durban.
Diga, K. (2013). Chapter 5: Access and usage of ICTs by the poor (part I). In L. Elder, H. Emdon, R. Fuchs, & B. Petrazzini (Eds.), *Connecting ICTs to development: The IDRC experience*. London: Anthem Press/IDRC.

Diga, K., Nwaiwu, F., & Plantinga, P. (2013). ICT policy and poverty reduction in Africa. *Info, 5*(5), 114–127.

Drèze, J., & Sen, A. (2013). *An uncertain glory: India and its contradictions.* Princeton: Princeton University Press.

Duncan, J. (2013). Mobile network society? Affordability and mobile phone usage in Grahamstown East. *Communicatio, 39*(1), 35–52. doi:10.1080/02500167.2013.766224.

Duncombe, R. (2011). Researching impact of mobile phones for development: Concepts, methods and lessons for practice. *Information Technology for Development, 17*(4), 268–288. doi:10.1080/02681102.2011.561279.

Galperin, H., & Mariscal, J. (2007). *Mobile opportunities: Poverty and mobile telephony in Latin American and the Caribbean.* Lima: DIRSI, IDRC.

Gillwald, A., & Stork, C. (2008). *Towards evidence based ICT policy and regulation: ICT access and usage in Africa (ICT adoption and diffusion)* (Vol. 1): Research ICT Africa and Johannesburg.

GiveWell. (2012, December). *Cash transfers in the developing world.* URL: http://www.givewell.org/international/technical/programs/cash-transfers/2012-version. Accessed Nov 2013.

Gomez, R. (2013). When you do not have a computer: Public-access computing in developing countries. *Information Technology for Development, 20*(3), 274–291.

Gordon, R. J. (2012). *Is US economic growth over? Faltering innovation confronts the six headwinds* (NBER Working Paper Series National Bureau of Economic Research). Cambridge. Retrieved from http://faculty-web.at.northwestern.edu/economics/gordon/Is%20US%20Economic%20Growth%20Over.pdf

Grimshaw, D. J., & Kala, S. (2011). *Strengthening rural livelihoods: The impact of information and communication technologies in Asia.* Ottawa: IDRC.

Harttgen, K., & Klasen, S. (2012). A household-based human development index. *World Development, 40*(5), 878–899. doi:http://dx.doi.org/10.1016/j.worlddev.2011.09.011.

Hulme, D. (2013). *Poverty and development thinking: Synthesis or uneasy compromise?* Manchester: University of Manchester.

infodev.(2012a). *Mobile usage at the base of the Pyramid in Kenya.* Washington, DC: The World Bank.

infodev. (2012b). *Mobile usage at the base of the Pyramid in South Africa.* Washington, DC: The World Bank.

ITU. (2011). *World telecommunication/ICT indicators database.* Retrieved from http://www.itu.int/ITU-D/ict/statistics/

ITU. (2012). *Measuring the information society 2012.* Geneva: International Telecommunications Union.

James, J. (n.d.). Product use and welfare: The case of mobile phones in Africa. *Telematics and Informatics* (0). doi:http://dx.doi.org/10.1016/j.tele.2013.08.007.

James, J., & Versteeg, M. (2007). Mobile phones in Africa: How much do we really know? *Social Indicators Research, 84*(1), 117–126.

Kaino, L. M. (2013). Information and Communication Technology (ICT) and Attainment of the Millennium Development Goals (MDGs): The interdependence between MDGs' educational and socioeconomic goals. *Journal of Communication, 4*(1), 33–40.

Kenny, C., & Sandefur, J. (2013, July/August). Can Silicon Valley save the world? *Foreign Policy.* http://foreignpolicy.com/2013/06/24/can-silicon-valley-save-the-world/

Leon, L., Rahim, F., & Chib, A. (2015). The impact of mFinance initiatives in the global south: A review of the literature. In A. Chib, J. May, & R. Barrantes (Eds.), *Impact of information society research in the global south.* Singapore: Springer.

Lipton, M. (1997). Editorial: Poverty—Are there holes in the consensus? *World Development, 25*(7), 1003–1007. doi:http://dx.doi.org/10.1016/S0305-750X(97)00031-4.

Mascarenhas, O. (2014). Impact of enhanced access to ICTs on small and micro enterprises in Tanzania. In E. O. Adera, T. M. Waema, J. May, O. Mascarenhas, & K. Diga (Eds.), *ICT pathways to poverty reduction: Empirical evidence from east and southern Africa.* Rugby: Practical Action Publishing/IDRC.

May, J. (2012a). Smoke and mirrors? The science of poverty measurement and its application. *Development Southern Africa, 29*(1), 63–75. doi:10.1080/0376835x.2012.645641.

May, J. (2012b). *Tweeting out of poverty: Access to information and communication technologies as a pathway from poverty.* Paper presented at the Towards Carnegie III: Strategies to Overcome Poverty & Inequality, University of Cape Town. http://www.carnegie3.org.za/docs/papers/175_May_Tweeting%20out%20of%20poverty%20-%20access%20to%20information%20and%20communication%20technologies%20as%20a%20pathway%20from%20poverty.pdf

May, J. D. (2012c). Digital and other poverties: Exploring the connection in four east African countries. *Information Technologies and International Development, 8*(2), 33–50.

May, J., Dutton, V., & Munyakazi, L. (2014). Chapter 2: Information and communication technologies as a pathway from poverty: Evidence from East Africa. In E. O. Adera, T. M. Waema, J. May, O. Mascarenhas, & K. Diga (Eds.), *ICT pathways to poverty reduction: Empirical evidence from East and Southern Africa.* Rugby: Practical Action Publishing/IDRC.

Meier, G. M. (1970). *Leading issues in economic development* (2nd ed.). Oxford: Oxford University Press.

Moser, C. O. N. (1998). The asset vulnerability framework: Reassessing urban poverty reduction strategies. *World Development, 26*(1), 1–19. doi:10.1016/s0305-750x(97)10015-8.

Muto, M. (2012). The impacts of mobile phones and personal networks on rural-to-urban migration: Evidence from Uganda. *Journal of African Economies, 21*(5), 787–807. doi:10.1093/jae/ejs009.

Nagavarapu, S., & Sekhri, S. (2013). *Role of ICT technologies in reforming TPDS: Information provision by SMS in Uttar Pradesh.* Brown University. Paper under review.

Oliner, S. D., & Sichel, D. E. (2000). The resurgence of growth in the late 1990s: Is information technology the story? *The Journal of Economic Perspectives, 14*(4), 3–22.

Rashid, A. T. (2011). A qualitative exploration of mobile phone use by non-owners in urban Bangladesh. *Contemporary South Asia, 19*(4), 395–408. doi:10.1080/09584935.2011.577206.

Ravallion, M. (2013). *How long will it take to lift one billion people out of poverty?* Washington, DC: The World Bank.

Rodrik, D. (2013). *The past, present, and future of economic growth.* Global Citizen Foundation.

Röller, L.-H., & Waverman, L. (2001). Telecommunications infrastructure and economic development: A simultaneous approach. *The American Economic Review, 91*(4), 909–923. doi:10.2307/2677818.

Sen, A. (1999). *Development as freedom.* Oxford: Oxford University Press.

Sivapragasam, N., & Kang, J. (2011). The future of the public payphone: Findings from a study on telecom use at the bottom of the Pyramid in South and Southeast Asia. *International Technologies and International Development, 7*(3), 33–44.

Soriano, C. R. R. (2007). Exploring the ICT and rural poverty reduction link: Community telecenters and rural livelihoods in Wu'an, China. *The Electronic Journal of Information Systems in Developing Countries, 32*.

Souter, D., Scott, D., Garforth, C., Jain, R., Mascarenhas, O., & McKemey, K. (2005). *The economic impact of telecommunications on rural livelihoods and poverty reduction.* Reading: Gamos.

Spence, R., & Smith, M. (2009). *Information and communication technologies, human development, growth and poverty reduction: A background paper.* Ottawa: IDRC.

Spence, R., & Smith, M. (2010). ICT, development, and poverty reduction: Five emerging stories. *Information Technologies and International Development, 6*(SE (Special Edition 2010)), 11–17.

Stork, C., Calandro, E., & Gillwald, A. N. (2013). Internet going mobile: Internet access and use in eleven African countries. *Info, 5*(5), 34–51.

Sumner, A. (2012). Where do the poor live? *World Development, 40*(5), 865–877. doi:http://dx.doi.org/10.1016/j.worlddev.2011.09.007.

Teltscher, S., & Korka, D. (2005). Macroeconomic impacts. In G. Sciadas (Ed.), *From the digital divide to digital opportunities: Measuring infostates for development* (pp. 45–55). Montreal: Orbicom.

Toyama, K. (2011). *Technology as amplifier in international development.* Paper presented at the proceedings of the 2011 iConference, Seattle.

Toyama, K. (2012). *Can technology end poverty?* Retrieved June 26, 2012, from http://bostonreview.net/forum/can-technology-end-poverty/kentaro-toyama-responds

United Nations. (2013). *A new global partnership: Eradicate poverty and transform economies through sustainable development.* New York. Retrieved from http://www.un.org/sg/management/pdf/HLP_P2015_Report.pdf

Waverman, L., Meschi, M., & Fuss, M. (2005). The impact of telecoms on economic growth in developing countries. *The Vodafone Policy Paper Series, 2*(03), 10–24.

Wolfensohn, J. D. (2000, October 3). *Poverty and development: The world development report 2000/2001.* Paper presented at the conference on World Poverty and Development: A Challenge for the Private Sector, Amsterdam.

World Bank. (1980). *WDR 1980: Poverty and human development.* Washington, DC: World Bank Group.

World Bank. (1990). *World development report 1990.* Washington, DC: World Bank Group.

World Bank. (2001). *World development report 2000/2001: Attacking poverty.* Washington, DC: World Bank Group.

World Bank. (n.d.). World Bank archives – Alden Winship ("Tom") Clausen. Retrieved June 26, 2013, from http://go.worldbank.org/DG1E29A900, Washington, DC.

Zainudeen, A., & Ratnadiwakara, D. (2011). Are the poor stuck in voice? Conditions for adoption of more-than-voice mobile services. *International Technologies and International Development, 7*(3), 45–59.

# The Impact of mFinance Initiatives in the Global South: A Review of the Literature

Arul Chib, Laura León, and Fouziah Rahim

## 1 Introduction

After more than two decades of research on technological interventions in the transition to information societies, the burgeoning of mobile phones in developing countries (ITU 2013) has shifted the information and communication technologies for development (ICTD) research lens to the different domains of mDevelopment. While advances have been made in domains of mHealth, mGovernment, mBusiness and mEducation, mFinance initiatives have had impressive adoption upon implementation in certain geographic locations. Services such as M-Pesa have been widely reported in the mainstream press and form the test beds for various scholarly investigations.

Due to these unique geographical successes (and less-reported failures), scholars have attempted to determine the factors behind the widespread adoption of mFinance applications. Prior reviews of the mFinance literature, largely reliant on studies conducted in industrialised nations, have focused primarily on technological and business-related success factors (Dahlberg et al. 2008; Dewan 2010; Ngai and Gunasekaran 2007). This review builds upon the work of Duncombe and Boateng (2009) in investigating the impact of these mFinance initiatives within a development context. Our aim, however, is to determine the relative focus of mFinance research, focusing on the bottom of the pyramid (Prahalad 2006). In this

---

A. Chib, Ph.D. (✉) • F. Rahim
Wee Kim Wee School for Communication and Information, Nanyang Technological University, Singapore
e-mail: arulchib@ntu.edu.sg

L. León
University of Lima, Lima, Peru
e-mail: leon.laura@gmail.com

© The Author(s) 2015
A. Chib et al. (eds.), *Impact of Information Society Research in the Global South*, DOI 10.1007/978-981-287-381-1_6

chapter, we contextualise the BoP within low- and middle-income countries and study mFinance initiatives in terms of technological inputs, mechanisms of adoption and the resultant outputs, or impact.

First, however, it is worthwhile to reflect upon existing, and propose alternative, definitions for the notion of impact of mFinance.[1] The traditional approach to measuring impact has relied on an economic perspective, measured in terms of increased productivity, income and savings. Developmental impact is alluded to multiple indirect indicators of financial effects at a variety of levels—structural, group and individual—rather than the resultant social, economic and cultural effects (Donner and Tellez 2008) of technology introduction, adoption and appropriation. As a result, scholars believe there is an issue assessing the broader development impacts of mFinance applications and interventions in a concise and coherent manner (Alampay and Bala 2010; Heeks and Molla 2009).

In this review, we focus on the development outcomes of mFinance initiatives. It is therefore important to provide our perspective on development. Development is understood as people achieving a better quality of life, meaning "being healthy, being well-nourished, being literate, etc." (Sen 1988: p. 16), and following Sen, freedom of choice. As Kleine (2010: p. 683) remarks, ICT are "multi-purpose technologies which offer far more significant changes to people's lives than the economic impact they might have". Relating and applying this notion of development with mFinance issues, we consider mFinance development outcomes not merely as economic, due to the nature of the assets that these services manage, but extending to other aspects, such as empowerment.

mFinance can be understood as "a set of applications that enable people to use their mobile telephones to manipulate their bank accounts, store value on an account linked to their handset, transfer funds, or even access credit or insurance products" (Donner 2008a: p. 3–4). Different transactions can be made through mFinance applications, as is being observed at a global scale, within the context of the industrialised economies. Typically, users can make person-to-person transfers of cash and airtime, make payments to retailers, receive bank statements, enquire about balances and top-up mobile phone credits (Casanova 2007; Wishart 2006). It is also possible in some cases to request and receive notifications about activity in the client's bank account, which has become a way to manage risk, for example, managing incoming transfers into bank accounts (Scornavacca and Hoehle 2007), and outgoing expenditures related to credit card use. In some countries, mFinance services allow customers to receive international remittances, pay bills or a loan, receive their monthly payroll or receive social security payments (Casanova 2007).

In developing countries, mFinance harnesses the rapid expansion of mobile phones among low-income users. In general, for this group, the benefits include

---

[1] The field uses various terms to describe the use of mobile networks to conduct financial transactions, including mBanking, mCommerce, mMoney, mPayments, mRemittances, mTransfers, etc. We choose to use the all-encompassing term mFinance and use the individual terms in the literature review and in the search methodology as appropriate.

faster and cheaper, and sometimes safer, banking transactions and payments. The most evident benefits are when transacting with social networks located in remote places where the lack of physical outlets is a limitation to accessing the formal financial system. Therefore, mFinance, by avoiding long journeys to bank branches, translates into savings in time and money (Datta et al. 2001; Donner 2007; Jones and Du Toit 2007; Rosemberg 2008; World Bank 2002) and avoidance of riskier informal routes (Jagun et al. 2007).

Besides these benefits, mFinance services "holds the prospect of offering a low cost, accessible transaction banking platform for currently unbanked and poorer customers" (Heeks and Jagun 2007; Porteous and Wishart 2006: p. 5). In other words, it appears that mFinance initiatives have the potential to expand financial services to those who have been previously systematically excluded (García et al. n.d.; Hughes and Lonie 2007; Mendes et al. 2007). mFinance offers to poor people, who normally belong to the informal sector, financial services such as "access to payments, transfers and stored value functionality without opening an actual banking account" (Donner 2008b: p. 8). The benefits in terms of economic development, or poverty reduction, include the power to access loans and insurance towards productive investment. The poor are thus potentially better able to take control of their own livelihoods (Donner 2007; Economist 2008). On the supply side, mFinance applications allow the banking sector to discover new business models targeting new segments at differential cost levels, leading to a disaggregation of the bank components (Klein and Mayer 2011). We definitely note that the scope of benefits that mFinance offers serves not only the poor but also the banks, hence strengthening their profits. This is just another example of the complex nature of the development initiatives.

Nonetheless, the question that arises is whether this ICT-based system is actually producing development impact for the bottom of the pyramid. Does the delivery of banking services via mobile phones lead to productive saving and investment, in turn translating into poverty reduction for the poor, or is it more beneficially suited for low-income customers? In developing countries, the regular banking system may not necessarily provide benefits for the poor, so there is little motivation to move from informal ways of economic transactions to more formal ones. A second idea focuses on foreign (both inter- and intra-national) worker remittances. Prior research conflates process improvements such as volume, frequency, speed and cost with development outcomes of remittances such as households "retaining a higher proportion of the money by paying lower fees" (Donner and Tellez 2008: p. 328; van Reijswoud 2007), leading to optimism about benefits (Heeks and Jagun 2007).

These arguments suggest that even though the potential of mFinance seems to be enormous, we believe that evidence of how these applications impact on the livelihoods of the poor has been prematurely assessed as having a net beneficial impact. Supporting this viewpoint, the prior literature reviews report gaps in the conceptualisation and measurement of the impact of mFinance (Dewan 2010; Duncombe and Boateng 2009; Ivatury and Mas 2008). Key issues highlighted by the literature related to adoption of mFinance systems include security and trust (Karunanayake et al. 2008; Mousumi and Jamil 2010).

```
┌─────────┐    ┌───────────┐    ┌──────────┐
│  Input  │ ➡  │ Mechanism │ ➡  │ Outcomes │
└─────────┘    └───────────┘    └──────────┘

•Feasibility       •Individual preferences   •Beneficiary outcomes
•Access and use    •Adoption factors         •Efficiency measures
•Response rate                               •Broader development
                                              impacts
```

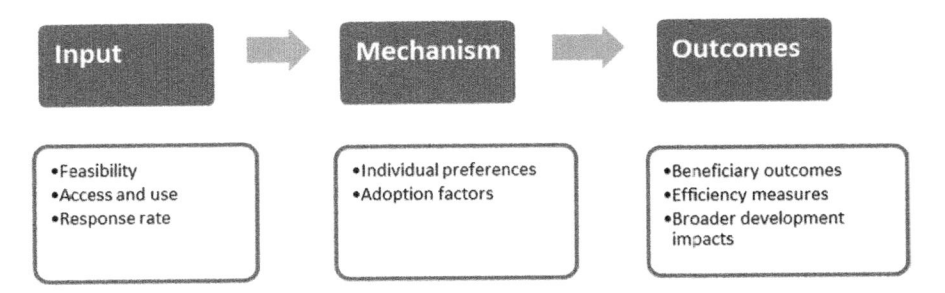

**Fig. 1** The Pathway Model

This chapter first interrogates the notion of impact of the mFinance initiatives. To do so, we approach the framework for measuring impact inspired by ICTD areas in which impact assessment is more advanced. The first research question we pose investigates how impact is conceptualised in the mFinance literature. What alternative definitions of impact can be proposed, beyond traditional notions of economic development (i.e. income and savings)?

In order to contribute to the discussion on impact measurement issues, this paper utilises the pathway of effects framework, the input-mechanism-output (IMO) model (Chib et al. 2014), as a framework for assessment. This framework is similar to the Duncombe and Boateng (2009) model with respect to a technical design and development phase and subsequent adoption and impact phases.[2]

The IMO model was chosen because it is one of the assessment frameworks that explicitly relate inputs and outcomes, adding rigour to the impact assessment, such as that provided by the Cost-Benefit Analysis (CBA) model (Heeks and Molla 2009). However, the IMO model offers a more flexible tool than the CBA framework, which is only based on a financial assessment.

Specifically, the terms in the IMO model (see Fig. 1) assist in the identification of the focus of the articles reviewed, where input refers to the access and use of technology being introduced. The second category, mechanism, relates to the process of user adoption and appropriation. The third category, outputs, comprises the process outcomes and end-user benefits, i.e. impact of mFinance initiatives.

The secondary research question posed interrogates the mFinance literature in terms of understanding technological inputs, mechanisms of adoption and outputs such as process improvements and end-user impact. The objective is to understand the focus of the mFinance research that has been conducted and, in doing so, to calibrate these efforts towards providing greater evidence of impact, using relevant theoretical frames and rigorous measures.

---

[2]The category of needs identification has been dropped for two reasons: the first being the focus of this chapter is impact and, second being, the low incidence of articles in this category found in the review of the literature.

The methodology utilises a secondary literature review of peer-reviewed and non-peer-reviewed sources, including the grey literature. Despite a significant amount of literature on mFinance contextualised in the developed world, we concentrate on underprivileged populations in the developing world and how mobile applications contribute to both livelihoods and a broader development perspective.

## 2 Methods

### 2.1 Inclusion and Exclusion Criteria

We included research papers fulfilling the following inclusion and exclusion criteria. Only research studies focusing on the application of mobile technologies to financial services in low- and middle-income countries (as categorised by the World Bank) have been included. We excluded research which studied mobile devices other than mobile phones and which focused on banking or commerce undertaken in high-income countries. Only English language papers were considered, although this may imply excluding valuable literature in other languages and therefore note that concentrating only on the production of English-speaking researchers is not representative of the developing world.

From a process perspective, we first demarcated the boundaries of the investigation to include mFinance research from a broad range of disciplines, with articles drawn from development studies, economics, banking and finance, technology and innovation, management and information systems, and information and communication technology for development (ICT4D). Because many of the studies have been disseminated by practitioners, published in the grey literature and focused on a consumer perspective, the scope of the present review includes peer-reviewed academic papers, published in scientific journals and/or conference proceedings, as well as non-peer-reviewed papers. As a result, commercial, government and international cooperation papers have also been included.

As the review focused on the impact of these mFinance applications on poor people's livelihoods, we examined primarily the individual and community levels (micro level) of analysis. Therefore, papers related to macro-level phenomenon, such as government policies related to the regulatory environment and market forces of demand and supply, and organisational perspectives of readiness assessments and business models were excluded.

### 2.2 Search Methods

The authors used mFinance related search terms such as mCommerce, mFinance, mBanking, mMoney, mPayment, mRemittances, mTransfers and MFS (and their m- and mobile- versions) to search the following electronic databases: Academic

Search Premier, ACN, Business Source Premier, Communication & Mass Media Complete, EBSCOhost, PsychINFO, Science Direct, SciVerse Scopus and Web of Science. Google Scholar was also used as a search tool under the mentioned terms. Reference lists of studies identified as relevant were also searched as a means of creating a snowball sample. Two co-authors were involved in the search process, so as to ensure maximum reach; specific search methodologies or analyses were not utilised other than as described. We note here that this review is not meant to be exhaustive and all-encompassing; it merely wishes to develop a sample sufficient enough to generalise trends for the adoption, use and impact of mFinance initiatives in developing countries.

## 2.3 Data Extraction and Analysis

The authors merged search results across databases, removed duplicates and screened citations against inclusion and exclusion criteria. Data were extracted using a standardised form created in Microsoft Excel including descriptive, inputs, mechanism factors and outputs (and can be made available upon request). Statistical pooling of results was not possible due to the extensive heterogeneity of the study methodologies. The papers were further categorised according to the type of main intervention. When studies focused on factors fell under more than one category, we chose to concentrate on the main intent of each study. Where studies exhibited more than one category in a significant manner, we examined the linkages.

# 3 Results

We found 41 studies addressing one of the three stages of the pathway, *input-mechanism-output*, as shown in Fig. 1. The majority of the studies 29 out of 41 studies analysed were in the mechanism stage, elaborating on factors leading to adoption of mFinance. We next elaborate on the specific studies constituting each stage.

## 3.1 Inputs

Eleven mFinance studies were concerned with input issues related to access and use and technological aspects, such as infrastructure requirements, and software and hardware issues. Key concepts relevant to developing countries that emerged relating to inputs included access; affordability; literacy, both textual and financial; security; and gender issues.

Unsurprisingly, mobile access was mentioned as the primary technological requirement (Duncombe 2009). Widespread coverage provides ubiquitous access to the user while high-speed SIM cards (e.g. 64 Kbps) were recommended to best utilise the mobile network (Mariscal and Flores-Roux 2011), while Hossain and Khandanker (2011) consider mobile handsets with advanced options. Access to technology was mentioned by Arora and Cummings (2010), in order to enable the mobile phone bank branch. Within a developmental context, affordability cannot be delinked from mobile access, especially for the low-income and poor segments of the population (Mariscal and Flores-Roux 2011; Zainudeen et al. 2011). From the user point of view, money is needed to access and use these systems, and the lack of affordability represents a constraint to access and use (Boadi et al. 2007; Duncombe 2009). Boadi et al. (2007) consider whether rural fishermen in Ghana could afford the needed investment to access both the fixed costs of equipment and the variable costs of mobile subscription services.

Studies of the optimal system infrastructure required for SMS-based mobile banking systems reaching not only urban but also remote rural areas of developing countries identify security as a key issue (Hossain and Khandanker 2011). A technological solution describes a push-pull system, wherein either the bank broadcasts information or the customer requests banking services (Mousumi and Jamil 2010). However, other studies point to human resource approaches to the issue of trust. One approach relies on banking agents for mobile ATM service (Karunanayake et al. 2008), while another mentioned cash-in, cash-out points (Singh 2009), conceptualised as partnerships with retail stores (Mariscal and Flores-Roux 2011).

A lack of literacy skills has been mentioned as a barrier to use of text-based services in Uganda (Duncombe 2009). A study of adequate mobile payment user interfaces for non-literate and semi-literate subjects concluded that non-text designs were strongly preferred over text-based designs (Medhi et al. 2009). Furthermore, while the use of rich multimedia user interfaces reported better task-completion rates, the spoken-dialogue system was faster and required less assistance.

Beyond textual and numerical literacy, financial literacy is considered an important factor in order to adopt the mobile financial systems. Financial literacy is understood by BCG (2011: p. 12) as the "advantages of becoming banked". Lack of financial literacy was a constraint to assimilating the required skills to interact effectively with mobile phones and mFinance technologies. Looking at financial literacy issues from a gender perspective, Singh (2009) examines issues of women empowerment via financial inclusion, proposing design principles that invoke a sociocultural perspective, including gender patterns of financial control and issues of privacy and trust in the transaction.

## 3.2 Mechanisms

The largest group of studies ($n=29$) investigated the reasons for technology adoption, with some using theoretical models for explanation or validation of the findings. Many researchers examined mobile banking as an emerging ICT artefact from the perspective of user adoption of information technology (Min et al. 2008; Zhou 2015). The most commonly used theories in information technology adoption included the theory of reasoned action and its extensions, the technology acceptance model (TAM), the extended TAM and the unified theory of use and acceptance of technology. Less studied explicitly, the importance of context and sociocultural factors in affecting or mediating mobile adoption was nonetheless emphasised (Crabbe 2009; Bankole et al. 2011; Donner and Tellez 2008; Najafabadi 2012; Zainudeen and Ratnadiwakara 2011; Berman 2011).

These findings are similar to an earlier review (Ha et al. 2012) which acknowledged the preponderance of TAM as an explanatory framework. Ha et al. (2012) identify key factors for the adoption of mobile banking as perceived cost, perceived risk, perceived usefulness and perceived compatibility. We elaborate on these factors found in the current review within the developmental context plus ease of use, as a key construct of the TAM.

The perceived cost of financial services had mixed evidence as a factor for adoption despite being identified as the main barrier in a number of studies (Alampay and Bala 2010; Bankole et al. 2011; Cruz et al. 2010; Joubert and Van Belle 2009; Lu et al. 2011; Medhi et al. 2009; Sripalawat et al. 2011; Tobbin 2012). It is important to note that perceived cost in this instance is defined in relationship to individual motivations to adoption of mFinance innovations rather than as an access barrier, as seen in the *Inputs* sections. Despite the lack of financial wherewithal being a barrier to adoption of mFinance services (Sripalawat et al. 2011; Tobbin 2012), Cruz et al. (2010) argue that cost is relevant only for specific groups, typically young, low-income males with high education levels. On the other hand, in the case of the EKO mBanking system, Nandhi (2012) reports that one-third of users became inactive following the introduction of transaction charges for deposits and withdrawals.

Lu et al. (2011) suggest that the perception of cost of mobile payment services exists only in the student group and not among salaried workers. On the other hand, low cost was perceived as a positive factor for adoption (Bankole et al. 2011; Joubert and Van Belle 2009), such as in the case of WIZZIT mBanking services (Ivatury and Pickens 2006). Extending the definition of costs as a determining factor in adoption of mFinance services beyond economic measures, Ha et al. (2012) propose that both tangible and intangible costs, such as transaction and switching costs,[3] should be taken into account.

Perceived risk was also found to be negatively associated with behavioural intention (Brown et al. 2003; Cruz et al. 2010; Joubert and Van Belle 2009; Lu et al. 2011; Morawczynski 2009; Rejikumar and Ravindran 2012; Sripalawat et al. 2011; Teo et al. 2012; with the exception of Bankole et al. 2011) while relative advantage

---

[3] Both transaction and switching costs are understood by Ha et al. (2012: p. 223) and are referred "to the efforts required by the user to adopt the service". An example of transaction cost is the time for performing a task, and an example of switching costs is the costs of changing from a platform to another.

(Cruz et al. 2010; Lu et al. 2011; Püschel et al. 2010; Zainudeen and Ratnadiwakara 2011), perceived usefulness (Bankole et al. 2011; Teo et al. 2012) and perceived ease of use (Cruz et al. 2010; Bankole et al. 2011; Püschel et al. 2010; Teo et al. 2012) were found to be positively associated with behavioural intention.

Perceived risk concerns security issues related to the disclosure of personal and financial information (Brown et al. 2003; Shen et al. 2010) and has been identified as a major negative factor to the adoption of mFinance services (Cruz et al. 2010; Ha et al. 2012; Teo et al. 2012) including theft and losses occurring during mRemittances (Medhi et al. 2009). In the same sense, the more positive perceived security and privacy, the more likely intention to use SMS banking was found by Amin and Ramayah (2010). Perceptions of the risks of mBanking had an adverse impact on perceptions of service quality and satisfaction (Rejikumar and Ravindran 2012). Further, risk was identified as a barrier in specific user groups such as women between the ages of 35 and 55 with higher income (Cruz et al. 2010) and not in other groups, as users experienced with online transactions minimised such risks (Sripalawat et al. 2011).

As a consequence of the risk involved in mFinance adoption, trust factors influenced the adoption and usage of mFinance (Joubert and Van Belle 2009; Medhi et al. 2009; Lu et al. 2011; Tobbin 2012). Joubert and Van Belle (2009) find that service provider risk exerted a significant negative influence on adoption. Lu et al. (2011) find that customers' initial trust in mPayment services positively affected their perception of relative advantage, which, in turn, increased their intention to use. Interestingly, similar to earlier findings, trust that had developed during prior Internet payment experiences transferred to mobile environments. Clearly, perceived credibility of the service provider is an important component of trust. This was seen in the case of adopters in Kenya, who acquired a great deal of trust in the new channel due to the marketing of the service by the provider Safaricom and strong pre-existing ties with local prepaid talk-time agents (Crabbe et al. 2009). Berman (2011) reports that M-Pesa users were afraid of losing their money, and in attention to this, Safaricom designed the paper logbook to provide users a feeling of safety, verifying the completion of the transaction.

Trialability, defined as "the extent to which users would like an opportunity to experiment with the innovation prior to committing to its usage" (Agarwal and Prasad 1997, as cited by Brown et al. 2003), has been identified as a factor that influences the initial adoption of cell phone banking (Brown et al. 2003; Brown and Molla 2005). Post-adoption, service quality was found to be a strong predictor for continuance intention (Rejikumar and Ravindran 2012; Zhou 2013).

Perceived compatibility was mentioned as a factor that influenced adoption of mFinance services by various authors (Brown and Molla 2005; Ha et al. 2012; Joubert and Van Belle 2009; Lu et al. 2011; Teo et al. 2012). It was defined as "the extent to which adopting the innovation is compatible with what people do" (Tobbin 2012: p. 3). Their findings also confirmed the effect of perceived ease of use on perceived compatibility, and at the same time, perceived usefulness and perceived compatibility played a mediator role in the relationship between perceived ease of use and behavioural intention. Related to the perception of compatibility, users' adoption was found to be determined by the task—technology fit, understood as "the fit between the technology characteristics and task requirements" (Zhou et al. 2015: p. 760).

Brown et al. (2003) broadly define perceived usefulness as "the variety of banking products and services required by an individual" (Tan and Teo 2000). Crabbe et al. (2009) find that it was the major factor that influences attitude of non-users while sustained usefulness played a minor role. Other studies also found perceived usefulness as an influencing factor (Sripalawat et al. 2011; Bankole et al. 2011; Ha et al. 2012). Dass and Pal (2011) find that the drivers for adoption of mFinance applications among the rural under-banked were the demand for banking and financial services and the difficulties of accessing them.

Ease of use includes convenience, understood as the "time saved by transacting at an m-banking agent store location" (Medhi et al. 2009), which was an important feature for interviewed subjects (Ivatury and Pickens 2006; Medhi et al. 2009). Perceived ease of use has been mentioned by Bankole et al. (2011), Sripalawat et al. (2011), Teo et al. (2012) and Tobbin (2012). Teo et al. (2012) find that the effect of perceived ease of use (PEOU) on perceived usefulness (PU) and perceived ease of use (PEOU) on perceived compatibility (PC) was the most influential determinant of mobile payment acceptance.

It should be noted that interviewed users associated banks with long queues and were impressed by the speed of the mobile payment systems, with which the transaction was completed, even in cases where there were delays. Tobbin (2012) points out that most participants emphasised time saving and convenience as a motivation to use mBanking.

From a cost-benefit perspective, convenience was seen as a key benefit of mobile banking (Shen et al. 2010) and a major influence on the adoption intention of the mobile banking systems. In early adopters who already had multiple mCommerce interactions, convenience was found to be more dominant than trust and risk in determining intention to use mCommerce (Joubert and Van Belle 2009). In addition, the perceived behavioural control, understood as the extent to which an individual perceives the situation is under his or her control, was found to positively influence the convenience perception.

Finally, image, defined as "the extent to which users of mobile payment systems have more prestige and a higher profile, where using these systems is considered a status symbol", is reported by Joubert and Van Belle (2009) as one of the three most significant factors that influence mobile payment systems adoption. Additionally, Lu et al. (2011) find that it strongly increases the intention to use such services and that it is also a strong determinant of behavioural intention. Related to image are subjective norms, understood as "a person's perception that most people who are important to her or him should or should not perform the behaviour in question" (Fishbein and Ajzen 1975, as cited by Amin and Ramayah 2010: p. 3). Subjective norms, in addition to attitudes, were significantly associated with intention to use banking via SMS.

Prospective mFinance users were studied with contrasting findings. Dewan and Dewan (2009) find that most respondents were interested to conduct banking via mobile phones, while an IMTFI study (2011) observes that unbanked respondents preferred banks as mMoney providers, although banked ones preferred mMoney network operators. At the same time, perceived service quality and satisfaction were essential prerequisites for continuance decisions of the customer with mobile banking in Kerala (Rejikumar and Ravindran 2012).

## 3.3 Outputs

The final set of outputs studies ($n = 8$) was most relevant to show actual transformational benefits of mBanking. Yet, only a handful of research studies was found in this category (Arora and Cummings 2010; BCG 2011; Berman 2011; Boadi et al. 2007; Morawczynski 2009; Nandhi 2012; Ndlovu and Ndlovu 2013; Zainudeen et al. 2011).

We found that studies based on theory (Morawczynski 2009) overlapped more with output studies than input studies (Zainudeen et al. 2011). Only three studies (Arora and Cummings 2010; BCG 2011; Morawczynski 2009) concluded with an emphasis on outputs such as financial inclusion, acceleration of the economic growth and employment. The rest of the studies did not address quantitative impacts on mBanking.

Five studies reported non-quantitative impacts (Berman 2011; Boadi et al. 2007; Nandhi 2012; Ndlovu and Ndlovu 2013; Zainudeen et al. 2011). The majority of these studies found economic impacts, specifically issues such as savings, increased income and financial inclusion. In terms of savings, Nandhi (2012) finds an increased ability to save in 90 % of interviewed users. Boadi et al. (2007) also notice cost savings for rural businesses.

Increased income was stated by Berman (2011) for business agents, although an increased business competition was also found. Ndlovu and Ndlovu (2013) report that mobile banking brings economic activity to rural communities. Finally, Zainudeen et al. (2011) consider that CellBazaar extends the market size of the business by connecting buyers and sellers. Aside from economic impacts, Boadi reports improvements in communication, as better information flows for rural businesses. It is important to note that most impact studies were not focused on the poor.

## 3.4 Links Between Stages: From Inputs and Mechanisms to Outputs

In this review, special attention was paid to articles that linked the various stages, inputs to mechanisms to outputs, such that some relationships could be established between the three stages. Seven articles fell under this category;[4] some of them linked inputs to outputs (Arora and Cummings 2010; BCG 2011; Boadi et al. 2007; Zainudeen et al. 2011), while others linked usage of different mBanking systems (mechanisms) to outputs (Berman 2011; Morawczynski 2009; Nandhi 2012). It is worth elaborating on these studies.

In assessing the direct impact of mobile phones on farming and fishing businesses in Ghana, Boadi et al. (2007) find adoption of mCommerce has brought about benefits in terms of better information flows, enhanced marketing activities, operational efficiencies and cost savings for rural businesses. They considered affordability to acquire the handsets and service as the main investment or input that enable this impact.

Arora and Cummings' (2010) Indian case study, focused on *A Little World*, provides insights about how the Zero technological platform created value for the different actors involved. The platform interacts with Near Field Communications (NFC27) technology-enabled mobile phones, contact-less Radio-Frequency Identification (RFID) Smart Cards, integrated biometrics authentication system and a transaction server, which resulted in convenient mobile transaction solutions for branchless banking and financial transactions.

Notably, the system also includes the human factor in the form of Customer Service Points, manned by village women appointed by the local village self-help groups (SHGs) (Arora and Cummings 2010: p. 9). These women were supported by the SHGs, who in turn ensured their trustworthiness and responsible behaviour.

The outputs mentioned include economic, social and environmental factors. In terms of economic impact, the article claims income enhancement from the women's employment for 16,000 individuals. Each customer service point earns Rs. 10 for every enrolment and Rs. 500 or 0.5 % of monthly transactions, which comes to about Rs. 1,000. This total amount is divided between the village (20 %) and the CSP, thus arriving at an average income for each woman of about Rs. 400 (approximately USD 6.30). Even though the calculation is precise, a comparison between the income of these women and a control group is needed to realise a clearer impact. In addition, villager's trips to and from the post office and bank branches are avoided to save money. The article also highlights the profits of the microfinance agents and other business actors.

Linked with the economic impact, some social and environmental effects are mentioned. These include the social recognition and status of employed women, a higher self-esteem which comes from a better control over one's own money, a reduction in rural–urban migration and the change of power dynamics within villages. Furthermore, the identification cards give a sense of identity and empowerment to villagers. The environmental impact is conceptualised, though not tested, as the waste saved when compared to that produced by a physical bank outlet.

---

[4]Four articles were excluded because they referred to potential outputs but failed to provide evidence of real effects.

Zainudeen et al. (2011) point to factors that contributed to emergence of *CellBazaar* and enabled it to reach a wider market, including high mobile penetration; affordability of access; association with Grameenphone, the largest mobile provider; and the entrepreneurial culture of Bangladeshis.

The success of CellBazaar demonstrated positive outputs for users of the service. For example, it enabled buyers and sellers of many kinds of goods and services in all parts of Bangladesh to connect with each other, extending the market size of the business. It also provided convenience for consumers by reducing the need to travel to buy a product and encouraged a thriving business environment.

The BCG (2011) report reviewed the preconditions for adoption of mobile finance services and linked them to the socioeconomic impact of these initiatives. Consumer education, understood as financial literacy, was mentioned as an input. The study quantified impact in terms of financial inclusion, finding that it ranged from a 20 %-point increase in Pakistan (from 21 to 41 %) to a 5 %-point increase in Malaysia (from 90 to 95 %). The other three countries were likely to experience an impact of around 10–12 % points.

Morawczynski (2009) studies the usage of M-Pesa through ethnographic fieldwork of an informal settlement near Nairobi and a farming village in Western Kenya. This study identified several factors (and events) that affected the use of mBanking. For example, the post-presidential election violence in December 2007 increased usage as mBanking was the only means to access cash. Seasonality and seasonal pricing also influenced how mBanking was used. During the harvesting and planting season, farmers solicited funds to pay for seeds and fertilisers. Due to seasonal pricing, some farmers bought stock when the prices were lower. From a gender perspective, the author found that women frequently used mBanking services to store their "secret savings" to decrease the risk of money being stolen or found by their domineering husbands.

These findings about uses of the application were interpreted and followed by findings about outputs. The most significant of the outputs generated by M-Pesa usage was a reduction in vulnerability, a measure little studied in other mFinance research. mFinance usage enabled urban migrants and subsistence farmers access to financial assets during the post-election period and "hunger months". It also provided a platform through which funds could be instantly sent to address urgent situations such as the onset of illness. A major determinant of adoption was ease of use, where recipients did not need to wait for the money to physically travel from the city. M-Pesa also helped rural–urban migrants to maintain their social networks by fostering money transfers between urban centres and rural areas. The gender perspective on beneficial outputs suggests that the application helped to reduce the vulnerability of women by providing them with a safe storage place, while simultaneously providing more financial autonomy and decision-making.

The M-Pesa application also facilitated another important outcome—it helped users to generate additional income. By sending money weekly or biweekly, the total amount remitted increased by 20–40 %. The recipients saved money on the transfer as they no longer needed to pay travel expenses when retrieving their cash. Finally, the application extended the network of potential remitters and lenders. Subsistence farmers found it easier to acquire small amounts of money from a larger base of contacts during the lean season. As a consequence, there was an increase in the gross remittance of inflows.

Departing from the techno-optimism and lack of negative results found in the literature review, this study took a critical perspective on the impact of mFinance usage. The author noted that, in some instances, M-Pesa usage engendered less than ideal outcomes—it weakened relations between urban migrants and their rural relatives because the former decreased their visits home.

Berman (2011), focusing on M-Pesa, describes the different uses across three distinct areas. In rural areas there were small withdrawals, with some flows back to urban areas to support family and students settling into new lives. Users included illiterate persons who relied on clerks to operate the system. In urban areas, however, the system was frequently used for business (mPayments), constituting greater amounts of money. In contrast, poor customers on the mainland were more regular in their usage of M-Pesa as savings accounts.

Evidence for outputs in the 2011 Berman study is mixed. On the one hand, the service created more business opportunities for agents, increasing their incomes. At the same time, it helped create a number of jobs. On the other hand, agents in Mombasa and Likoni referred that the proliferation of M-Pesa agencies brought about increased competition. Whether this resulted in better prices and services for marginalised consumers is unclear.

Nandhi (2012) investigates the usage of EKO, India's Simplibank mBanking system, finding that a high percentage of users save in EKO for emergencies, which is considered a robust substitute to many informal savings mechanisms, as well as a bank account. At the same time, the service is used in conjunction with, or as complementary to, existing saving practices.

The outputs reported in EKO were diverse. Ninety percent of users stated that their ability to save had increased after opening an EKO account due to the following reasons: (1) the service was much safer than keeping cash on hand; (2) the mobile account helped users to avoid wasteful expenses and to save, thus improving saving habits; and (3) small amounts and more frequent savings were more feasible.

## 4 Discussion and Conclusion

The analysis of 41 research studies on mFinance in developing countries is illuminating not only in what it reveals but in the shadows that permeate the field. The findings indicate that most of the studies were in the mechanism stage, suggesting that there is sufficient academic interest in investigating the factors that lead to success of mFinance initiatives. It is however heartening to note that the emphasis has moved from technological inputs and measures access. However, the fact that success within this mechanism's frame limits itself to adoption suggests that we are at a fairly nascent phase of understanding the deep impacts that such a technological revolution could offer marginalised populations.

This is not to suggest that there is a sophistication to the unique study of mFinance. The theoretical models applied to garner understanding are borrowed from other disciplines, and the lack of consensus about the key factors may lead one to imagine that a number of factors are a major influence underlying adoption of mFinance. While such approaches may be an inherent characteristic of the academic approach, there is the potential and fairly serious risk of confusing practitioners and policymakers seeking to guide investment to initiatives that produce impactful results.

From this review we consider two issues as significant. First, trust is highlighted as a mechanism factor that leads to adoption. As the poor relies more on physical money and face-to-face relationships and mediations to exchange money, issues of trust may be important for future research, to understand more of its functions and how to manage it. It is important to note how the literature reports that trust can be transferred, being this a remarkable feature to be applied by mFinance practitioners. Second, the review has come through interesting literature, discussing and deepening issues of affordability. Even though it is considered a main barrier for adoption, the literature shows that cost is relevant only for some groups, showing that the poor should be thought of as a heterogeneous group. That being said, it is the rare study that focused on the poor.

Indeed, despite the rhetoric (*potential* is a term much associated in the literature with mFinance) and the related optimism, there is little evidence that mFinance has made a substantive impact on the well-being and empowerment of the poor. Certainly there is evidence that mobile phones are being widely accessed and used by the poor; the uptake of mFinance services is sketchy at best, with little conclusive evidence of their benefits to those at the bottom of the pyramid.

Assessing the few papers that focus on outputs from the perspective of the ICT4D Value Chain (Heeks and Molla 2009), it appears to be that almost all the outputs may be categorised as outcomes of financial inclusion, cost savings and improved communication. Nonetheless, there is no study bringing evidence of net beneficial financial impacts, comparing regular finance services and transformational mFinance ones and balancing invisible costs to access the latter. In our opinion the next aims to be assessed in mFinance research projects should focus on development impacts as outlined previously in terms of capabilities and empowerment.

Towards this end, the definition of impact has largely been examined within a financial lens, ignoring the changes that are wrought in individual and community contexts. As one example, broadening the impact of mFinance to encompass gender perspectives may contain revelations about the pressures being exerted upon traditional patriarchal structures by the access to savings, investment and entrepreneurship by women with mobile phones and the resultant fissures and power struggles that accompany such transformational social change.

Alternative approaches may understand contributions to well-being, including the capabilities approach (Sen 2001) and sustainable livelihood framework, stressing social inclusion and development domains such as improvements in poverty alleviation, living conditions, education, health (World Economic Forum 2011) and gender equality, among others. We hope that this contribution to the field of mFinance may lead to both broader and deeper investigations that shed light on this complex yet potentially extremely rewarding field.

**Open Access** This chapter is distributed under the terms of the Creative Commons Attribution Noncommercial License, which permits any noncommercial use, distribution, and reproduction in any medium, provided the original author(s) and source are credited.

# References

Agarwal, R., & Prasad, J. (1997). The role of innovation characteristics, and perceived voluntariness in the acceptance of information technologies. *Decision Sciences, 28*(3), 557–582.

Alampay, E., & Bala, G. (2010). Mobile 2.0: M-money for the BoP in the Philippines. *Information Technologies and International Development, 6*(4), 77–92.

Amin, H., & Ramayah, T. (2010). SMS banking: Explaining the effects of attitude, social norms and perceived security and privacy. *The Electronic Journal of Information Systems in Developing Countries, 41*(2), 1–15.

Arora, B., & Cummings, A. M. (2010). *A little world: Facilitating safe and efficient M-banking in rural India.* New York: The United Nations Development Programme.

Bankole, F. O., Bankole, O. O., & Brown, I. (2011). Mobile banking adoption in Nigeria. *The Electronic Journal of Information Systems in Developing Countries, 47*(2), 1–23.

Berman, M. (2011). The development, use and cultural context of MPESA in Costal Kenya. *Independent Study Project (ISP) Collection.* Paper 1197 [Online]. Available: http://digitalcollections.sit.edu/isp_collection/1197. Accessed 1 Oct 2013.

Boadi, R. A., Boateng, R., Hinson, R., & Opoku, R. A. (2007). Preliminary insights into m-commerce adoption in Ghana. *Information Development, 23*(4), 253–265.

Boston Consulting Group [BCG]. (2011). *The socio-economic impact of mobile financial services analysis of Pakistan, Bangladesh, India, Serbia and Malaysia.* BCG. Available at http://www.telenor.com/wp-content/uploads/2012/03/The-Socio-Economic-Impact-of-Mobile-Financial-Services-BCG-Telenor-Group-2011.pdf

Brown, I., & Molla, A. (2005). Determinants of Internet and cell phone banking adoption in South Africa. *Age, 20*, 20–29.

Brown, I., Cajee, Z., Davies, D., & Stroebel, S. (2003). Cell phone banking: Predictors of adoption in South Africa—an exploratory study. *International Journal of Information Management, 23*(5), 381–394.

Casanova, A. (2007). Remittances and mobile banking. *Migrant Remittances Newsletter, 4*, 3.

Chib, A., van Velthoven, M., & Car, J. (2014). mHealth adoption in low-resource environments: A review of the use of mobile healthcare in developing countries. *Journal of Health Communication: International Perspectives,* 1–31. doi:10.1080/10810730.2013.864735.

Crabbe, M., Standing, C., Standing, S., & Karjaluoto, H. (2009). An adoption model for mobile banking in Ghana. *International Journal of Mobile Communications, 7*(5), 515–543.

Cruz, P., Neto, L. B. F., Muñoz-Gallego, P., & Laukkanen, T. (2010). Mobile banking rollout in emerging markets: Evidence from Brazil. *International Journal of Bank Marketing, 28*(5), 342–371.

Dahlberg, T., Mallat, N., Ondrus, J., & Zmijewska, A. (2008). Past, present and future of mobile payments research: A literature review. *Electronic Commerce Research and Applications, 7*(2), 165–181.

Dass, R., & Pal, S. (2011). *Exploring the factors affecting the adoption of mobile financial services among the rural under-banked* (W.P. No. 2011-02-02). Ahmedabad: Indian Institute of Management.

Datta, A., Pasa, M., & Schnitker, T. (2001). Could mobile banking go global? *The McKinsey Quarterly, 4*, 71–80.

Dewan, S. (2010). *Past, present and future of m-Banking research: A literature review*. ACIS 2010 proceedings, Paper No. 84 [Online]. Available: http://aisel.aisnet.org/acis2010/84. Accessed 6 Oct 2013.

Dewan, S. M., & Dewan, A. M. (2009, December). Young consumers' m-banking choice in urban Bangladesh: Preliminary indication. In *12th international conference on Computers and Information Technology*, (pp. 121–126). Dhaka.

Donner, J. (2007). M-banking. Extending financial services to poor people. *id21 insights*, IDS, University of Sussex [Online]. Available: http://r4d.dfid.gov.uk/PDF/Articles/insights69.pdf. Accessed 17 Oct 2013.

Donner, J. (2008a). Research approaches to mobile use in the developing world: A review of the literature. *The Information Society, 24*(3), 140–159.

Donner, J. (2008b). A couple of very good m-banking review articles. *Most mobiles...*, [Online]. Available: http://jonathandonner.com/archives/41. Accessed 24 May 2008.

Donner, J., & Tellez, C. A. (2008). Mobile banking and economic development: Linking adoption, impact, and use. *Asian Journal of Communication, 18*(4), 318–332.

Duncombe, R. (2009). *Assessing the potential for mobile payments in Africa: Approaches and evidence from Uganda* (Development Informatics Working Paper No, 41). Manchester: Centre for Development Informatics.

Duncombe, R., & Boateng, R. (2009). Mobile phones and financial services in developing countries: A review of concepts, methods, issues, evidence and future research directions. *Third World Quarterly, 30*(7), 1237–1258.

Economist. (2008). Send me a number. *The Economist* [Online]. Available: http://www.economist.com/node/10286141. Accessed 18 Oct 2013.

Fishbein, M., & Ajzen, I. (1975). *Belief, attitude, intention, and behavior: An introduction to theory and research*. Reading: Addison-Wesley.

García, J., Wilke R., & Navajas, S. (n.d.). *M-Banking. Extending the reach of financial service through mobile payment systems* (The Multilateral Investment Fund Concept Note). Available at http://idbdocs.iadb.org/wsdocs/getDocument.aspx?DOCNUM=1328322

Ha, K. H., Canedoli, A., Baur, A. W., & Bick, M. (2012). Mobile banking—insights on its increasing relevance and most common drivers of adoption. *Electronic Markets, 22*(4), 217–227.

Heeks, R., & Jagun, A. (2007). *m-Development: Current issues and research priorities*. Short paper by Development Informatics, IPDM, University of Manchester [Online]. Available: http://www.sed.manchester.ac.uk/research/events/conferences/documents/mobiles/mDevelWorkshopReport.pdf. Accessed 18 Oct 2013.

Heeks, R., & Molla, A. (2009). *Impact assessment of ICT-for-development projects: A compendium of approaches*. Manchester: University of Manchester Institute for Development Policy and Management (IDPM).

Hossain, M. S., & Khandanker, M. R. A. (2011, December). Implementation challenges of mobile commerce in developing countries-Bangladesh perspective. In *14th International Conference on Computer and Information Technology*, (pp. 399–404). Dhaka.

Hughes, N., & Lonie, S. (2007). M-PESA: Mobile money for the "unbanked" turning cellphones into 24-hour tellers in Kenya. *Innovations, 2*(1–2), 63–81.

IMTFI. (2011). *Nigerian mobile money knowledge and preferences: Highlights of findings from a recent mobile money survey in Nigeria*. Institute for Money, Technology and Financial Inclusion Working Paper 2011-1.

International Telecommunication Union [ITU]. (2013). *The world in 2013: ICT facts and figures*. ITU. Available at http://www.itu.int/en/ITU-D/Statistics/Documents/facts/ICTFactsFigures2013-e.pdf

Ivatury, G., & Mas, I. (2008). The early experience with branchless banking. *CGAP Focus Note*, (46), 1–16. Available at http://papers.ssrn.com/sol3/papers.cfm?abstract_id=1655257

Ivatury, G., & Pickens, M. (2006). *Mobile phone banking and low-income customers: Evidence from South Africa*. Washington, DC: Consultative Group to Assist the Poor. Available at http://www.cgap.org/sites/default/files/CGAP-Mobile-Phone-Banking-and-Low-Income-Customers-Evidence-from-South-Africa-Jan-2006.pdf

Jagun, A., Heeks, R., & Whalley, J. (2007). *Mobile telephony and developing country microenterprise: A Nigerian case study* (Developing Informatics Working Paper Series, No. 29). IDPM, University of Manchester, Manchester.

Jones, C., & Du Toit, C. (2007). Boom time for e-banking. *IT Web* [Online]. Available: http://www.itweb.co.za/sections/computing/2007/0707121040.asp?S=IT%20in%20Banking&A=ITB&O=FRGN. Accessed 16 Oct 2013.

Joubert, J., & Belle, J. P. V. (2009). The importance of trust and risk in M-commerce: A South African perspective. *PACIS 2009 Proceedings* (Vol. 96). Hyderabad.

Karunanayake, A., De Zoysa, K., & Muftic, S. (2008, August). Mobile ATM for developing countries. In *Proceedings of the 3rd international workshop on Mobility in the evolving internet architecture* (pp. 25–30). Seattle.

Klein, M. U., & Mayer, C. (2011). *Mobile banking and financial inclusion: The regulatory lessons* (World Bank Policy Research Working Paper Series No. 5664). Washington, DC: World Bank.

Kleine, D. (2010). ICT4WHAT?—Using the choice framework to operationalise the capability approach to development. *Journal of International Development, 22*(5), 674–692.

Lu, Y., Yang, S., Chau, P. Y., & Cao, Y. (2011). Dynamics between the trust transfer process and intention to use mobile payment services: A cross-environment perspective. *Information and Management, 48*(8), 393–403.

Mariscal, J., & Flores-Roux, E. (2011). *The development of mobile money systems* (Documento de Trabajo No. 256). México: CIDE.

Medhi, I., Gautama, S. N., & Toyama, K. (2009, April). A comparison of mobile money-transfer UIs for non-literate and semi-literate users. In *Proceedings of the SIGCHI Conference on Human Factors in Computing Systems* (pp. 1741–1750). Boston.

Mendes, S., Alampay, E., Soriano, E., & Soriano, C. (2007). *The innovative use of mobile applications in the Philippines – Lessons for Africa*. Sida, Department for Infrastructure and Economic Development [Online]. Available: http://siteresources.worldbank.org/EXTEDEVELOPMENT/Resources/20071129-Mobiles_PH_Lessons_for_Africa.pdf. Accessed 18 Oct 2013.

Min, Q., Ji, S., & Qu, G. (2008). Mobile commerce user acceptance study in China: A revised UTAUT model. *Tsinghua Science and Technology, 13*(3), 257–264.

Morawczynski, O. (2009). Examining the usage and impact of transformational m-Banking in Kenya. In *Internationalization, design and global development* (pp. 495–504). Berlin/Heidelberg: Springer.

Mousumi, F., & Jamil, S. (2010). Push pull services offering SMS based m-Banking system in context of Bangladesh. *International Arab Journal of e-Technology, 1*(3), 79–88.

Najafabadi, M. (2012). Identifying barriers of mobile marketing in the agricultural section: A case study in Iran. *International Journal of Mobile Marketing, 7*(2), 78–85.

Nandhi, M. A. (2012). *Effects of mobile banking on the savings practices of low income users – The Indian experience* (Working Paper 2012-7). Irvine: Institute of Money, Technology and Financial Inclusion.

Ndlovu, I., & Ndlovu, M. (2013). Mobile banking the future to rural financial inclusion: Case study of Zimbabwe. *IOSR Journal Of Humanities And Social Science, 9*(4), 70–75.

Ngai, E. W., & Gunasekaran, A. (2007). A review for mobile commerce research and applications. *Decision Support Systems, 43*(1), 3–15.

Porteous, D., & Wishart, N. (2006). m-Banking: A knowledge map. *Information for Development Program (infoDev)* [Online]. Available: http://www.infodev.org/en/Publication.169.html

Prahalad, C. K. (2006). *The fortune at the bottom of the pyramid*. Delhi: Pearson Education India.

Püschel, J., Mazzon, J. A., & Hernandez, J. M. C. (2010). Mobile banking: Proposition of an integrated adoption intention framework. *International Journal of Bank Marketing, 28*(5), 389–409.

Rejikumar, G., & Ravindran, D. S. (2012). An empirical study on service quality perceptions and continuance intention in mobile banking context in India. *Journal of Internet Banking and Commerce, 17*(1), 1–22.

Rosemberg, J. (2008). Mobile banking needs "standardized innovation. *Mobile Banking Blog*. CGAP Technology Program [Online]. Available: http://technology.cgap.org/2008/05/15/mobile-banking-needs-standardized-innovation/. Accessed 26 May 2008.

Scornavacca, E., & Hoehle, H. (2007). Mobile banking in Germany: A strategic perspective. *International Journal of Electronic Finance, 1*(3), 304–320.

Sen, A. (1988). The concept of development. *Handbook of Development Economics, 1*, 9–26.

Sen, A. (2001). *Development as freedom*. Oxford: Oxford University Press.

Shen, Y. C., Huang, C. Y., Chu, C. H., & Hsu, C. T. (2010). A benefit–cost perspective of the consumer adoption of the mobile banking system. *Behaviour & Information Technology, 29*(5), 497–511.

Singh, S. (2009). Mobile remittances: Design for financial inclusion. In *Internationalization, design and global development* (pp. 515–524). Berlin/Heidelberg: Springer.

Sripalawat, J., Thongmak, M., & Ngarmyarn, A. (2011). M-banking in metropolitan Bangkok and a comparison with other countries. *The Journal of Computer Information Systems, 51*(3), 67–76.

Tan, M., & Teo, T. S. (2000). Factors influencing the adoption of Internet banking. *Journal of the Accounting Information Systems, 1*(1es), 5.

Teo, A. C., Cheah, C. M., Leong, L. Y., Hew, T. S., & Shum, Y. L. (2012). What matters most in mobile payment acceptance? A structural analysis. *International Journal of Network and Mobile Technologies, 3*(3), 49–69.

Tobbin, P. (2012). The adoption of 'Transformational Mobile Banking' by the unbanked: An exploratory field study. *Communications and Strategies, 86*, 103.

van Reijswoud, V. (2007). Mobile banking: An African perspective. *World Dialogue on Regulation for Network Economies*. Available at http://www.regulateonline.org/content/view/948/63/

Wishart, N. (2006). *Micro-payment systems and their application to mobile networks*. Washington, DC: infoDev/World Bank.

World Bank (2002) *WBI courses and events: New technologies for small and medium-size enterprise finance* [Online]. Available: http://info.worldbank.org/etools/docs/library/159695/smetech/overview.html. Accessed 17 Oct 2013.

World Economic Forum. (2011). *The mobile financial services development report 2011*. Available at http://www3.weforum.org/docs/WEF_MFSD_Report_2011.pdf

Zainudeen, A., & Ratnadiwakara, D. (2011). Are the poor stuck in voice? Conditions for adoption of more-than-voice mobile services. *Information Technologies and International Development, 7*(3), 45.

Zainudeen, A., Samarajiva, R., & Sivapragasam, N. (2011). Cell Bazaar: Enabling m-commerce in Bangladesh. *Information Technologies and International Development, 7*(3), 61.

Zhou, T. (2013). An empirical examination of continuance intention of mobile payment services. *Decision Support Systems, 54*(2), 1085–1091.

Zhou, B. (2015). ICTs and opinion expression: An empirical study of new- generation migrant workers in Shanghai. In A. Chib, J. May, & R. Barrantes (Eds.), *Impact of information society research in the global south*. New York: Springer.

# An Analytical Framework to Incorporate ICT as an Independent Variable

Matías Dodel

This chapter presents an analytical framework to guide the assessment of information and communications technologies' (ICTs) impact on individual-level development (or wellbeing). Based on the content analysis methodology, we argue that the amount of polysemy and lack of common basic guidelines in ICT's research fields constitute one of the main barriers both to the incorporation of ICT into a broader research problems spectrum (outside the ICT researchers' communities) and, consequently, to widen ICT's impact research. After a synthesis of the historical development of the digital divide concept (a framework for the analysis for digital inequalities), we discuss and select some plausible analytical models to assess ICT's impact on wellbeing. Based on Selwyn's approach, we advocate the idea that every researcher testing an ICT-related hypothesis should analyse at least three stages of hierarchical digital achievements (access, usage and appropriation) plus one last divide stage: ICT's outcomes (measured by the effect of previous stages on the dependent wellbeing variable). Finally, we propose some guidelines for the applications of this framework and present an actual case of use, showing how this framework guided the research design of this author's SIRCA II's project, which tested the effect of digital skills on education-to-work transition.

## 1 Introduction

From the richest to the poorest countries, ICT has already shaped the way we live in contemporary societies: in the Information Age, knowledge is a critical resource and information is a primary commodity (Flor 2001: p. 3). And, thus, the way we

---

M. Dodel (✉)
Department of Social & Political Sciences, Catholic University of Uruguay, Montevideo, Uruguay
e-mail: matias.dodel@ucu.edu.uy

work and research, talk and communicate with each other (Castells 2005: p. 88) and also the way we participate in our societies have changed dramatically.

Taking these ideas as foundations, we hypothesize that from a sociological point of view, the centrality of ICT in our lives has achieved a similar level of relevance as, e.g. progress education or gender, and thus it should be understood as a key independent variable for studying an array of diverse social phenomena.

This chapter attempts to address the problem that although most social scientists share the general idea of ICT relevance to micro- or individual-level development achievement (from now on referred to as wellbeing[1]), there are relatively few studies that consider the effects of ICT outside the fields of information society or ICT for Development (ICTD). There are probably multiple and complex causes for this, but based on the experiences acquired from collaborating with scientists and colleagues who were not specialized on the topic, we propose that at least one cause arises from some characteristics of ICT studies' field itself.

As ICT inequalities (a concept we have embedded in the digital divide) are a multidimensional phenomenon that involves several theoretical levels, conceptualization and operationalization are particularly difficult to integrate into nonspecific ICT research. Furthermore, we believe that the main cause of this problem comprised of a combination of (1) polysemy regarding key terminology in the information society field and (2) the lack of consensus on basic analytical and methodological measurement guidelines, consequently raising the barriers to entry in the field.

From this perspective, this chapter aims to present an analytical framework that links ICT inequalities to wellbeing, a digital divide analytical model that will be useful to assess ICT's effect as an independent variable or dimension on any non-ICT wellbeing-dependent variable. We argue that this model or tool is crucial because it aids to reduce some of the barriers to entry mentioned above, for a particular audience outside our field.

With this goal in sight, the document is structured into three sections:

1. An overview of the conceptual development and evolution of the digital divide (understood as different levels of inequality in ICT). After stating the consequences of the lack of basic analytical consensuses, we emphasize the importance of addressing multiple hierarchical and coexisting levels of divide (access, usage, appropriation), as well as the assessment of a "final" level or dimension focusing on the impacts of ICT on the wellbeing.
2. The proposal of two plausible analytical models (one using a capabilities approach and another based on Selwyn's model of digital divide) for addressing ICT inequalities, reasonable but practical enough to be adopted by researchers outside specific ICT studies. Selwyn's model (2004, 2010) not only successfully

---

[1] For example, quality of life, social equity, education, health and income levels of individuals and families.

complies with the key issues reviewed in the first section but is widely operational. Therefore, we advise the adoption of this model.
3. The description of a case of use of the analytical framework, presenting some basic guidelines for its application while showing how it guided the authors' research design in testing the effect of digital skills on education-to-work transition.

## 2 The Evolution of ICT Inequalities' Conceptualization and Their Link with Development

### 2.1 The Problem of "ICT4D Polysemy"

It is highly likely that whoever tries to enter the study of information society and ICT4D faces a first great barrier typical of fields with relative novelty: the lack of major agreements about main conceptual categories and operative terminologies. The relative time proximity of the phenomenon, as well as the complexity and acceleration of ICT developments, makes it extremely difficult to reach a point of maturity of the field which allows generating minimum agreements and consensus.

This characteristic of the field is considered to be of relative seriousness, to such an extent that specialized bibliography of more specific subtopics (e.g. digital literacy) begins its works stating explicitly the difficulty of this polysemy and the lack of conceptual maturity causes. As Lanksher and Knobel propose: "the most immediately obvious facts about accounts of digital literacy are that there are many of them and that there are significantly different kinds of concepts on offer" (Lankshear and Knobel 2008: p. 2).

This makes it difficult to select a suitable set of relevant background and findings, as well as to establish a proper analytical conceptual framework that, in terms of Bunge, possesses both a pertinent *range* to achieve a theoretical level that allows its operation and connection with other theories (Bunge 1999: p. 176) and *depth* to give account of its components and mechanisms ("translucent box" or at least "grey box" models in contrast to "black box" models; Bunge 1999: p. 178–180).

Furthermore, Peña-López (2009: p. 42) states that this conceptual ambiguity has severe sociopolitical consequences: without clear conceptual and analytical frameworks, it is difficult to evaluate the impact and reduction of the divide. This opens the path for political discretional ICT strategies and policies to the detriment of technical and social criteria.

### 2.2 Digital Divide

Despite the current extension of the term, the first governmental enunciation of "digital divide" is recent and can be attributed to the 1990s Clinton Administration in the United States (Peña-López 2009: p. 42). The fact is, as Rivoir et al. (2010: p. 1) state, the concept is complex and has suffered diverse mutations in the course of time.

The recent popularization and increasing relevance of the term has produced a higher inclination of national states to influence ICT development, but it did not generate major agreements in the conceptualization of the phenomenon: "Yet, while substantial policies are being put into place to combat the digital divide, much of the surrounding debate remains conceptually oversimplified and theoretically underdeveloped" (Selwyn 2004: p. 343).

Moreover, beyond the vague common idea of a division generated or caused by ICT, the diversity of uses and conceptualization of the divide is huge.[2] As a way of presenting a brief summary, we have reviewed empirical studies dedicated to studying divides related to ICT (e.g. ITU 2010; Sunkel et al. 2010), benchmarking exercises (Cobo 2009; Peña-López 2009) and conceptual frameworks (i.e. Kaztman 2010; Selwyn 2004, 2010).

Regarding discipline approaches, the wide majority of the reviewed literature has been mainly dedicated to social or socioeconomic dimensions that cause or are consequences of the divide or both. Although due to the characteristics of the issue themselves, the field is intrinsically cross-disciplinary.[3]

## 2.3 Digital Divide's Conceptual Development

We have identified at least four stages considered as key factors in the historical and conceptual development of the divide: (1) discarding excessive technological optimism, (2) criticism to the dichotomous conception of access, (3) studying simultaneous but different levels or stages at which divides exist and (4) the divide's conceptualization model that must include ICT's impact on wellbeing as a final stage. These stages are briefly characterized below.

The overly optimistic beginnings of the divide's studies proposed that the mere introduction of ICT on country or household levels or both would practically revert poverty and inequality historical conditions. These technological deterministic origins, which stood on the initial potentials of technological breakthroughs, have almost no current serious adherents. Arguments for leaving behind this initial optimism of ICT4D are supported by recent empirical evidence at both international

---

[2] It is worth stating explicitly that this position is not new or innovative on this document. Already in 2003, Fink and Kenny faced the same dilemma: "The term of digital divide came to prominence more for its alliterative potential than for its inherent terminological exactitude. In another world we might have had the 'silicon split,' the 'gigabyte gap' or the 'pentium partition.' As such, it would be wrong to ponder for too long on what, exactly, should be meant by the term" (Fink and Kenny 2003: p. 2).

[3] For example, texts from education (OECD 2010; Prado et al. 2009), economics/business (e.g. White et al. 2011) and psychology (Thatcher and Ndabeni 2011; Reig 2012) have been revised.

(e.g. Peña-López 2009; ITU 2010) and national country levels (e.g. in Uruguay, Rivoir et al. 2010; Moreira 2010; in Brazil, Cetic.br 2009, among most other countries). In general terms, the literature proposes that if there are no intentional or planned interventions of men through ICT public policies or digital inclusion programmes, the effects of technologies on the society will be more regressive than redistributive due to current or prior socioeconomic inequalities (Hargittai 2008: p.942–943; PNUD 2009: p. 211). In the prophetic words of one of the "fathers" of informational society: "The information age does not have to be the age of stepped-up inequality, polarization and social exclusion. But for the moment it is" (Castells 2005: p. 403, as cited in Selwyn 2004: p. 342).

The next factor is related to the binary or simplistic[4] conceptualizations of the divide. The criticism to this is sustained by authors like Hargittai (2008), Selwyn (2010) and Van Dijk and Van Deursen (2010) who have considered insufficient the idea that the mere access to technology will end inequalities. Consequently, the study of the issue has been refined to much more diverse and complex dimensions: quality of access (e.g. characteristics of equipment, connection speed), ICT effective uses (different types of usage), presence of social support networks, digital literacy or ICT skills and notions related to appropriation (Hargittai 2008: p. 937, based on a revision of several authors).[5] Nonetheless, increasing the complexity of the concept started an explosion of approaches on new informational society inequalities, and today, there are almost as many perspectives as authors studying the subject (e.g. PNUD 2009; Peña-López 2009; Hargittai 2002).

In a later stage of development, some scholars have suggested that the divide must be further refined by studying simultaneous but different levels or stages at which divides exist (Hargittai 2002): Norris (2001: p. 4) signalled three levels of inequalities (between countries, within countries and in participation within countries; taken from Hargittai 2002), while DiMaggio and Hargittai (2001) proposed more complex approaches by suggesting five dimensions of possible divides (technical means, autonomy of use, use patterns, social support networks and digital skill). Van Dijk's (2005) *model of successive kinds of access to digital technologies* (motivational, material, skills and usage accesses) described four successive stages or kinds of access that are supposed to be cumulative (Van Dijk 2005: p. 21). Not only do we agree with this perspective, we suggest that this is the only way the divide can be conceptually and empirically addressed adequately.

---

[4] For example, as Selwyn (2004: p. 344–345) states, a position addressed by Devine (2001: p. 28) or Edwards-Johnson (2000: p. 899) at the start of divide's studies.

[5] Mark Warschauer (2002) was one of the first scholars to systematically promote the necessity to evolve the conceptualization of the divide, including digital capabilities and digital literacy as key factors in it (Peña-López 2009: p. 79). Peña-López argues the role played by Paul DiMaggio and Eszter Hargittai (2001a, b) was central, as they contributed enormously to the shift in focus from the dichotomist divide to digital inequalities in a more comprehensive way, including also their concern about the divide on skills (Peña-López 2009: p. 79). For a more complete synthesis, see Peña-López (2009: p. 7).

Finally, we argue that there is a last stage which is still unanswered: *any divide conceptualization model must also include ICT's impact on wellbeing or development* by integrating the concept of "development" to the name of the subject field itself. In fact, only a few studies include this dimension on an individual or personal level.

From the approaches that could be categorized in this final stage, we have opted to present here two of the plausible hierarchical divide's conceptualizations which clearly point to the impact or outcome dimension as the last stage in the hierarchy: (1) a comprehensive (but thus complex) adaptation of Amartya Sen's capabilities approach (from now on referred to as CA) to ICT (in several texts as Alampay 2006, Zheng 2007 or Forester and Handy 2008) and (2) a more concise but nevertheless coherent and easier to operationalize divide's model based on Selwyn's approach (2004, 2010) of four hierarchical and coexisting stages of the divide (access, use, appropriation and results/outcomes).

However, before concisely describing these two alternatives, it is important to emphasize the criticism proposed by Hargittai about the oversimplification of the discussion regarding the relationship between ICT and social reproduction or mobility: even after recognizing ICT's potential effect on equity, it is naïve to suppose that ICT will nullify the pernicious effect of previous inequalities such as background social class, gender or socioeconomic status (Hargittai 2008: p. 942).

## 2.4 Two Plausible Analytical Approaches: Sen's and Selwyn's Frameworks

The selection or recommendation of a divide's analytical framework to adopt is not an easy task. There are many alternatives (as reviewed in the former sections), some classified as good, others classified as bad and probably many reasonable ones. Between all the plausible digital divide models, the most comprehensive are generally too complex or have huge barriers to entry to introduce in non-ICTD researches, and the most basic ones tend to lack a proper theoretical coherence or a wide range of application (i.e. social inequality, participation, education and health studies).

In this section, we will present and briefly discuss two different analytical frameworks in compliance with the conceptual development previously identified. As the discussion of each model will reflect, both have strengths and weaknesses, making the final selection both a matter of practicality and personal preferences.

## 2.5 A Capabilities Approach-Based Analytical Framework: Exhaustive but Not Cost-Efficient

Widely adopted in economics and poverty studies, the CA is far from an ICT-specific conceptualization. CA is a more general human development paradigm that considers the expansion of individual freedoms (or agency in CA terms) as a main development goal (Heeks and Molla 2009: p. 33). As argued by Zheng, Sen proposes that reaching a substantive level of individual freedom is the only means of objective and social development (Sen 1999, as cited in Zheng 2007: p. 2). In this sense, agency is the goal and central concept of the CA, understood as the ability to pursue and achieve goals that people value or have reason to value (Alkire 2005: p. 1–2).

Due to this general approach, CA can be applied to almost any area of social research, particularly at individual or household levels; e-development, ICTD or information society is no exception.

The driving idea of CA's application to ICT, in accordance to the mentioned hierarchical logic we adhere to, is that the possession of ICT does not irreducibly result in an increase of wellbeing. While access to these goods is necessary, it is insufficient for assuring ICT's impact on the capabilities and performances of people. For ICTs to have a positive effect on wellbeing, CA's authors argue that a mediation or simultaneous presence of several other factors is required (Alampay 2006: p. 9).

Although Fig. 1 tries to present a condensed representation of CA's analytical framework to assess ICT's impact, complexities of the approach itself and its glossary render it important to further explain some of its core concepts. According to the objectives of this section, there are three key (functioning, capabilities and commodities) and two subsidiary (characteristics, conversion factors) concepts of CA to work with.

ICT's goods and services are the first entry of the process. Although the concept of commodities is probably more mainstream, its interaction with capabilities is not as direct and has major consequences for the CA and scope of this document. Zheng (2007: p. 2) explains that for Sen, commodities become relevant in so far as their *characteristics* enable the individuals to generate capabilities from their properties.

In the opposite corner of the figure, functionings refer to the huge number of activities and states that make the actual wellbeing of individuals; they are the "beings and doings" of people (Zheng 2007: p. 2). On the other hand, capabilities reflect the concept of freedom that was previously emphasized by the notion of

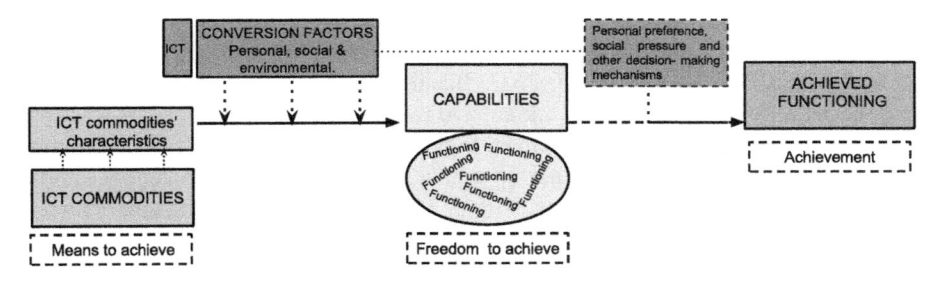

**Fig. 1** Adaptation of capabilities approach to ICT and wellbeing (Source: Based on Zheng (2007) (Adapted from Robeyns (2005)) and Heeks and Molla (2009) (based also on Zheng's work))

*agency*: these refer to a "pool" of performances to which an individual can access at any given time (Alkire 2005: p. 1; Zheng 2007: p. 2). Nevertheless, effective functionings are the only things researchers would probably be able to assess in the majority of studies: well-paid jobs, formal education achievement, income, etc. This would be the area or dependent variable in which ICT may have an impact.

However, Fig. 1 shows that CA's framework is even more complex, in the way that not all individuals are able to convert or generate capabilities in equal rates from the same features. This is due to differences in their *conversion factors*, which may be personal (i.e. literacy, cognitive ability, gender), social (i.e. culture, norms, values) and environmental (Zheng 2007: p. 2).

Conversion factors may also be considered capabilities themselves, which mediate the conversion of ICT's characteristics (Alampay 2006: p. 9; Garnham 1997: p. 32): literacies, knowledge on the use of ICT and the understanding of the implications of using information as a resource, to name just a few of them. Moreover, ICT commodities may also act as conversion factors or conversion factor enablers (Heeks and Molla 2009: p. 34).

At this point, it is probably clear to the reader that the problem with CA is that as strong and comprehensive its theoretical and philosophical framework is, it is a very difficult framework to understand and apply, especially to a non-ICT researcher. As Heeks and Molla (2008: p. 33) argue, CA is "quite a dense set of ideas that can be hard to understand and translate into practical evaluation terms". For the sake of lowering barriers to entry to the ICT field, while CA provides a strong theoretical model on ICT's impact on wellbeing, it is not the best candidate to use as a first-entry analytical framework.

## *2.6 A Basic Analytical Model: Selwyn's Hierarchical Approach*

Selwyn's digital divide model (2004, 2010) is not only comprehensive, but also specific to the ICT4D field, and presents adequate alternatives and solutions for the already mentioned problems of the conceptual development of the term.

Having a more direct and specific ICT's impact focus, Selwyn considers that it is essential to conceptualize the divide "as a hierarchy of access to various forms of technology in various contexts, resulting in differing levels of engagement and consequences" (Selwyn 2004: p. 351). Basing an analytical framework on this approach seems at least logically correct and reasonable.

As Fig. 2 illustrates, the hierarchical logic is expressed in the plausible but not assured progression from one stage to another, culminating with potential short-term or long-term benefits (Selwyn 2010: p. 351). However, following authors who adopt this approach to the Latin American context, we believe that "more than distinguishing development phases, it is necessary to think in gap levels that occur simultaneously" (Sunkel et al. 2010: p. 12).

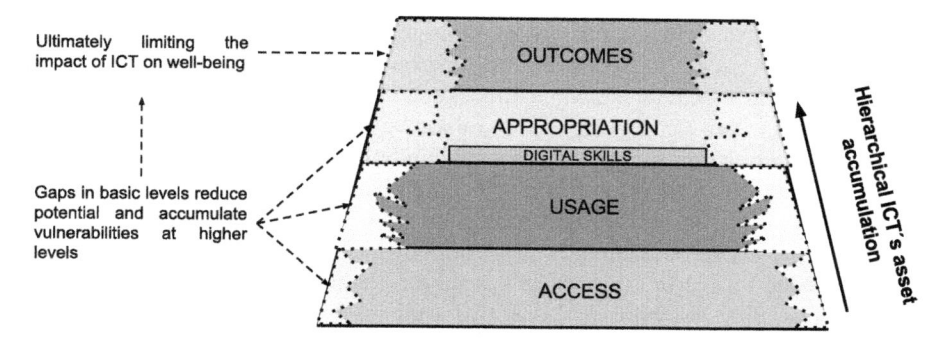

**Fig. 2** Adaptation of Selwyn's digital divide theoretical model (Source: Dodel (2013), based on Selwyn (2010) and Sunkel et al. (2010))

We think that not only assets accumulate, but also disadvantages or liabilities. As Fig. 2 shows, the gaps in lower levels of the divide have consequences to the whole chain of digital achievements and, thus, to the chances of ICT having a positive impact on wellbeing.

As obvious as it may seem at first glance, this disadvantage accumulation is a key point of the framework: ICT's effect is non-assured; it could be null, positive or even some combinations of assets, i.e. of low access and poor usage; and it may have negative effects on several aspects of wellbeing. We think it is useful to remember that we are assessing ICT's effect and not blindly preaching about its potential; it is not only logical but scientifically desirable, we think, to find empirical evidence of situations in which ICT's effect is not clearly positive.

## 2.7 Divide Stages

Going back to the use of the approach as an ICT's impact assessment framework, we have opted to simplify Selwyn's proposed five stages (2004, 2010) into four (collapsing *formal* and *effective* access into one category of access only).[6] We suggest assessing four levels of digital achievements: *access, use, appropriation* and *outcomes*.

The denominations of Selwyn's proposed stages have a secondary advantage: when compared to other models, the simplicity of the terms used makes the framework almost instantly understandable. Obviously, there are some technicalities and nuances, but mostly, this accessibility is one of the strengths of this approach.

A second basic component of this model is that even after taking into account the specific but shareable conceptualizations of each stage, they should not be

---

[6]Following Sunkel et al. strategy (2010: p. 12).

regarded as static and dichotomic. Besides this general definition, each stage's most appropriate indicator differs in relation to the outcome or impact in which ICT's effect must be assessed: the outcome dimension.

Having these two main guidelines in mind, we will address each dimension in some detail in order to discuss the approach limitations. The access stage refers to the availability of the "hard" components of the divide: from general infrastructure and connectivity to specific hardware and software with different purposes.

Selwyn gives a special emphasis on the importance of avoiding falling in analysis centred on the formal or "theoretical" aspect of the concept.[7] Access assessment should focus on accessibility and effective availability of ICT goods. As Selwyn states, we believe that any realistic notion of access must be defined from the demand's or individual's perspective (Selwyn 2004: p. 347).

Usage refers not only to a use or not use dichotomy but also to the usage frequency, places of use, the kind of activities conducted through ICT and the amount of content generation and consumption, among others (e.g. AGESIC-INE 2010; CETIC.br 2009).

There is an enormous amount of usage measurement possibilities, probably even more than on the access stage, making the selection of one or a small quantity of indicators more arbitrary. As we will develop the idea in the next section, the final stage or dependent variable in which the study tries to assess the impact of ICT should be used as a guideline for the selection of the most relevant usage indicators.

In turn, appropriation is a bit more complex and difficult to define and carry out due to its predominantly subjective character. According to Selwyn, appropriation can be understood as "meaningful use of ICT... where the 'user' exerts a degree of control and choice over the technology and its content, thus leading to a meaning, significance and utility for the individual concerned" (Selwyn 2004: p. 349).

In this particular case, alternative definitions of appropriation are very useful, not only due to the ambiguity of the notion but also as they provide more clues about the application of the concept in the literature.

As an example, Prado et al. (2009: p. 87) propose that appropriation must be understood as "the integration and adoption process within user's daily life". They emphasize the fact that technology appropriation results from routine and stand on this idea to measure the concept.[8]

Finally, as we previously hinted, the last stage of the framework refers to what we believe is the sociologically key issue: the outcome, impact and consequences of accessing and using ICT—the ends of engagement of ICT use (Selwyn 2004: p. 349).

---

[7] Many studies about the education gap tend to use excessively gross indicators of digital access, such as "percentage of educational institutions with PC access", which do not take into account the PC/student ratio or the possibility of use or effective use of this ICT by students (Claro et al. 2011).

[8] For them, it is necessary to inquire about the conditions facilitating the technology, perception of the technology as an object, the simplicity of use, perception of usefulness, auto-efficiency, technology in use and satisfaction with the same (Prado et al. 2009: p. 87).

Moreover, this is the area of expertise of the non-ICT researcher: the dependent variable in which he or she is much more skilled than the actual ICTD and information society veterans. He or she will construct better dependent variables, studying not only ICT's effect but several other dimensions that affect these outcomes, controlling confounding effects. For ICTD studies though, this is crucial as this logic enormously strengthens the methodological design and testing of ICT's impact.

Another aspect in which this stage is crucial is related to the choice of the operationalization of the first 3 stages of the divide. As we will elaborate, this task not only depends on the conceptual definition of each stage but also on the dependent variable. This variable should be taken as the main guide to set the bar in what we would like to study about access, usage or appropriation: they should have a logical connection with the sought outcome. The last section of this chapter will expand on this topic and present a fully developed example/case of use of this approach.

## 2.8 Limitations: Motivational Stage and Linkage with General Socioeconomic Inequality

Aside from the (maybe too) general approach (which some may consider as a limitation, but we consider it one of its main strengths), the main weakness of Selwyn's model refers both to the lack of a motivational component and to the absence of a direct theoretical linkage with non-ICT inequalities.

Despite some people conceptualizing a motivational stage as the first level of the divide (Van Dijk J. 2005), motivations or attitudes towards ICT are complex, multicausal and affect all of the three ICT stages of the proposed divide model in circular ways. From the household perspective, positive attitudes towards ICT could encourage the purchase of digital commodities, but on the other side, without the possibility of access, individuals will not be able to get to the usage stage even if absolutely motivated. Also, motivation could increase the chances of usage (Van Dijk 2005), but usage could lead to the acknowledgement of what can be achieved with ICT and, thus, motivation or appropriation or both.

The second weakness of the version we adopted from Selwyn's approach is the lack of any direct linkage within ICT inequalities and socioeconomic-based ones (e.g. social class, Bourdieu's capitals). It is not that Selwyn forgets to include these subjects (Selwyn 2004), but for the sake of simplification, we excluded them in the specific framework.

Obviously, differences in access, usage, appropriation and also motivation related to ICT are determined, at least in part, by socioeconomic inequity (Dodel 2013), and at least some kind of theoretical linkage between the two was already stated in this chapter (the regressive effect of ICT without public policy intervention).

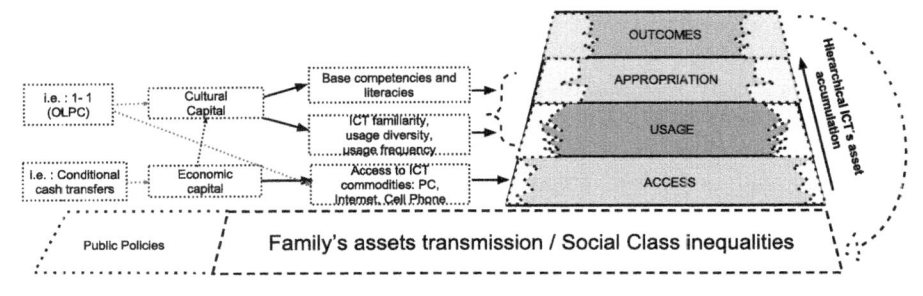

**Fig. 3** Adaptation of Selwyn's digital divide theoretical model taking into account previous socioeconomic inequalities (Source: Dodel (2013), based on Selwyn (2010) and Sunkel et al. (2010))

We think these two limitations are not fatal flaws to the analytical framework, and as Fig. 3 shows, we propose a broader version of the model, taking into account both socioeconomic inequality and motivation/attitudes by including a first and parallel source of inequity: family's asset transmission and, in a lesser amount, public policies.

Borrowing some of Bourdieu's notions (Bourdieu 1986), we propose that similar mechanisms that determine the social class or status prior to any digital divide affect economic, cultural and other capitals of the families, which thus affect both the means to access ICT and the attitudes and base competencies needed to use and engage with technology.

Nevertheless, Fig. 3 points to a third problem related to ICT's impact assessment in general. As ICT inequalities are caused by prior socioeconomic disparities, without the proper confounding controls, ICT's relation with any wellbeing-dependent variable could be overestimated or even completely spurious. One example of this problematic phenomenon would arise if a researcher heads towards the assessment of the effect of a household's Internet connection on educational or health achievements without controlling any measure of income inequality: even if there is a marginal ICT effect, most of the observed relationships would probably fall under the effect of income inequalities. Thus, contextualizing ICT and wellbeing relationship in other sociological issues, both as theoretical and confounding control, stands as crucial as considered in any serious ICT's impact study.

Summarizing, we think Selwyn's approach limitations need to be taken into consideration, but they do not disable its application. Moreover, even if we think Fig. 3 provides a more comprehensive framework, in most of the cases which this chapter addresses (non-ICT-specific research), the basic model (Fig. 2) should be chosen by taking the parsimony criteria.

In the next section, we will present a specific case of use and discuss some criteria on how to apply this framework.

## 3 Operationalization: A Case of Use on the Study of Digital Skills' Effect on Education-to-Work Transitions

### 3.1 Research Problem

Important cumulative experience exists regarding the increasing diversification of formal education and labour market pathways (from now on referred to as pathways) of contemporary Uruguayan society's young people and their serious consequences on the processes of social inequality reproduction and the beginning of social mobility. As the effects of several socioeconomic variables on these pathways have already been stated (socioeconomic origin inequalities, gender, educational achievements, early labour market pathways), we suggest the existence of another determining key factor for the opportunities of social welfare in informational contemporary societies, which has not yet been addressed in the education-to-work transition field: the digital skills (e-skills).

The preceding paragraph summarizes the main hypothesis of the SIRCA II's quantitative research on which the author of this chapter has participated: "ICT and Welfare policies: Digital skills' impact on formal education and labour market pathways of young Uruguayans evaluated by Programme for International Student Assessment (PISA) 2003 (Panel study)". The dependent variable in this research was young Uruguayans' early occupational achievements (if they were able to access a white-collar job at or before the age of 19).

### 3.2 Application Based on the Analytical Framework

We propose a short but useful guideline for the operational process, which can be summarized in a number of steps, with a general, overarching guideline to use an up-to-down strategy: i.e. start from the dependent variable, and then go to the appropriation stage, then usage and finally access (when available):

1. Starting from a theoretical relevance perspective: which ICT dimension is the one which could have a direct impact on the dependent variable?
2. If you are producing primary data, think ahead on how you are going to measure such data. If you are using secondary data, browse your questionnaires or interview guides in order to search for potential variable candidates.
3. In order to select or create a specific variable in a stage, choose wisely based on the characteristic of the study's universe: the level of dissemination of the selected ICT's achievement on the population could be too scarce or almost universal. "Set the bar" using an indicator of reasonable spread.
4. Repeat steps 1 to 3 through lower levels of the divide.

### 3.2.1 Linkage Between ICT and the Dependent Variable

Step one is the most complex but also crucial in an adequate operationalization process. As we were researching education-to-work transition and conceived ICT as a key factor in them, we first needed to establish the theoretical links between our research problem (early labour market achievements) and ICT field; we will briefly present this conceptualization.

As Mills and Blossfeld argue (2006: p. 1), young people in industrialized nations have experienced significant changes in the transition to adulthood in the past decades, particularly due to the rapid dissemination of knowledge networks and the expansion of the new technologies. The new technological or informational paradigm (Castells 2005: p. 88) will severely affect the chances of getting a job or accessing quality occupations or both, especially in the contexts of great socioeconomic inequalities (e.g. Latin America). Thus, further increasing the importance of ICT achievements due to *skill-biased technological change*, Cobo (2009: p. 3) explains that "The acquisition of ICT competencies is increasingly becoming a key requirement for employability".

The way we conceive the problem, it is neither the hardware nor the mere use that is relevant for the desired outcomes. Achievements on a higher stage of the divide are necessary to have skills or abilities related to ICT that could impact on occupational outcomes. We propose that these skills correspond to a low level of the third divide stage: e-competencies and e-skills are very basic types of ICT appropriation (Dodel 2013).

### 3.2.2 The Data

The microdata which constitutes the research's empirical base comes from PISA-L Uruguay (Boado and Fernández 2010), the first panel follow-up survey for PISA (2003) carried out in 2007 to a sample of 2,201 Uruguayan young people (between 19 and 20 years old at that time).

Application of the different divide's dimensions was carried out based on the variables gathered by PISA's original ICT questionnaire and some ICT variables from the student's questionnaire (both from 2003). It is relevant to stress that this data was collected when respondents were 15 years old, giving us strong arguments (temporal precedence) to talk about causality in their effect on dependent variable at the age of 19.

The first of these questionnaires contained several specific ICT variables regarding access (availability of ICT goods for use at home, school and other places), usage (from the year of first use and general frequency to the frequency of specific activities performed within certain programmes), e-skills (perception or confidence on his or her ability to perform certain tasks) as well as attitude towards ICT, among others.[9]

---

[9] For further details, see the PISA 2003 Information Communication Technology Questionnaire.

Although it would be possible to develop an exhaustive analysis based on this whole information, given the research design choices (we chose to prioritize confounding control with an already large amount of variables), we opted to select only four ICT indicators as proxy of achievements at the three first levels of the divide.

The main criterion for selecting these indicators was to focus on a reduced number of important digital achievements, identifying key aspects of each dimension of the divide without overloading the already broad group of variables to be included in multivariate models.

### 3.2.3 The Proper Application

Appropriation: Office User E-Skills

Not addressing here all the complexities of the final construct used in the research (see Dodel 2013 for a similar discussion), we opted to create an e-skills variable based on the already stated theoretical conceptualization: a measure of the e-skills (probably) required on a white-collar job.

Therefore, we decided to build a dichotomous variable indicating if the young individual can perform on his or her own the vast majority of skills considered as part of the solid core of ICT tools required *to participate in a socially and economically valuable activity*.

We choose to "set the bar" on at least eight out of nine of the skills related (as a whole) to a standard office-like user, a cluster of skills shared by 27 % of the survey's population: *opening a file, editing a file, saving a file, printing a file, downloading a file from the Internet, sending e-mails, attaching a file to an e-mail, creating graphs in excel or similar programmes and creating a PowerPoint (or similar programme) presentation.*

Usage: Early ICT Socialization and Frequent Persistent Use

In turn, usage achievement indicators involve a much greater number of alternatives (20 questions). However, in line with the conceptual framework, we reduced the number of variables in this field to two: according to Cobo (2009) and our own analytical model in Fig. 3, it is possible to state that (1) an early socialization in ICT and (2) ICT frequent use are key as a platform for the subsequent e-skills achievement.

In this sense, we choose two relatively high usage achievements, shared by 25 and 38 % of the population, respectively: (1) years using a PC (5 years or more) as an "early" ICT socialization proxy and (2) the most frequent use of the PC (almost daily) as a regularity or routine indicator of use.

Access: A Significant Kind

Having several technologies available to use as indicators on this stage, we dismissed connectivity because of the low penetration rate at Uruguay in 2003 but also because PISA's questionnaire conceived the Internet more like an activity or use of the PC rather than as a commodity by itself.

Then, we chose an access indicator based on two criteria. On the one hand, according to Selwyn (2004) and Dodel (2013), we stated the democratizing effect of PC access at home for the subsequent levels of the divide. On the other hand, also in accordance with Selwyn (2004), we wanted to emphasize the effective provision of ICT goods that enables the subject to use them for the desired or required activities. In this sense, we opted for a relatively common (38 %) but rather "high" indicator of access' achievement: the availability of a PC at home where the young individual can do schoolwork if needed.[10]

Figure 4 shows a visual synthesis of the application of ICT's impact analytical framework on a more complete representation of the education-to-work transition research design. Obviously, the design has some limitations and improvements could be done to the research, but we think the case of use exemplifies more than adequately a reasonable conceptualization and application of an ICT's impact research.

Based on this framework and operationalization and fitting strategies presented in Fig. 4 (strong control of the ICT's working hypothesis under several hypothesis blocks), the research was able to assess an "e-skills effect" on pathways. E-skills constitute a significant part of the explanatory component in the variance of occupational achievements at 19–20 years old. Despite the fact that their effect is not the strongest, having a quantum of e-skills (an office-like level) at 15 years old compared to not having them increases (ceteris paribus) the chances of getting a white-collar occupational achievement at 19–20 years old by 60 %.

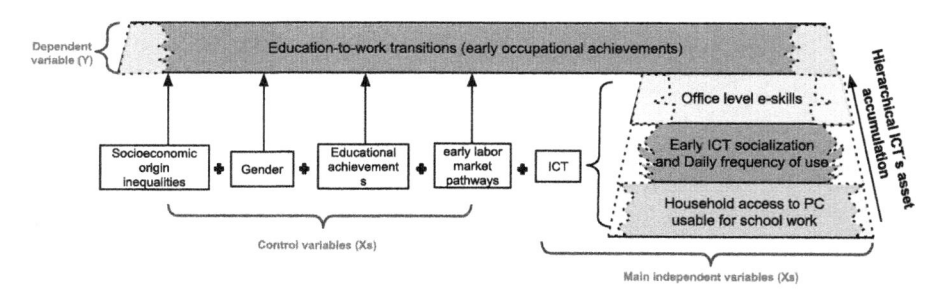

**Fig. 4** Visual representation of "ICT and Welfare policies: Digital skills' impact on formal education and labour market pathways of young Uruguayans evaluated by Programme for International Student Assessment (PISA) 2003 (Panel study)" using the proposed analytical framework

---

[10]The original question is: "Which of the following do you have in your home? A computer you can use for schoolwork"

## 4 Synthesis and Implications

This chapter proposed an analytical framework in order to guide the assessment of ICT's impact on wellbeing. We argued that the amount of polysemy and lack of common basic guidelines in ICT fields constitute one of the main barriers for the incorporation of ICT into a broader research problem spectrum (outside the ICT researchers' communities) and, consequently, to widen ICT's impact research. After a synthesis of the historical development of the digital divide concept, we discussed and developed two plausible analytical models to assess ICT's impact on wellbeing, finally opting for a framework based on Selwyn's approach. We supported the idea that the testing of any ICT-related hypothesis should analyse, at least and when possible, three stages of hierarchical digital achievements (access, usage and appropriation) plus one last divide stage: ICT's outcomes (measured by the effect of previous stages on the dependent wellbeing variable). Finally, we propose five guidelines for the applications of this framework and present an actual case of use, showing how this framework guided the research design in a study of the effect of e-skill on education-to-work transitions: *use an up-to-down strategy from higher to lower stages of the divide, start from a theoretical relevant perspective, assess the possibilities the data or research instruments enable for each stage and repeat this strategy for the lower levels of digital inequalities.*

To conclude, we would like to emphasize that this document does not aim to become a final analytical framework to research ICT's impact on wellbeing as it is only a first and basic guide for orientation purposes.

We conceive this document as a first step in the task of expanding ICT's impact research and discourse outside the information society and ICTD community. It is our belief that there is a need to construct quality and cost-efficient research instruments that will enable a wider spectrum of social researchers, those not focused on information society, to introduce ICT assets (access, usage, appropriation) as independent variables to their studies. This would also enhance our opportunity to study ICT's impact in a much wider and more diverse array of subjects.

**Open Access** This chapter is distributed under the terms of the Creative Commons Attribution Noncommercial License, which permits any noncommercial use, distribution, and reproduction in any medium, provided the original author(s) and source are credited.

## References

AGESIC-INE. (2010). *Encuesta de Usos de las Tecnologías de la Información y Comunicación.* Retrieved from http://www.ine.gub.uy/encuestas%20finalizadas/tics2011/tics2011.asp

Alampay, E. (2006). Beyond access to ICTs: Measuring capabilities in the information society. *International Journal of Education and Development using ICT, 2*(3), 4–22. ijedict.dec.uwi.edu/include/getdoc.php?id=1343

Alkire, S. (2005). *Capability and functionings: Definition & justification.* Human Development and Capability Association. Retrieved from http://www.capabilityapproach.com/pubs/HDCA_Briefing_Concepts.pdf

Boado, M., & Fernández, T. (2010). *Trayectorias académicas y laborales de los jóvenes en Uruguay. El panel PISA 2003–2007*. Montevideo: FCS. Retrieved from http://www.fcs.edu.uy/pagina.php?PagId=869

Bourdieu, P. (1986). The forms of capital. In J. Richardson (Ed.), *Handbook of theory and research for the sociology of education* (pp. 241–258). Westport: Greenwood.

Bunge, M. (1999). *Buscar la filosofía en las ciencias sociales*. México: Siglo XXI.

Castells, M. (2005). *La era de la información. Economía, Sociedad y Cultura* (La Sociedad en Red, Vol. 1). Madrid: Alianza.

Cetic.br. (2009, 2011). *Pesquisa sobre o Uso das TICs no Brasil – 2009/Survey on the use of ICTs in Brazil* [On line]. Retrieved from http://www.cetic.br/tic/2009/

Claro, M., Espejo, A., Jara, I., & Trucco, D. (2011). *Aporte del sistema educativo a la reducción de las brechas digitales. Una mirada desde las mediciones PISA* (Documento de proyecto LC/W.456). Retrieved July 21, 2013, from http://www.eclac.org/publicaciones/xml/4/45634/Aporte_del_sistema_DCTO_W__NR__con_ultimas_indicaciones_editx.pdf

Cobo, R. (2009). *Strategies to promote the development of e-competencies in the next generation of professionals: European and International trends* (SKOPE Issues Paper Series, Monograph No. 13). Retrieved July 21, 2013, from www.skope.ox.ac.uk/sites/default/files/Monograph%2013.pdf

Devine, K. (2001). Bridging the digital divide'. *Scientist, 15*(1), 28.

Dimaggio, P., & Hargittai, E. (2001). *From the digital divide to digital inequality: Studying Internet use as penetration increases* (Working Paper Series Number 15). Princeton: Princeton University Center for Arts and Cultural Policy Studies. (http://www.princeton.edu/~artspol/workpap/WP15%20-%20DiMaggio+Hargittai.pdf)

Dodel, M. (2013). *Las tecnologías de la información y comunicación como determinantes del bienestar: el papel de las habilidades digitales en la transición al empleo en la cohorte PISA* (Sociology M.A. dissertation). Montevideo: Universidad de la República.

Edwards-Johnson, A. (2000). Closing the digital divide. *Journal of Government Information, 27*(6), 898–900.

Fink, C. & Kenny, J. (2003). Whither the digital divide? *Info, 5*(6), 15–24. Retrieved July 21, 2013, from http://www.itu.int/wsis/docs/background/themes/digital_divide/fink-kenny.pdf

Flor, A. G. (2001). *ICT and poverty: The indisputable link*. In SEARCA, paper for the third Asian development forum on "Regional Economic Cooperation in Asia and the Pacific" (pp. 11–14). Bangok: Asian Development Bank. http://www.academia.edu/1110532/ICT_and_poverty_the_indisputable_link. http://ftp.unpad.ac.id/orari/library/library-ref-ind/ref-ind-1/application/poverty-reduction/!%20ICT4PR/ICT%20and%20poverty%20-%20the%20link%20(WB%20Report).pdf

Foster, J., & Handy, C. (2008). *External capabilities* (OPHI Working Paper 8). Oxford: University of Oxford.

Garnham, N. (1997). Amartya Sen's capabilities' approach to the evaluation of welfare: Its application to communication. *Javnost – The Public Journal of the European Institute for Communication and Culture, 4*(4), 25–34.

Hargittai, E. (2002). Second-level digital divide: Differences in people's online skills. *First Monday, 7*(4). http://firstmonday.org/htbin/cgiwrap/bin/ojs/index.php/fm/article/view/942/864. URL: http://firstmonday.org/issues/issue7_4/hargittai/index.html

Hargittai, E. (2008). The digital reproduction on inequality. In D. Grusky (Ed.), *Social stratification: Class, race, and gender in sociological perspective* (pp. 936–944). Boulder: Westview Press.

Heeks, R. B., & Molla, A. (2009). *Impact assessment of ICT-for-development projects: Compendium of approaches* (Development Informatics Working Paper No. 36). Manchester: University of Manchester. http://www.sed.manchester.ac.uk/idpm/researcpublications/wp/di/di_wp36.htm

ITU-International Telecommunications Union. (2010). *Informe sobre el Desarrollo Mundial de las Telecomunicaciones/TIC de 2010: Verificación de los objetivos de la CMSI [Resumen ejecutivo]*. CH-1211. Ginebra: UIT

Kaztman, R. (2010). *Impacto social de la incorporación de las nuevas tecnologías de información y comunicación en el sistema educativo* (Serie Políticas Sociales N°166. CEPAL). Retrieved July 21, 2013, from http://www.eclac.cl/publicaciones/xml/4/41364/sps166-kaztman-gsunkel-alis-2010.pdf

Lankshear, C., & Knobel, M. (Eds.). (2008). *Digital literacies: Concepts, policies and practices.* New York: Peter Lang.

Mills, M., & Blossfeld, H. P. (2006). Globalization, uncertainty and changes in early life courses. A theoretical framework. In H.-P. Blossfeld, E. Klijzing, M. Mills & K. Kurz (Eds.), *Globalization, uncertainty and Youth in society.* New York: Routledge.

Moreira, N. (2010). *Acceso y uso de las Tecnologías de la Información y Comunicación en los jóvenes evaluados por PISA 2003-2006 en Uruguay* (Sociology M.A. dissertation). Montevideo: Universidad de la República.

Norris, P. (2001). *Digital divide: Civic engagement, information poverty, and the internet worldwide.* Cambridge: Cambridge University Press.

OECD. (2010). *Are the new millennium learners making the grade? Technology use and educational performance in PISA.* Paris: Educational Research and Innovation.

Peña-López, I. (2009). *Measuring digital development for policy-making: Models, stages, characteristics and causes.* Information and Knowledge (Doctoral dissertation). Barcelona: Universitat Oberta de Catalunya.

PNUD – Programa de las Naciones Unidas para el desarrollo. (2009). *Informe sobre desarrollo humano para Mercosur, Innovar para incluir: jóvenes y desarrollo humano, 2009–2010.* Tucuman: Libros del Zorzal.

Prado, C., Romero, S., & Ramírez, M. (2009). Relaciones entre los estándares tecnológicos y apropiación tecnológica. *Enseñanza & Teaching, 27*(2), 77–101.

Reig, D. (2012). Disonancia cognitiva y apropiación de las TIC. *Telos, 90,* 9–10.

Rivoir, A., Baldizan, S., & Escuder, S. (2010, septiembre). Plan Ceibal: acceso, uso y reducción de la brecha digital según las percepcio-nes de los beneficiarios. In *IX Jornadas de Investigación de la Facultad de Ciencias Sociales.* Montevide: Universidad de la República.

Robeyns, I. (2005). The capability approach: A theoretical survey. *Journal of Human Development, 6*(1), 93–117.

Selwyn, N. (2004). Reconsidering political and popular understandings of the digital divide. *New Media & Society, 6*(3), 341–362.

Selwyn, N. (2010). Degrees of digital division: Reconsidering digital inequalities and contemporary higher education. In Redefining the digital divide in higher education [online monograph]. *Revista de Universidad y Sociedad del Conocimiento (RUSC), 7*(1). UOC. ISSN 1698-580X. http://rusc.uoc.edu/ojs/index.php/rusc/article/view/v7n1_selwyn/v7n1_selwyn. Accessed dd/mm/yy.

Sunkel, G., Trucco, D., & Möller, S. (2010). *Aprender y enseñar con las tecnologías de la información y las comunicaciones en América Latina: potenciales beneficios* (Serie Políticas Sociales 169). Retrieved July 21, 2013, from http://www.eclac.org/publicaciones/xml/9/42669/sps-169-tics-aprendizajes.pdf

Thatcher, A., & Ndabeni, M. (2011). A psychological model to understand e-adoption in the context of the digital divide. In J. Steyn & M. Johanson (Eds.), *ICTs and sustainable solutions for the digital divide: Theory and perspectives.* Hershey: Information Science Reference.

Van Dijk, J. (2005). *Deepening digital divide: Inequality in the information society.* Thousand Oaks: Sage.

Van Dijk, J., & Van Deursen, A. (2010). Inequalities of digital skills and how to overcome them. In E. Ferro, Y. K. Dwivedi, J. R. Gil-Garcia, & M. D. Williams (Eds.), *Handbook of research on overcoming digital divides: Constructing an equitable and competitive information society* (pp. 278–291). New York: Information science Reference.

Warschauer, M. (2002). Reconceptualizing the digital divide. *First Monday, 7*(7). doi:http://dx.doi.org/10.5210/fm.v7i4.942.

White, D., Gunasekaran, A., Shea, T., & Ariguzo, G. (2011). Mapping the global digital divide. *International Journal of Business Information Systems, 7*(2), 207–219.

Zheng, Y. (2007, May). *Exploring the value of the capability approach for e-development.* Proceedings of the 9th international conference on Social Implications of Computers in Developing Countries, San Pablo.

# Part II
# Research on Impact

# (Un)Balanced Conversations: Participatory Action Research in Technology Development in Peruvian Primary Schools

Paz Olivera, Komathi Ale, and Arul Chib

## 1 Introduction

Scholars have argued that information and communication technologies (ICTs) offer many benefits for educational systems (Alexander 1999; Charoula and Nicos 2004, 2009; Wang 2002). These include innovative instruction (Becker 1991), enhanced access to educational materials that allows for the reformulation of teaching through more flexible and interactive methods (Yavuz and Coskun 2008) and facilitation of communication between stakeholders (Sprague 1995). ICTs have been increasingly recognized as catalysts in improving education in developing countries (Cristia et al. 2012), with significant efforts being channelled towards development of educational applications for classrooms with the goal of targeting marginalized populations (infoDev 2010). The belief is that educational systems are instrumental in expanding capabilities and empowering individuals economically, socially and psychologically (Khan and Ghadially 2009).

It is often expected that the introduction of new technologies in schools would necessarily result in better educational outcomes among students. It is further argued that technological applications, when implemented with due consideration for

---

P. Olivera
Instituto para la Sociedad de la Información, Lima, Peru
e-mail: paz.olivera@pucp.pe

K. Ale (✉)
Annenberg School for Communication and Journalism, University of Southern California,
Los Angeles, CA, USA
e-mail: komathia@usc.edu

A. Chib, Ph.D.
Wee Kim Wee School for Communication and Information, Nanyang Technological University,
Singapore
e-mail: ArulChib@ntu.edu.sg

© The Author(s) 2015
A. Chib et al. (eds.), *Impact of Information Society Research in the Global South*,
DOI 10.1007/978-981-287-381-1_8

unique pedagogical practices along with the constraints and motivations of specific agents, such as teachers (Martínez and Olivera 2012), "amplifies the pedagogical capacity of educational systems" (Toyama 2011: p. 1). However, there is a tendency to concentrate either on the technological inputs or educational outcomes without necessarily understanding how the space where innovations are being introduced is functioning and evolving around the expectations of various stakeholders in the process. The same is true for the development of educational software where cultural assumptions that embody application design and subsequently impact institutional use of the applications are often overlooked in the equation (McIntyre et al. 2007).

In the education domain, this has meant that elements like school management, teachers' beliefs and power relations between the different actors are overlooked (Hosman 2010; Mercer 2005). The introduction of ICTs has the potential to alter the workings of the institutions and the relationships between actors at different levels (Gobbo and Girardi 2002; Lim and Khine 2006).

Within schools, the role of teachers is crucial for ICT tools to achieve positive impacts on students (Wang 2002). To be successful in computer use and integration, teachers need "to engage in conceptual change regarding their beliefs about the nature of learning, the role of the student, and their role as teacher" (Niederhauser et al. 1999: p. 157). Reaping the potential benefits of a technology-enabled teaching environment requires a shift in the learning and teaching paradigm, including recognizing the inordinate role teachers play in this process (Afshari et al. 2009). Indeed, learning with educational applications has been proven to be more effective when used in the presence and guidance of teachers (Rouet and Puustinen 2009).

Within the realm of ICTs for educational improvement in developing countries, the focus of analysis and impact has largely been on the educational attainment of students (Habgood and Ainsworth 2011; Wood et al. 2010). Impact of ICT in education programmes has rarely been conceptualized at the level of teacher attitudes, capabilities and practices (Drent and Meelissen 2008; Judson 2006; Yavuz and Coskun 2008). The study attempts to fill this gap by identifying teachers' expectations not only in technology usage but also in software interface and interaction design.

## 2 Barriers to Technology Integration by Teachers

Ertmer et al. (1999) offer a conceptual basis to understand the barriers to technology introduction at the level of teachers. Building upon earlier work (Brickner 1995; Cuban 1993; Fullan and Stiegelbauer 1991), they propose that there are external (first-order) and internal (second-order) barriers to the successful incorporation of technology into educational curricula by teachers. First-order barriers focus on practical impediments to effectiveness and efficiency, while second-order barriers pertain to fundamental beliefs, such as attitudes towards engrained and novel practices, and the capacity for change of stakeholders such as teachers.

Within the developmental context for technology projects, first-order barriers have been recognized as basic economic, sociocultural, infrastructural and technological issues (Chib et al. 2008; United Nations for Development Programme 2005). Much of this analysis has focused on availability of hardware, financial resources, physical infrastructure, training sessions, etc. However, the debate has largely moved on from that of access and adoption of technological inputs to their effective appropriation and translation into educational objectives. In this chapter, we focus on teachers as key stakeholders, whose interests and needs need to be understood in order to achieve viable, sustainable and coherent impact (Ale and Chib 2011; European Commission 2001). It is important to understand the specific characteristics unique to this important stakeholder community, such as perceptions and practices concerning technological resources (Hollow and Masperi 2009; Misuraca et al. 2011) as second-order barriers to successful integration of technology into the educational milieu.

## 3 Attitudinal Barriers of Teachers

Successful implementation of technological resources in education has relied on teachers' attitudes towards ICT (Buabeng-Andoh 2012; Hismanoğlu 2011; Khan et al. 2012), which are expected to predict their uses of technology (Bai and Ertmer 2008). In the same sense, Cavas et al. (2009) claim that "during the process of combining ICT with education, teachers' attitude(s) towards using knowledge besides their talent and desire will be a crucial point affecting the results of application" (p. 21). On the other hand, Snoeyink and Ertmer (2001) and Buabeng-Andoh (2012) find that positive attitudes of teachers towards ICTs were related to perceptions of usefulness to the teaching and learning processes.

Khan et al. (2012) claim that "less technologically capable teachers who possess positive attitudes towards ICT, require less effort and encouragement to learn the skills necessary for the implementation of ICT in their design activities into the classroom" (p. 71). From these reflections, we consider attitude a key factor for ICT integration in education. In order to provide further granularity to the all-encompassing attitudinal factor, we next conceptualize teacher attitudes as competence, confidence and capacity for change.

Pelgrum (2001) suggests that competence is related to skills and knowledge about ICTs, while Rychen and Salganik (2003) claim that it involves not only resources of a component but also the ability to mobilize these resources appropriately in complex learning situations. While teacher training as a response to a first-order barrier of inadequate competencies has been identified as crucial (Kirkwood et al. 2000), teachers have sometimes been unable to convert this information into effective teaching practices in classrooms (Goktas et al. 2009).

Teachers' lack of confidence is revealed by an anxiety to adopt and incorporate technology in classes and the fear of encountering situations where they noticeably display a lack of proficiency in the technology use, hence significantly undermining their authority as experts in the classrooms (Guha 2000; Larner and Timberlake 1995; Olivera 2012). Peralta and Costa (2007) find that control of the technology usage is an important factor in determining their self-efficacy. Likewise, we note that these attitudinal measures can be interlinked. Teachers consider competence a key factor to increase confidence in technology use (Peralta and Costa 2007).

The capacity for change is related to the ability to adopt, and adapt to, new teaching environments and practices (Abdullah 2009; Balanskat et al. 2006; BECTA 2004; Buabeng-Andoh 2012). Levin and Wadmany (2008) maintain that strongly held pedagogical beliefs and practices would deter the adoption of a novel technological approach. Gomes (2005) points out that integration of ICTs in education implies the advancement of new pedagogical strategies that teachers could render ineffective. Integration of computers in education by teachers is a complex process and implies learning new roles (Van den Akker et al. 1992) that usually generate a resistance in teachers because they have to leave old ones.

## 4 Institutional Impediments to Addressing Attitudinal Barriers of Teachers

Teachers are one key stakeholder among many within the context of an ICT in education (ICTE) project implementation; the others constituting government officials, school administrators, technology developers, students and parents, and often the implementing agency or researchers or both as the change agency. Each of these actors may have different perceptions and opinions about the role of ICTs in teaching and learning. We argue that recognizing dynamic interests of special stakeholder groups in practice would not only influence impact but may in fact constitute an impact in itself. Thus, addressing teacher attitudes, in terms of competencies, confidence and capacity for change, as deeply engrained internalized values could be understood as meaningful to the impact of a project.

The inclusion of teacher empowerment and agency as a potential positive outcome of an ICTE project has a basis in the capabilities approach (Sen 1985). This in itself is a radical departure from the individual blame bias (Rogers 2003) that has often held teachers as individuals responsible for project failure, rather than recognizing the success or failure of the system or the innovation. In other words, stakeholders (technology developers, change agencies) involved in the management of an innovation have tended to hold users (teachers) responsible for lack of project success rather than the innovation itself or the system of which these users are part of. When stakeholders involved in the management of ICT initiatives are cognizant of the limitations of a beneficiary group, in terms of their knowledge capabilities or attitudes towards adopting a technology, it could either work for or against the eventual impact of the initiative (Armstrong et al. 2005).

The identification of teachers' capabilities and values as key to the success or failure of ICTE implementation often fails to ascribe a role to teachers in the design of the technology. We hope to understand the point of view of individual adopters, or the community of teachers as users, by identifying their attitudes towards an innovation alongside the unique problems and needs they encounter. This study maintains that addressing barriers would not only build teachers' capacities needed to support local infrastructure but also equip them to be better involved in the development of educational applications used for teaching and learning. The design of educational applications is often left to software developers who get limited interaction or feedback from the actual users of the application (Buzhardt and Heitzman-Powell 2005). Hence, it becomes necessary to align to stakeholder-user considerations, as part of a participatory process of technology development. We therefore adopt the participatory action research paradigm to create an iterative technology development process that incorporates teachers' feedback as an important component.

Berman and Allara (2007: p. 115) define participatory action research (PAR) as "an approach to an alternative system of knowledge production, based on the community's role on setting the agendas, participating in data gathering and analysis, and in controlling the use of its outcomes". In allowing participating stakeholders to adopt a critical perspective in the research process, key principles of PAR include "participation, action and reflection, the empowerment and emancipation of individuals and groups...and the production of various forms of knowledge" (Lennie and Tacchi 2007: p. 3).

A deeper understanding of the realities of teachers' ICT usage in classrooms, particularly in terms of the barriers that they face, allows for customized ICT initiatives that can be catered to address the unique challenges that teachers face. In doing so, the process not only aids teachers in utilizing ICTs in education but also puts them in better positions to actively participate in the creation of applications that they would subsequently use to teach.

We recognize that the PAR approach has rarely been adopted in ICT application design (Kajtazi et al. 2011; Lennie and Tacchi 2007) and when incorporated, has tended to focus on the technology, rather than the dynamics of participation (Kalra et al. 2007). Teacher involvement in application design could either be a one-off affair or they could be continually sought out for feedback in order to repeatedly improve an application. An iterative approach would empower teachers to engage in the development of educational materials prior to use. Prior research reveals that collecting first-hand information from users themselves, in this case teachers, resulted not only in a better final product but also increased their sense of ownership of the finished application (Chin 1996; Morris 2003). Additionally, the PAR process facilitates a collective reflection on macro-level cultural assumptions (McIntyre et al. 2007; Mohatt et al. 2004), as well as in eliciting individual-centric changes in teachers' attitudes and beliefs (Vaino et al. 2013). Following this line of reasoning, our research question investigates the influence of participatory action research approach in educational technology development on teacher attitudes of competence, confidence and capacity to change as conceptualizations of impact.

## 5 Context

In Peru, initiatives focusing on the integration of ICTs in classrooms began in the 1960s as a solution to geographical problems faced by the country (Trinidad 2005). Subsequently, significant resources have been devoted to the implementation of

ICTs in education such as INFOESCUELA (Info-school), Educación a Distancia (Distance Education), Plan Huascarán and Una Laptop por Niño (One Laptop per Child) (Marcone 2009). It is argued that, despite the investment, these projects allowed access to technological resources but had little impact on children's learning, not taking into account scholastic infrastructure, teacher training and more specifically, the needs and challenges of teachers and students (Derndorfer 2010; Martínez and Olivera 2012; Montero et al. 2009; Trinidad 2005).

The recent National Curriculum Design (Ministerio de Educación 2009) plan which encourages ICT use in schools resulted in a rapid uptake of computers in Peru's schools. As a result, teachers have been exposed to extensive computer training workshops that were initiated in response to their requests and encouraged to integrate technological tools into teaching practices. Therefore, Peruvian teachers have become generally technologically knowledgeable, though in varying degrees, in terms of operating basic computer programmes and using the Internet. Teachers, however, are very much at a nascent stage of trying to incorporate computers for academic subject-related purposes, using them primarily for general web surfing and information gathering.

According to PISA evaluations in 2001 and 2009, Peru occupies last place in reading comprehension and mathematical reasoning (Trathemberg 2010). Likewise, Montero et al. (2009) identify that Peruvian students reach very low levels of achievement in language and mathematics. Further, these authors also found that there were significant differences between state and private schools and that learning gaps were bigger between urban and rural areas. In a nationwide effort to advance pedagogy in local schools, the "*Programa Leer es Estar Adelante*" was initiated to improve reading comprehension among third to sixth graders in urban, suburban and peri-urban schools.

The Leer Programme has been active in select schools (993) for over 6 years across eight regions, involving over 2,000 teachers and 50,000 students, and has resulted in increases in the level of reading comprehension (BBVA Foundation 2011). Following an increased recognition of the positive impact that ICTs could potentially bring to educational outcomes, there was interest in utilizing digital resources to supplement the existing reading comprehension programme. The Instituto de Estudios Peruanos (IEP), the central institution responsible for the coordination and management of the programme, aimed to develop an e-learning element using the established Leer curriculum. The proposed innovation, Leer Digital, comprised software developed from a programme textbook for dissemination to computer classes and laboratories in chosen schools, for use by children guided by teachers. The aim was to investigate the impact of the digital textbooks on education, specifically on the reading comprehension of children. It was proposed that the software would allow students to interact in innovative ways with the current materials developed by the programme. We studied the educational software development as a participatory process, focusing on teachers as key stakeholders.

**Fig. 1** Theorized PAR process in development of educational software

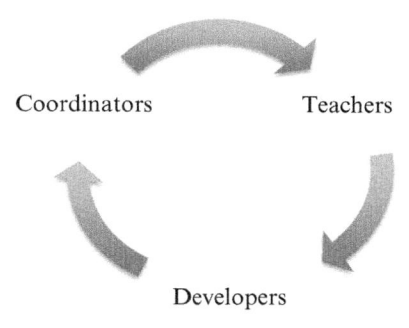

Coordinators → Teachers → Developers → (cycle back to Coordinators)

## 6 Methodology

Key aspects of participatory research that recommend it as a methodology for ICT projects include a focus on simultaneous data gathering and analysis, a flexible inquiry process that focuses on the lived experiences of participants as well as a cyclical approach to planning, acting, observing and reflecting on findings (Coghlan and Brannick 2001). As illustrated in Fig. 1, the theorized PAR process assumes a balanced conversation between the varied groups of teachers, coordinators and developers. This includes a variety of ethnographic approaches such as semi-structured and in-depth interviews and participant observations. Further, it has been noted that the PAR method should be adopted from the onset of application design and must involve constant monitoring (Tufte and Mefalopulos 2009). Towards this end, teachers' opinions and perceptions of the application were gathered at every stage of technology development prior to actual implementation.

## 7 Participants

Fieldwork was conducted in two primary schools in the urban and peri-urban districts of Lima—Nuestra Señora del Carmen in Cercado de Lima and Javier Pérez de Cuéllar in San Juan de Lurigancho—from May 2012 to December 2012, with three cycles of teacher interviews during the PAR process (at the beginning, during implementation of software and at the end of the school year). Schools were selected based on prior participation in the Leer Programme, existence of functional computer laboratories and a willing administration.

Seven fifth-grade teachers along with two computer laboratory staffs, comprising a total of four males and five females, participated in the study. Teachers, aged between 46 and 51, belonged to similar socioeconomic backgrounds. They had an average of 25 years of teaching experience in various Peruvian schools, of which they had spent about 14 years teaching in the respective two study schools and had used computing devices for an average of 4 years. The school children were familiar with computer usage and expressed basic proficiency in a variety of computer skills, from typing on the keyboard to searching for information online. Ethical considerations included informed consent from teachers and the right to withdrawal at any stage of the project.

## 8 Procedure

Data were collected through teacher interviews and participant observations during weekly computer classes. An interview guide was designed based on the three components of attitudes defined in the framework: competence, confidence and capacity for change of teachers, with specific questions related to each component. Competence questions related to "integration of ICT in class", "abilities and competence" and "use of more applications for more purposes". Confidence questions were related to "kids know more than teachers" and "teachers are afraid to make mistakes in class". Capacity for change questions related to "teacher training received", "gaps in teacher training" and "capacity to handle classroom teaching using computers".

Interviews were audio recorded for transcription and translation purposes. The translations and transcriptions for analysis were conducted by a research assistant bilingual in English and Spanish. Thematic coding involving the interpretation and categorization of information was used for data analysis. Data were collected and reviewed concurrently, with transcripts reviewed and topics catalogued based on recurring issues. A triangulation of methods such as interviews and observations allowed for the cross-validation of the findings and to add to the emerging descriptions. Respondent comments and interpretations of observation notes are provided in the next section to highlight the analysis.

## 9 Findings and Discussion

The PAR process comprised of the collective commitment of various stakeholders—researchers, digital developers, Leer Programme coordinators and teachers, involved in the iterative technology development process. The process was studied as a baseline study then studied again during the first, second and third cycles of PAR. Analysis of the data reveals two broad themes worth reflecting upon. The first observation in viewing the impact of participatory methods in technology development suggests that there are both cumulative effects and diminishing returns to teacher empowerment and participation. The second observation is that PAR, beyond merely being a teacher-focused intervention, inherently involves negotiations among multiple stakeholders, each with varying powers of influence. These observations suggest a certain value for multi-stakeholder involvement, yet on the other hand, point to the limited impacts of the participatory model for any single group of users (teachers).

We conclude with comments on the internalizations of changes in values, attitudes and practices of individual participants and reflections on the role of researchers in the participatory process. It is worth noting that gender and age variables were not items of analysis; hence, we refer directly to teachers' comments.

Due to the dynamic nature of the educational context, it was necessary to equally recognize varying uses of technology in schools, where each translates to different thresholds of attitudes and skill levels. For that reason, we limited our analysis to teacher-level adoption and reaction to the development of an educational software within a formal education space that consists of interactions between teachers and children.

## 10 PAR Viewed as a Longitudinal Process: Cumulative Effects and Diminishing Returns

At the beginning of the research study, teachers were generally supportive of the strategy of technology introduction in education, yet exhibited a wide heterogeneity of attitudes and practices. Teachers' positive attitudes stemmed from a belief that the technology would benefit them as well as student learning, yet betrayed a sense of having bought into an ideology. Teacher 1 said that ICTs are "a tool that the government is trying to implement in order to improve quality of education". Teacher 2 echoed this sentiment, "we have to unite education and technology in order for children to learn more and be updated with advances".

While there was consensus on the potential benefits of technology introduction, in practical terms, however, teachers integrated ICTs into teaching in different ways, with some exhibiting proficiency and others at the opposite end of the spectrum, displaying a lack of confidence about their competence. Teacher 3 proposed visiting an online resource to review the water cycle, a topic previously reviewed in regular class, while Teacher 1 stated that students were required to search for information online using search engines and present their work using PowerPoint. Further, some teachers were willing to resolve ICT-related doubts raised by students. On the other hand, others faced difficulties in using technology in the classroom and depended largely on the assistance of computer laboratory assistants.

Most reluctant teachers admitted to a lack of confidence related to computing technologies, partially related to competence, as exemplified by Teacher 4, who claimed that "when I used a computer for the first time, I was lost and confused. I tried to handle the mouse but I could not". Despite gradually being able to manage to adopt the technology, Teacher 3 recognized that she continues to face technical trouble, claiming that "I still have problems with boxes in MS Excel. Sometimes I want to move or erase a line but I cannot, and in these aspects I need a little help". The lack of competence and confidence led to a limited capacity for change, with Teacher 5 sharing that during her first experience with ICT, she was "very afraid to spoil (the computer) or make a mistake". Beyond confidence and competence, the capacity for change was also seen to be related to age, with older teachers pointing to a lack of experience. Teacher 3 said that "for me, it was very difficult and still is. Maybe it is due to (the) age that I am (in); we are not entirely involved into computers. We use these when it is necessary.... When I need, I use a computer but I am honest, I delay a lot. I spend many hours trying". For others, their relative computing knowledge versus that of their students was a reason to resist change. While the teachers articulated their role as a guide to students, the presence of computers acted as a source questioning their credibility. Teacher 8 said "...the intention is to use them (computers) but these are scary because sometimes it is complicated to keep order...".

The PAR process involved observation, training and solicitation for improvement areas in the software from the respondents. Teachers expressed interest in receiving more ICT training, with Teacher 4 stating that "we work with students, but we have to be trained". All teachers attended a training workshop in each school to familiarize with the functional aspects of the first two software units (of five). During this training, some teachers had difficulties in using computers that ran the software. In both schools, teachers were nervous when they had to independently run the programme, while some faced difficulties in handling the mouse. Overall, however, it was found that teachers were interested in learning to use the digital software. Reflecting this sentiment, Teacher 3 said, "but without training we cannot develop this class in a right way".

Over time, the PAR process led teachers to take charge of their own learning and teaching approaches and consequently depended less on the laboratory assistants. To aid self-directed learning, teachers received USB flash drives containing a tutorial guide and the Leer Digital software. Teacher 3 indicated that this was crucial "because researchers gave USBs that helped us in addition to training...to know how the software is, and how we should use... it helped us". By this stage, a majority of the teachers expressed that they were confident in being able to use the software in classes. According to Teacher 5, "a group (of researchers) had come and taught us, a partner (of investigators) guided with tutorial and the rest, in addition I know something, I know how to use a computer... and practice. And by practicing we learn". Similarly, Teacher 2 mentioned that "having the portable version of the software in USB file, I can review it in my house using the tutorial card that they gave us to read. This way, we can familiarize with the software that we have (in school)".

It was found that teachers integrated software into their teaching practices in different ways, yet found noticeable impacts on students. Teacher 8 said "software Adelante is providing many advantages to our educational labour as teachers because it lets us motivate students,...children learn through play,...because they like to manipulate the computer a lot,...they learn easily (through computers)...they love coming to class". Although all teachers emphasized skills and strategies of Leer Digital sessions, they subsequently progressed with classes in different ways. While variances in teaching approaches are understandable, differences in teachers' competence to both incorporate the software to teaching and to control the class in a new ICT environment influenced the way in which learning with the software complemented Leer objectives.

With repeated practice and training, teachers maintained that they were knowledgeable about appropriate software usage. For instance, Teacher 5 said, "well, until now I had no problem....I know how to use this educational software". Teacher 4 stated that he knew how to use the software, explaining that "because, more or less, we are following the instructions given to us since beginning". In other words, from the perspective of teachers, knowing how to use this software was related to the initial training received.

However, due to this perceived familiarity with the software, the feedback that was obtained from teachers declined. Teachers linked their improvements in competence and confidence with the training and supplementary materials received and subsequent iterations of the software. As the PAR process proceeded, it became increasingly difficult to get teachers to be critical about the software in terms of providing feedback to make significant changes to it.

After numerous iterative cycles of PAR, beyond technological competencies, teachers exhibited a nascent confidence in leading the technology intervention. Teacher 5 mentioned that "the role that teacher assume(s) is to be a guide...generates learning of students, but guides them". On the other hand, Teacher 1 mused that "...I have been a guide because I created conditions to bring something that was established, this is established because software is all planned, this has a sequence...". Overall, it can be argued that being a part of the software development process gave teachers a sense of ownership over the technology and also to the general sense of responsibility over students' learning, hence prompting them to assume leadership positions in the learning process.

## 11 PAR as a Negotiated Process

The second observation is that PAR, beyond merely being a teacher-focused intervention, inherently involves negotiations among multiple stakeholders, each with varying powers of influence. This study was conducted on the premise that there is value to including teacher inputs in the technology development process. However, we realize that the motivations and influence of a variety of other stakeholders meant that teacher influence on the process was deeply contested and that the researchers played an important mediating role.

Teacher inputs garnered during the PAR process had varying degrees of success in being incorporated into the software and were largely dependent on the inputs of external stakeholders. The resultant negotiations led to situations in which teachers were able to influence the software development, those wherein teachers had to form alliances with other groups to make their opinion count, as well as instances where teachers' inputs were ignored completely.

During PAR feedback, Teachers 2 and 6 suggested that the software should include an exam at the end of each section to assess students. In response, researchers highlighted that the software was equipped with the option for generating student work as PDF files for assessment. However, Teacher 1 mentioned difficulties in saving student work in the PDF format, which made it hard for her to review students' performance in each section. Other stakeholders, namely, coordinators of the Leer Programme, agreed that addressing problems with generating PDF versions was pertinent to improving the software. This glitch in the software was communicated to the digital developers who rectified it for subsequent units.

Another idea from teachers gained less traction, with different groups of stakeholders uniting in their opposition to it. Teacher 5's suggestion was to add an element of play, such as educational games, to effectively engage students. Coordinators of the Leer Programme rejected the proposal, feeling strongly that a play element "contradicted with the pedagogical approach of the original *Leer Adelante* book". Later in the process, the teacher repeated the idea, but the researchers, heeding the position of the coordinators, did not take this recommendation under consideration.

Similarly, Teacher 2 suggested that the software should include a sign or image that indicated to students if they got the right answer to questions. While researchers considered this a valid input, coordinators of the Leer Programme felt strongly that the software should not include this suggestion since correcting students was deemed to be the role of teachers rather than that of the software. In addition, Teacher 6 mentioned that at the end of each unit, the software should include a conceptual map that summed up progress made in class. This suggestion too was not incorporated in the software because researchers and Leer coordinators concluded that an existing feature of the programme "Me Autoevalúo" summarized students' progress to a similar extent. Researchers later clarified this option to teachers, who nonetheless considered it an insufficient response compared to the original request.

On the other hand, teachers sometimes formed alliances with other groups to push through their ideas. Teacher 4 felt that the software should include audio of readings of each unit, demonstrating to children reading techniques such as correct pronunciation. Researchers and coordinators of Leer Programme agreed that these changes were pertinent to improving the software. However, digital developers rejected these modifications as being too difficult to incorporate to the existing version of the software. After much persuasion, the developers agreed to make the changes in the version that was due in the following academic year.

It must be noted that participation occurred across the various stakeholders involved in the project, sometimes excluding the teachers. While teachers were engaged as participants, other groups were engaged in heated negotiations as well. Coordinators of the Leer Programme suggested that the software reflect the colour scheme adopted by the physical textbook, wherein various stages of the learning process for reading comprehension are uniquely colour coded. Developers were expected to incorporate the initial feedback to the development of unit 3. However, similar to the problems faced at the baseline phase when delays in deliverables were encountered, developers were not able to meet the expected deadlines for delivering subsequent units to schools, hence, once again, building tensions between researchers and digital developers.

The differences in work requirements and resultant schedules of the various groups of stakeholders were another source of tension for the participatory process. Delays in software delivery by the developers and national-level strike by teachers not only delayed commencement of the PAR process with teachers but also interrupted the PAR schedule as agreed with schools. Hence, the majority of the feedback and interaction regarding the initial development of the software was limited to researchers and developers.

Further along in the PAR process, teachers did not review or reflect on modifications made, taking their role in the process for granted. It was observed that teachers tended to repeat their suggestions on improvements to software during the PAR. For example, Teacher 1 repeatedly brought up issues she faced with saving PDF files to assess student work, but this problem was solved in the first cycle. In addition, this teacher highlighted that the software include a feature to review student achievements over the weeks. However, when asked if she had checked the software upgrade recently, she mentioned not having the time to do so. Such instances demonstrated that teachers might not necessarily keep track of the improvements made to the software, where in some cases, they might merely repeat previously mentioned suggestions without much interest in recognizing the progress made to the software.

## 12 Challenges of Externalities to PAR

Externalities in the form of maintenance support and differential syllabi hampered the effectiveness of PAR as a tool to empower teachers in the adoption and usage of technology. Generally, teachers found it easy to use the software in classes but had issues with the lack of maintenance and inoperative computers in their schools. It was found that this dissatisfaction with the poor condition of computer hardware translated to teachers' lack of trust in being able to progress in classes with the software.

A secondary issue arose when teachers from different schools used the software during respective reading comprehension classes. While all teachers started with section 1 of unit 1, some of them chose to proceed without necessarily following the syllabus order as proposed by Leer. This could be because teachers advanced with the coursework at different levels. Another reason was that individual factors dictated the pace of teaching; school Teachers 5 and 6 did not conduct their classes because of "personal reasons". This lag between the PAR software iterations and proceedings in the class schedule adversely affected teacher participation in the programme, where they were both unprepared and unwilling to provide feedback on the Leer software.

## 13 Conclusion

This research study set out to interrogate the assumption that participatory processes in technology introduction in the domain of education would yield beneficial results. The PAR process may apply different definitions of impact to developmental domains beyond that of education. Therefore in this study, impact was defined as empowerment of teachers in terms of their competences, confidence level and attitudes towards change. The approach assumed that an inclusive approach with a hitherto absent stakeholder, the teachers, as the focal point would lead to improvements in second-order benefits, as framed by Ertmer et al. (1999, 2012).

Defining impact in a development project related to education in terms of second-order benefits such as attitudinal and competence change of key stakeholder (teachers), rather than educational attainment of the ultimate beneficiary (students), creates problematic assessments. The objective of the project may begin with the premise that community dialogue and debate are the immediate objectives, leading into collective action (Figueroa et al. 2002), thus diverting attention and resources to the creation of participatory activities rather than keeping the end goals in sight. We caution against such conceptualizations of participation as an end in itself and a recognition that the community of stakeholders is diverse, both in its groupings and the motivations that guide it to action.

We note that over the course of time, teacher's attitudes improved with increases in technological capacities, a confidence to provide technological leadership in the classroom and a capacity to act as engaged participants in the technology development process. While these second-order benefits were being realized, we noticed increasingly diminished returns as the process went along. Teachers' identities regressed from technology development partners to their core identities as guides and leaders in the classroom. This behaviour may be considered a sign of success, as teachers were less reliant on the participatory process as source of support, but achieved a self-reliance in the use of the technology. This perspective on the iterative participatory process that recognizes a natural dynamic of ebbs and flow echoes Huesca's (2001) calls for attention to temporal dimensions of social movements.

It is important to note that the participatory process needs to acknowledge the varied motivations and skill sets of the engaged constituencies, rather than elevating a formerly ignored albeit important community into a focal position. The developers and coordinators were unwilling to accept the teachers' inputs as given, rather weaving these demands into their own objectives, leading to a back-and-forth negotiated process. These unforeseen realities of working with specific communities for PAR were not taken into consideration when planning the PAR process yet revealed a dynamic that could allow for multiple voices to create an output that is potentially better than one decided by a single group. Further research is required to judge whether educational outcomes at both student and teacher levels respond to such a collaborative technology development process.

Finally, we reflect upon the strange looking-glass dynamic of the PAR process because we, as researchers, cannot avoid being central characters. While the theorized process assumed an equal conversation between the varied groups, in reality (see Fig. 2), the researchers became the conduit for relaying and moderating the exchanges. It would be interesting to understand how such dynamics would be practiced in reality without the presence of a research group. Further, we acknowledge that the subsequent dissemination, usage and adoption of the

**Fig. 2** Observed PAR process in development of educational software

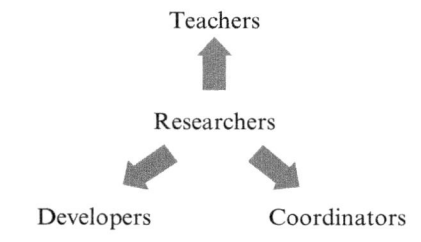

```
                        Teachers
                           ↑
                       Researchers
                       ↙         ↘
               Developers       Coordinators
```

digital Leer innovation by Peruvian teachers will occur without a technological development process. New questions shall necessarily arise—whether the principles of participatory democratic engagement in technology design and development translate into measurable impacts at the student level—and to be answered in the future.

**Open Access** This chapter is distributed under the terms of the Creative Commons Attribution Noncommercial License, which permits any noncommercial use, distribution, and reproduction in any medium, provided the original author(s) and source are credited.

# References

Abdullah, K. (2009). Barriers to the successful Integration of ICT in teaching and learning environments: A review of the literature. *Eurasia Journal of Mathematics, Science and Technology Education, 5*(3), 235–245.

Afshari, M., Bakar, K. A., Luan, W. S., Samah, B. A., & Fooi, F. S. (2009). Factors affecting teachers' use of information and communication technology. *International Journal of Instruction, 2*(1), 77–104.

Ale, K., & Chib, A. (2011). Community factors in technology adoption in primary education: Perspectives from rural India. *Information Technologies and International Development, 7*(4), 53–68.

Alexander, S. (1999). An evaluation of innovative projects involving communication and information technology in higher education. *Higher Education Research and Development, 18*(2), 173–184.

Armstrong, V., Barnes, S., Sutherland, R., Curran, S., Mills, S., & Thompson, I. (2005). Collaborative research methodology for investigating teaching and learning: The use of interactive whiteboard technology. *Educational Review, 57*(4), 455–467.

Bai, H., & Ertmer, P. A. (2008). Teacher educators' beliefs and technology uses as predictors of preservice teachers' beliefs and technology attitudes. *Journal of Technology and Teacher Education, 16*(1), 93–112.

Balanskat, A., Blaimire, R., & Kafal, S. (2006). *A review of studies of ICT impact on schools in Europe*. Retrieved from http://ec.europa.eu/education/pdf/doc254_en.pdf

BBVA Foundation. (2011). *El Programa* [The program]. Retrieved from http://www.leer.pe/ique-hacemos

Becker, H. J. (1991). When powerful tools meet conventional beliefs and instructional constraints. *The Computing Teacher, 18*(8), 6–9.

BECTA. (2004). *A review of the research literature on barriers to the uptake of ICT by teachers*. Retrieved from http://dera.ioe.ac.uk/1603/1/becta_2004_barrierstouptake_litrev.pdf

Berman, K., & Allara, P. (2007). Transformational practices in community learning: A South African case study. *International Journal of Learning, 14*(8), 113–127.

Brickner, D. (1995). *The effects of first and second order barriers to change on the degree and nature of computer usage of secondary mathematics teachers: A case study.* Unpublished doctoral dissertation, Purdue University, West Lafayette.

Buabeng-Andoh, C. (2012). Factors influencing teacher's adoption and integration of information and communication technology into teaching: A review of the literature. *International Journal of Education and Development Using Information and Communication Technology, 8*(1), 136–155.

Buzhardt, J., & Heitzman-Powell, L. (2005). Stop blaming the teachers: The role of usability testing in bridging the gap between educators and technology. *Electronic Journal for the Integration of Technology in Education, 4*(1), 13–29.

Cavas, B., Cavas, P., Karaoglan, B., & Kisla, T. (2009). A study on science teachers' attitudes toward information and communication technologies in education. *The Turkish Online Journal of Educational Technology, 8*(2), 20–32.

Charoula, A., & Nicos, V. (2004). The effect of electronic scaffolding for technology integration on perceived task effort and confidence of primary student teachers. *Journal of Research on Technology in Education, 37*(1), 29–43.

Charoula, A., & Nicos, V. (2009). Epistemological and methodological issues for the conceptualization, development, and assessment of ICT-TPCK: Advances in technological pedagogical content knowledge. *Computers & Education, 52*(1), 154–168.

Chib, A., Lwin, M. O., Ang, J., Lin, H., & Santoso, F. (2008). Midwives and mobiles: Using ICTs to improve healthcare in Aceh Besar, Indonesia. *Asian Journal of Communication, 18*(4), 348–364.

Chin, S. (1996). *Participatory communication for development.* Retrieved from http://www.southbound.com.my/communication/parcom.htm

Coghlan, D., & Brannick, T. (2001). *Doing action research in your own organization.* London: Sage Publications.

Cristia, J., Ibarrarán, P., Cueto, S., Santiago, A., & Severín, E. (2012). *Technology and child development: Evidence from the One Laptop per Child program.* Retrieved from http://www.academica.mx/sites/default/files/xo_evaluacion_peru.pdf

Cuban, L. (1993). *How teachers taught: Constancy and change in American classrooms, 1880–1990* (2nd ed.). New York: Teachers College Press.

Derndorfer, C. (2010). *OLPC in Peru: A problematic One Laptop per Child program for children.* Retrieved from http://edutechdebate.org/olpc-in-south-america/olpc-in-peru-one-laptop-per-child-problems/

Drent, M., & Meelissen, M. (2008). Which factors obstruct or stimulate teacher educators to use ICT innovatively? *Computers & Education, 51*(1), 187–199.

Ertmer, P., Addison, P., Lane, M., Ross, E., & Woods, D. (1999). Examining teachers' beliefs about the role of technology in the elementary classroom. *Journal of Research on Computing in Education, 32*(1), 54–72.

Ertmer, P. A., Ottenbreit-Leftwich, A. T., Sadik, O., Sendurur, E., & Sendurur, P. (2012). Teacher beliefs and technology integration practices: A critical relationship. *Computers & Education, 59*(2), 423–435.

Figueroa, M. E., Kincaid, D. L., Rani, M., & Lewis, G. (2002). *Communication for social change: An integrated model for measuring the process and its outcomes.* New York: Rockefeller Foundation and John Hopkins. Retrieved from http://www.communicationforsocialchange.org/pdf/socialchange.pdf

Fullan, M., & Stiegelbauer, S. (1991). *The new meaning of educational change.* New York: Teachers College Press.

Gobbo, C., & Girardi, G. (2002). Teachers' beliefs and the integration of ICT in Italian schools. *Journal of Information Technology for Teacher Education, 10*(1), 63–87.

Goktas, Y., Yildirim, S., & Yildirim, Z. (2009). Main barriers and possible enablers of ICTs integration into preservice teacher education programs. *Educational Technology & Society, 12*(1), 193–204.

Gomes, C. (2005). Integration of ICT in science teaching: A study performed in Azores, Portugal. In A. M. Vilas, B. G. Pereira, J. M. González, & J. A. González (Eds.), *Recent research developments in learning technologies* (Vol. III, pp. 165–168). Badajoz: Formatex Research Center.

Guha, S. (2000, November 8–11). *Are we all technically prepared? Teachers' perspective on the causes of comfort or discomfort in using computers at elementary grade teaching.* Paper presented at the annual meeting of the National Association for the Education of Young Children, Atlanta.

Habgood, M. P., & Ainsworth, S. E. (2011). Motivating children to learn effectively: Exploring the value of intrinsic integration in educational games. *Journal of the Learning Sciences, 20*(2), 169–206.

Hismanoğlu, M. (2011). The integration of information and communication technology into current ELT coursebooks: A critical analysis. *Procedia-Social and Behavioral Sciences, 15*, 37–45.

Hollow, D., & Masperi, P. (2009). *An evaluation of the use of ICT within primary education in Malawi.* Proceeding of the third international conference on Information and Communication Technologies and Development, Qatar, ISBN 978-1-4244-4662-9. doi:10.1109/ICTD.2009.5426707.

Hosman, I. (2010). Policies, partnerships, and pragmatism: Lessons from an ICT-in-Education Project in rural Uganda. *Information Technology for International Development, 6*(1), 48–64.

Huesca, R. (2001). Conceptual contributions of new social movements to development communication research. *Communication Theory, 11*(4), 415–436.

infoDev. (2010). *Quick guide: Low-cost computing devices and initiatives for the developing world.* Retrieved from http://www.infodev.org/en/Publication.107.html

Judson, E. (2006). How teachers integrate technology and their beliefs about learning: Is there a connection? *Journal of Technology and Teacher Education, 14*(3), 581–597.

Kajtazi, M., Haftor, D., & Mirijamdotter, A. (2011). Information inadequacy: Some causes of failures in human and social affairs. *The Electronic Journal of Information Systems Evaluation, 14*(1), 63–72.

Kalra, N., Lauwers, T., Dewey, D., Stepleton, T., & Dias, M. B. (2007). *Iterative design of a Braille writing tutor to combat illiteracy.* Proceedings of the 2nd IEEE/ACM international conference on Information and Communication Technologies and Development, Bangalore, ISBN 978-1-4244-1990-6. doi: 10.1109/ICTD.2007.4937386.

Khan, F., & Ghadially, R. (2009). *Empowering Muslim youth through computer education, access, use: A gender analysis.* Proceedings of the 3rd IEEE/ACM international conference on Information and Communication Technologies and Development, Qatar, ISBN 978-1-4244-4662-9. doi:10.1109/ICTD.2009.5426696.

Khan, M. S., Hasan, M., & Clement, C. K. (2012). Barriers to the introduction of ICT into education in developing countries: The example of Bangladesh. *International Journal of Instruction, 5*(2), 61–80.

Kirkwood, M., Van Der Kuyl, T., Parton, N., & Grant, R. (2000, September 20–23). *The New Opportunities Fund (NOF) ICT training for teachers programme: Designing a powerful online learning environment.* Paper presented at the European conference on Educational Research, Edinburgh.

Larner, D. K., & Timberlake, L. M. (1995). *Teachers with limited computer knowledge: Variables affecting use and hints to increase use* (ERIC Document Reproduction Service No. ED 384 595). Charlottesville: Curry School of Education, University of Virginia.

Lennie, J., & Tacchi, J. A. (2007). *The value of participatory action research for managing a collaborative ICT impact assessment project in Nepal.* Retrieved from http://eprints.qut.edu.au/10623/1/10623.pdf

Levin, T., & Wadmany, R. (2008). Teachers' views on factors affecting effective integration of information technology in the classroom: Developmental scenery. *Journal of Technology and Teacher Education, 16*(2), 233–236.

Lim, C. P., & Khine, M. S. (2006). Managing teachers' barriers to ICT integration in Singapore schools. *Journal of Technology and Teacher Education, 14*(1), 97–125.

Marcone, S. (2009). *Perspectivas de desarrollo de las TIC en el Perú, con especial incidencia en la educación* [Prospects for the development of ICT in Peru, with special emphasis on education]. En Las TIC en la educación. Lima: Grupo Santillana, Consejo Nacional de Educación. Retrieved from http://www.cne.gob.pe/docs/cne-publicaciones/Las_TIC_en_la_educacion.pdf

Martínez, M., & Olivera, P. (2012). *Qué ha pasado para que el programa Una Laptop por Niño en el Perú sea (hasta ahora) un fracaso?* [What happened to the One Laptop per Child program in Peru is (so far) a failure?]. Lima: FELAFACS.

McIntyre, A., Chatzopoulos, N., Politi, A., & Roz, J. (2007). Participatory action research: Collective reflections on gender, culture, and language. *Teaching and Teacher Education, 23*(5), 748–756.

Mercer, C. (2005). Telecenters and transformations: Modernizing Tanzania through the Internet. *African Affairs, 105*(419), 243–264.

Ministerio de Educación. (2009). *Diseño curricular nacional* [National curriculum]. Lima: MINEDU.

Misuraca, G., Alfano, G., & Viscusi, G. (2011). Interoperability challenges for ICT-enabled governance: Towards a pan-European conceptual framework. *Journal of Theoretical and Applied Electronic Commerce Research, 6*(1), 95–111.

Mohatt, G. V., Hazel, K. L., Allen, J., Stachelrodt, M., Hensel, C., & Fath, R. (2004). Unheard Alaska: Culturally anchored participatory action research on sobriety with Alaska Natives. *American Journal of Community Psychology, 33*(3–4), 263–273.

Montero, C., Eguren, M., Gonzales, N., Uccelli, F., & De Belaunde, C. (2009). *El estado de la educación. Estudios sobre políticas, programas y burocracias del sector* [The state of education. Studies on policies, programs and bureaucracies sector]. Lima: IEP.

Morris, N. (2003). A comparative analysis of the diffusion and participatory models in development communication. *Communication Theory, 13*(2), 125–248.

Niederhauser, D. S., Salem, D. J., & Fields, M. (1999). Exploring teaching, learning, and instructional reform in and introductory technology course. *Journal of Technology in Teacher Education, 7*(2), 153–172.

Olivera, P. (2012). *Usos y percepciones de los niños y niñas respecto a la laptop XO del Programa "Una laptop por niño"* [Uses and perceptions of children regarding the XO laptop program "One laptop per child"]. Retrieved from http://tesis.pucp.edu.pe/repositorio/bitstream/handle/123456789/4751/OLIVERA_RODRIGUEZ_MARIA_LAPTOP_NI%C3%91O.pdf?sequence=1

Pelgrum, W. J. (2001). Obstacles to the integration of ICT in education: Results from a worldwide educational assessment. *Computers & Education, 37*(2), 163–178.

Peralta, H., & Costa, F. (2007). Teachers' competence and confidence regarding the use of ICT. *Educational Sciences Journal, 3*(1), 75–84.

Rogers, E. (2003). *Diffusion of innovations* (5th ed.). New York: Free Press.

Rouet, M., & Puustinen, F. (2009). Learning with new technologies: Help seeking and information searching revisited. *Computers & Education, 53*(4), 1014–1019.

Rychen, D. S., & Salganik, L. H. (2003). *Defining and selecting key competencies*. Gottingen: Hogrefe & Huber.

Sen, A. K. (1985). *Commodities and capabilities*. Oxford: Oxford University Press.

Snoeyink, R., & Ertmer, P. (2001). Thrust into technology: How veteran teachers respond. *Journal of Educational Technology Systems, 30*(1), 85–111.

Sprague, D. (1995). ITS changing teachers' paradigms. In J. Willis, B. Robin, & D. Willis (Eds.), *Technology and teacher education annual* (pp. 352–356). Charlottesville: Association for the Advancement of Computing in Education.

Toyama, K. (2011). *There is not technology shortcuts to good education: Educational technology debate*. Retrieved from https://edutechdebate.org/ict-in-schools/there-are-no-technologyshortcuts-to-good-education/

Trathemberg, L. (2010). *Perú en las pruebas PISA 2009* [Peru in the PISA 2009]. Retrieved from http://www.trahtemberg.com/articulos/1684-peru-en-las-pruebas-pisa-2009.html

Trinidad, R. (2005). *Entre la ilusión y la realidad. Las nuevas tecnologías en dos proyectos educativos del estado [Between illusion and reality. New technologies in two educational projects of the state]*. Lima: Instituto de Estudios Peruanos.

Tufte, T., & Mefalopulos, P. (2009). *Participatory communication. A practical guide*. Washington, DC: World Bank. Retrieved from http://siteresources.worldbank.org/EXTDEVCOMMENG/Resources/Participatorycommunication.pdf

United Nations for Development Programme. (2005). *Promoting ICT for human development in Asia*. New Delhi: Elsevier.

Vaino, K., Holbrook, J., & Rannikmae, M. (2013). A case study examining change in teacher beliefs through collaborative action research. *International Journal of Science Education, 35*(1), 1–30.

Van den Akker, J., Keursten, P., & Plomp, T. (1992). The integration of computer use in education. *International Journal of Educational Research, 17*(1), 65–76.

Wang, Y. (2002). When technology meets beliefs: Preservice teachers' perception of the teacher's role in the classroom with computers. *Journal of Research on Technology in Education, 35*(1), 150–161.

Wood, C., Pillinger, C., & Jackson, E. (2010). Understanding the nature and impact of young children's learning interactions with talking books and during adult reading support. *Computers & Education, 54*(1), 190–198.

Yavuz, S., & Coskun, A. E. (2008). Attitudes and perceptions of elementary teaching through the use of technology in education. *The Journal of Education, 34*(1), 276–286.

# The Institutional Dynamics Perspective of ICT for Health Initiatives in India

Rajesh Chandwani and Rahul De

As there has been a considerable investment in ICT for development (ICT4D) initiatives, policymakers, practitioners and academics are calling for a more comprehensive and meaningful assessment of the impact of such initiatives. While the impact assessment of ICT4D can be carried out from multiple perspectives, the institutional lens is opportune in examining the softer aspects of the impact such as the behavioural, cultural, and social dimensions. ICT4D interventions juxtapose two institutional logics, that of designer and of the users, which may or may not align with each other. The impact of the initiative depends on how the interplay between the logics unfolds. We exemplify the importance of institutional context in impact assessment of ICT4D initiatives by examining the interplay of the institutional logics in the healthcare system. We conceptualise the healthcare system in terms of the logic of choice, perpetuated by the ICT for health initiative, and the logic of care which is embedded in the core of the health system. The interaction between the two logics, in turn, determines how the intervention evolves. We arrive at a framework outlining the tensions arising from the interplay of the logic of choice and logic of care in the healthcare system when an ICT4D intervention is introduced.

R. Chandwani (✉)
Personnel and Industrial Relations, Indian Institute of Management, Ahmedabad, India
e-mail: rajeshc@iimahd.ernet.in

R. De
Quantitative Methods and Information Systems, Indian Institute of Management, Bangalore, India
e-mail: rahul@iimb.ernet.in

# 1 Introduction

Information and communication technologies for development (ICT4D) projects are not implemented in a vacuum. Rather, ICT4D interventions can potentially influence the existing sociocultural–technical systems. In turn, the systems themselves can influence the evolution and adoption of the technology. Scholars have emphasised that human agency and technology have a bidirectional relationship and that evolution of technological intervention depends upon the interaction between human agency and technology (Orlikowski 1992). The interaction between human agency and technology is determined by institutional forces. "Institutions are the rules of the game in a society or, more formally, are the humanly devised constraints that shape human interaction" (North 1990: p. 3). The institutional perspective is opportune in highlighting the sociocultural–technical aspects of the ICT4D interventions. Introduction of an ICT4D initiative juxtaposes two different institutional systems, that of the project designers and implementers and of project users. The outcome of the project is then determined by the evolution of this *institutional dualism* (Heeks and Santos 2007).

Previous research has highlighted the importance of sociocultural factors as critical determinants for realisation of benefits from ICT usage (Chib et al. 2008; Miscione 2007; Walsham et al. 2007). The holistic assessment of impact of ICT4D initiatives, however, needs deliberation on how the ICT4D initiatives shape the existing institutional systems and how the institutional systems shape the technology usage and adaption. Scholars have called for research on the interaction of technology itself with *specific aspects of social, economic, and cultural contexts* (Walsham et al. 2007: p. 323). While most of the assessment frameworks used to measure the impact of ICT4D projects highlight the economic aspects of "impact" (such as in Arul Chib's introductory chapter and Kathleen Diga and Julian's May chapter later in this book), the institutional perspective is opportune in assessing the softer aspects of impact of ICT4D initiatives, such as the behavioural, cultural and social dimensions of the impact (Heeks and Molla 2009). Figure 1 below

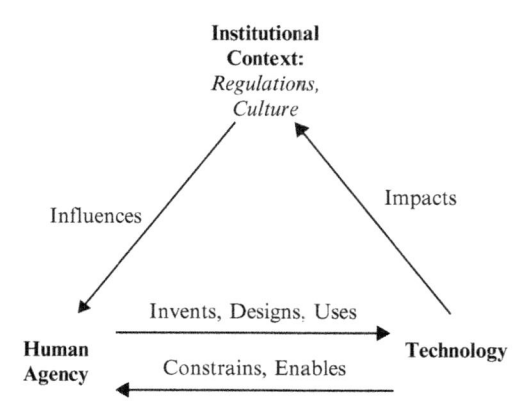

**Fig. 1** Relationship of technology and human behaviour (Adapted from Heeks and Molla 2009)

highlights the complex interaction between human agency, technology, and the institutional environment, both formal and informal, in shaping the impact of ICT4D interventions.

In this chapter, we focus on understanding how the implementation of an innovation in the form of an ICT4D initiative creates tension in the institutional dynamics and how these tensions affect the adaptation of the intervention. To highlight the institutional dualism (Heeks and Santos 2007), we refer to the "institutional logics" perspective. Institutional logics are the sociocultural norms, beliefs, and rules that shape the actors' cognition and behaviour; that is, how they make sense of the issues and how they act (Friedland and Alford 1991; Thornton 2004; Lounsbury 2007). Institutional logics provide a "stream of discourse that promulgates, however unwittingly, a set of assumptions" (Barley and Kunda 1992: p. 363). But institutional logics are seldom unidimensional and coherent. Rather, institutions, especially those which involve multiple and diverse stakeholders, such as healthcare systems, are characterised by coexistence of multiple, and sometimes conflicting, logics (Dunn and Jones 2010).

We exemplify the interplay of institutional dynamics by situating our discussion in the context of healthcare systems. The system of healthcare delivery can be conceptualised as an institution that is governed by logics that determine behaviour of stakeholders of health systems (Dunn and Jones 2010). Healthcare systems should be seen as institutions that are socio-technical systems with multiple stakeholders interacting with each other, such as the public and private healthcare organisations, political bodies, local community, regulatory bodies, financial institutions and so on (Arora 2010). The stakeholder behaviour and the socio-technical system are determined by the institutional context of healthcare. It is argued that ICT interventions can enable innovations in healthcare service delivery to extend the provision of affordable and quality healthcare for all (Sosa-Iudicissa et al. 1995). There has been substantial interest and investment in ICT interventions to enhance efficiency and effectiveness of healthcare delivery in developing countries. Facilitated by the increasing penetration of ICT, governments have made heavy investments in e-health initiatives (Blaya et al. 2010). E-health refers to the "use of information and communications technologies (ICT) in support of health and health-related fields, including healthcare services, health surveillance, health literature, and health education, knowledge and research" (World Health Organization 2005). The broad area of e-health includes several types of ICT for health initiatives such as m-health which involves use of mobile technologies, e.g. cell phones, SMS, etc. for strengthening healthcare delivery (Kahn et al. 2010), telemedicine which refers to use of telecommunication for connecting patients and doctors across geographies (Zolfo et al. 2011), Electronic Medical Records (EMR) which refers to creation and storage of health-related information of individuals in an electronic form that can be used by clinical and analytical purposes (Fraser et al. 2005), and so on. ICT interventions in the form of e-health initiatives can potentially influence the institutional system of healthcare delivery, and in turn the system can influence the evolution and adoption of the technology itself (Nicolini 2006). The posited

**Fig. 2** Interplay between ICT for health interventions and the institutional context

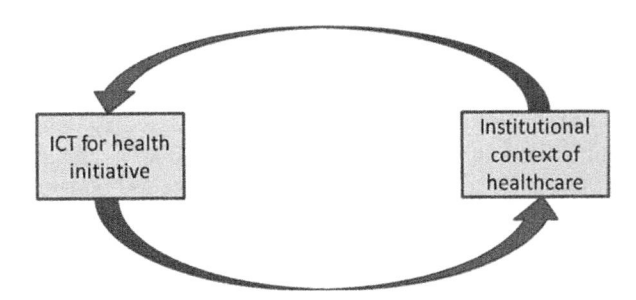

interplay between the ICT for health initiative and the existing institutional forces that shape healthcare delivery is represented in Fig. 2.

The literature on ICT for development, especially the dominant discourse on ICT for health initiatives, examines the economic dimension of interventions which highlights the efficiency-related aspects of the intervention (Blaya et al. 2010). There is a need to understand the adoption and evolution of ICT4D interventions from the sociocultural perspective and to explore how the system as an institution undergoes change, if any, from the intervention. The sociocultural and the institutional aspects of ICT interventions are more relevant for the effectiveness and sustainability of the intervention (Heeks and Molla 2009). In this paper, we attempt to address this gap by examining the evolution of ICT for health initiatives from the institutional logics perspective. ICT for health initiatives can be regarded as innovations that provide citizens with an alternative to their usual health-seeking avenues and that which can potentially alter the balance between conflicting logics prevalent in healthcare institutions where the interventions are attempted. The dominant rational view that ICT can act as a conduit for information transfer and hence knowledge transfer, making it possible for extending the access of medical knowledge for marginalised population and geographies, should be critically examined by investigating its effect on the basic assumptions and values that characterise the system (Arora 2010; Miscione 2007). In other words, innovations in healthcare can be understood in the backdrop of the logics that govern the healthcare system. Hence, we situate our discussion in the broad domain of logics in healthcare service institutions, specifically highlighting the trade-off between the logic of choice and logic of care (Mol 2008; van Schie and Seedhouse 1997).

The rest of the chapter is structured as follows. In the next section, we dwell on the concepts of logic of choice and logic of care in healthcare. Next, based on the extant literature, we highlight the emerging themes that arise from the interplay of the two logics in the healthcare domain. In the discussion and conclusion section, we arrive at a theoretical model explicating important dilemmas and tensions occurring due to institutional dynamics when an ICT initiative is introduced in the healthcare system. The chapter concludes by outlining the agendas for future research.

## 2 The Logic of Choice and Logic of Care

The logic of choice in healthcare represents the libertarian conception of healthcare systems, emphasising that market-driven competition-enhancing policy measures can enhance the efficiency and effectiveness of the services provided to the patient (Fotaki 2010; van Schie and Seedhouse 1997). This assumption, however, does not take into account the complex relationships between the diverse stakeholders, the socio-economic, and the cultural norms which determine the aspect of "care" in the healthcare context. The logic of care refers to practices such as support, advice, encouragement and consolation, thus including both medical and social dimensions. The logic of care broadens the scope of healthcare by regarding patients as individuals being embedded in a social milieu rather than just diseased bodies and entails collaborative attempts to understand and *attune diseased bodies and complex lives* (Mol 2008). The logic of care takes into account the practices, "what they do", while the logic of choice refers to the possibilities presented to the stakeholders, "what are the choices available and what they choose to do" (Fotaki 2010; van Schie and Seedhouse 1997). The "they" in the above discussion could represent any stakeholder in the healthcare system. However, most of the studies, and rightly so, conceptualise the variables in terms of "they" as patients. For example, when a telemedicine programme is implemented in rural areas, the patient has a choice to consult a remote specialist on telemedicine or to continue seeking medical services from the local practitioners, usually practising complementary and alternative medicine (CAM). Thus, the technology drives the logic of choice. However, studies have shown that the patients continue to consult their local health service providers and come for telemedicine only if there is no relief from their primary recourse (See Miscione 2007). One of the important reasons for this behaviour is that the existing network of the health system provides the environment of care (Miscione 2007).

Health-seeking behaviour, and choice, is determined by the prevalent norms which in turn are determined by the logic of care. Franckel and Lalou (2009) studied the health-seeking behaviour for childhood malaria in rural Senegal. In the community, the child care-taking was a collective process involving mother, father, friends, and relatives, and treatment decision was a collective one. The collective management favoured home care and resulted in delayed recourse to health facilities. The above study highlights that logic of care is embedded in the relational and sociocultural and economic web. Indeed, it is argued that "care" is an integral and central part of healthcare systems and "choice" operates in a milieu of a broader "care".

It should be noted that the policy interventions and innovations in healthcare domain ultimately aim at improving the distal outcome variables such as decreased morbidity and mortality and enhanced quality of life. For example, the Millennium Development Goals (MDG), adopted by the UN General Assembly in 2000, targets poverty alleviation and improvement in health by 2015 as their ultimate distal outcome through international development programmes. The three MDGs directly relating to health aim at more proximal and measurable outcomes—reducing child mortality (MDG 4), improving maternal health (MDG 5), and controlling HIV, malaria, tuberculosis and other diseases (MDG 6). The inherent pluralism of logics in the healthcare domain highlights an important dilemma faced by the studies examining the impact of ICT for health interventions to identify a proximal variable that is relatively easy to assess (e.g. enhanced patient choice and incidence of malaria) or to examine a distal variable that is more difficult to measure, although it is more desirable (e.g. enhanced quality of life for inhabitants and that of care facilities). However, the above distinction is not very compartmental as it can be argued that the intermediate or proximal variables can be regarded as an end in themselves, for example, enhancing patient autonomy or patient choice. Further, there may be conflict between variables that can be categorised in a single domain, for example, increasing the lifespan of elderly patients (albeit with associated morbidities) may not align with the goal of enhancing the quality of life for the elderly. The policymakers' dilemma of emphasising proximal versus distal variables is discussed in the next section. Here, we would posit that in the context of ICT for health interventions, generally, the proximal variables such as patient choice, patient autonomy, patient centredness, and adoption of technology represent the logic of choice while the distal variables such as patient satisfaction, equity of care, and quality of life derive more from the care perspective.

The extant literature highlights the following aspects of the interaction between the logic of choice and the logic of care: (1) the complex relationships between the proximal and the distal variables, (2) the contextual nuances that affect how the interaction between the logics unfold, (3) the overemphasis on the "expert patient" in the logic of choice, and (4) the issues arising from the coexistence of a formal system of choice and a predominantly informal system emphasising the logic of care.

## 3  Policymaker's Dilemma: Proximal or Distal Variables

The assumption that intermediate variables such as patient autonomy relate to distal variables such as improved health outcomes should be understood in relation to the contextual and individual level variables. For example, if the patient is unable to make a choice, or for that matter if he or she does not wish to make a choice, or if the choice involves gathering and assimilating loads of information that is not readily available, emphasis on the patient autonomy and choice can be detrimental to long-term quality of life. Lee and Lin (2010) in their study on diabetic patients highlight that patient's autonomy is not directly related to favourable outcomes in the form of glycemic control. The relation is contingent upon various factors such as high decisional and high informational preferences. Further, there may be a conflict between choice and autonomy. Aune and Möller (2010), for example, demonstrated that women welcomed the option of getting an early ultrasound for detecting chromosomal anomalies in the fetus but did not want to take a decision regarding getting the ultrasound done themselves. Rather they preferred that their doctor should prescribe the investigation. Thus, they welcomed choice but not autonomy. Pilnick and Dingwall (2011) problematise patient centredness which entails patient autonomy as a universal good. They highlight that asymmetry is engrained in the institution of medicine and hence in the doctor–patient relationship and implementing autonomy that may be beneficial for some patients who prefer decision-making on their own. Dixon-Woods et al. (2006: p. 2742) similarly highlight how women provided "informed consent" for surgery though they were ambiguous about the decision as the decision-making process was "enmeshed in the hospital structure of tacit, socially imposed rules of conduct". "Informed decision-making", yet another concept emphasising patient autonomy and choice, thus, reinforced passivity rather than autonomy. Arguably, the complexities of the relationship between proximal and distal variables would be more pronounced in the context of healthcare for poor populations in developing countries, which is characterised by a high level of health illiteracy.

Isolated emphasis by policymakers on some intermediate outcomes, such as patient autonomy and choice, may negatively impact the universal values that a healthcare system envisages, for example, equity of care. The individualistic paradigm that forms the fundamental basis of patient choice and autonomy is diametrically opposite to the collectivistic and welfare paradigm that emphasises solidarity and equity of care (Fotaki 2010). Scholars call for de-emphasising the implicit incorporation of independence in autonomy, arguing for a relational understanding of autonomy while recognising its embeddedness in a web of relationships, and emphasise incorporating logic of care in doctor–patient relationships (Entwistle et al. 2010). To summarise, the logic of choice, while having merit, emphasises more on the proximal variables and may not resonate with the long-term distal outcomes. Thus, arguably, ICT for health interventions that solely emphasise choice without taking into consideration the distal variables determined by care is more likely to face resistance in their adoption.

## 4 Contextual Aspects Can Affect the Logics

The logic of care emphasises that contextual nuances should be taken into account in designing and modifying an intervention design to suit the context. It is the context that determines the environmental factors involved in the delivery of care. Hardon et al. (2011) problematise the mono-dimensional view, emphasising a patient's autonomy without taking contextual factors into consideration. They found that contrary to the choice logic, "provider-initiated tests" for HIV were more acceptable than voluntary testing in HIV centres in Uganda and Kenya. An in-depth analysis revealed that the sociocultural aspects of the society made voluntary testing, based on the principle of patient autonomy and choice, less attractive. The patients going for voluntary testing were considered to have a loose character, having "slept around", and hence were more comfortable when the tests were initiated by the healthcare provider. Further, the design of the interventions should pay attention to the existing health-seeking behaviour of the patients. Adoption of any intervention that undermines the existing channels or patterns of health-seeking behaviour is less likely. Chandler et al. (2011), for example, examined the introduction of a diagnostic test for malaria through a drugstore in Uganda. The intervention, however, did not result in expected increase in use of the test before taking treatment for malaria. The drug shops, which were an important source of healthcare services for community and were an important stakeholder in the established network of care and health-seeking, considered the diagnosis and treatment of malaria as synonymous. Thus the rational "choice" of having a diagnosis before the treatment was not deemed necessary in the existing network of care (i.e. drug shops). In both the examples above, the logic of care contrasted with the logic of choice.

Contextual factors shape the evolution of an intervention, determining whether the intervention will be adopted or not, or will be adopted fully or partially, or not at all, and the possible intended and unintended consequences (Orlikowski 1993). The dominant logics of healthcare systems may, thus, shape the introduction of interventions emphasising the logic of choice or logic of care. Robertson et al. (2011) explicate how the phenomenon of "shared decision-making" highlighted the role of a general practitioner as an expert rather than a partner in decision-making. "Shared decision-making" was used in minimising resistance to treatment solutions rather than in involving patients in their treatment decisions. Thus the characteristic of the context (i.e. power distance in doctor–patient interaction in healthcare) shaped the adoption and use of intervention (i.e. emphasis on shared decision-making) and, in fact, ossified the existing power distance between the provider and the receiver of service (Greenfield et al. 2012; Robertson et al. 2011). Kaufman et al. (2011), on the other hand, examine how a technology with indefinite or indeterminate effects found universal acceptance among the stakeholders. Their study demonstrates how the "technology imperative" drove the physicians, patients, relatives, and other stakeholders, such as manufacturers of the instruments and the insurance companies, to adopt novel technologies (e.g. implantable cardioverter defibrillator for elderly patients) that have ambiguous results in terms of choice as well as care (e.g. postponement of death but prolongs morbidity).

The existing health-seeking behaviour and the sociocultural and economic milieu in which the health-seeking behaviour is embedded form an integral part of the logic of care. Any intervention that supports the logic of choice should be designed in a manner that is supported by the contextual factors that determine the logic of care. Stoopendaal and Bal (2012) explicate how a "sociomaterial" setup was used in organisations for providing care to the elderly to enhance the quality of care provided by facilitating the choice of food for the inhabitants. The attempts for quality improvement recognised the situatedness of the phenomenon, thus providing alignment between the logic of care and logic of choice.

The above discussion highlights the importance of contextual dimensions such as the sociocultural, economic, and political aspects, which determine the logic of care. While ICT for health initiatives such as telemedicine or m-health can be assumed to promote the logic of choice by making modern medical knowledge accessible to remote and rural populations, how the technology is adopted and how it evolves will be determined by the existing logic of care in which the intervention is embedded.

## 5 Assumption of an "Expert Patient"

Changing lifestyles and demographics have resulted in epidemics of chronic lifestyle-related illnesses such as diabetes and hypertension across the developed and the developing world. In case of such illnesses, where lifestyle modification forms an important aspect of treatment, it is often assumed that effective management involves converting a patient into an "expert patient" (Greenhalgh 2009). Driven by the logic of choice, the concept of the expert patient is based on the assumption that "teaching and training" of the patient in self-management will equip the patient with adequate knowledge and motivation to adhere to the prescribed treatment protocols (Mol 2008). The logic of choice emphasises that the patient is a rational individual who, once acquainted with the benefits of self-management, will indulge in actions that would maximise his or her wellbeing as an individual, that is, adherence to the treatment protocols (Gomersall et al. 2012). However, the studies emphasise that equipping patients with self-management may deprive them of the care environment and put the "blame" of any mismanagement onto the patients themselves (Mol 2008). Indeed, some patients consider self-management of diabetes at home as a demanding work (Hinder and Greenhalgh 2012). The success of self-management depends not only upon individual factors such as knowledge and motivation but also upon the family support and socio-economic contexts (Hinder and Greenhalgh 2012). The latter form a part of the care environment. Henwood et al. (2011) describe how a citizen patient, who is "nudged to adapt" the choice of healthy living habits in everyday life, negotiates between this logic of choice and the alternative logic of care in adopting health-promoting practices in their daily life. The sense-making that occurs in the process of negotiation is determined by the logic of care.

Patient expertise has three aspects: managing illness, managing everyday tasks with illness and enhancing the valuable sense of self. While the first aspect relates to logic of choice, the third aspect relates to logic of care, to feel secured and connected, and developing a sense of meaning and coherence (Aujoulat et al. 2012). Thus, the proper care of patients with chronic illnesses requires looking beyond the patient as an individual and laying isolated emphasis on making the patient "expert" in managing his or her illness (Greaney et al. 2012). The logic of care emphasises building more holistic models of care with the patient embedded in the family, society and the political contexts (Gomersall et al. 2012; Greenhalgh 2009).

Potentially, ICT for health interventions such as telemedicine, m-health, etc. can have a differential impact on patient empowerment and autonomy. An "expert" patient, who is thoroughly conversant with the use of technology and has the knowledge about his or her illness and management, may feel empowered by the use of ICT for health as it will result in "freeing" up of the abilities of the "expert" patient. However, the naïve patient, who has a limited knowledge of his illness and management or who is unable to utilise that knowledge effectively, may feel abandoned as the logic of care gets de-emphasised. A large majority of the poor population, which is the focus of most of the ICT for health interventions, has limited health literacy (Bhattacharyya et al. 2010) to fully utilise the capabilities emphasised by telemedicine. Further, ICT for health introduces a new dimension to the "expertise", technological expertise which refers to being conversant and comfortable with technology, thus complicating the issues arising from the assumption of an "expert" patient.

## 6 Formal Versus Informal Systems

An important aspect related to the logic of choice and logic of care is the interrelationship between the formal and informal systems that coexist within the healthcare context. ICT for health initiatives, largely driven by the governments or funding agencies, emphasises changes within the formal healthcare delivery system to enhance efficiency or effectiveness of the delivery process. For example, telemedicine initiatives in the developing countries, which seek to make "expert" specialist knowledge available to remote rural populations through the use of ICT, are usually implemented in the existing public health infrastructure in the remote areas. However, the informal systems in healthcare, such as the sociocultural aspects, play a crucial role in determining delivery and perception of "care". ICT for health initiatives, which provide an efficient alternative "choice" of healthcare delivery to the patients, generally highlight the formal aspects of the health system. In other words, ICT for health initiatives driven by the logic of choice largely emphasises the formal healthcare system. The logic of care, on the other hand, concerns the informal system of healthcare delivery. Empirical studies have shown that patients frequently resort to the informal systems that are driven predominantly by logic of care. For example, Stenner et al. (2011) find that patients preferred a nurse practitioner over a specialist in accessing primary care for chronic illnesses as they valued the "non-hurried" approach adopted by the nurses, the involvement of nurses in providing care and showing concern for the patients, their higher degree of approachability and length of the interactions, and better interpersonal skills, which raised the satisfaction level of patients' interaction with nurse practitioners. The above aspects highlighted the awareness about embeddedness and care in increasing patient satisfaction from the interaction. Similarly, studies have highlighted the preference of the informal over formal channels in case of doctors. Birk and Henriksen (2012) explicate that general practitioners, when referring a patient to a particular hospital, rely on the informal channels for gathering information about quality of care and services offered by the hospitals, for example, feedback from previously referred patients, recommendations from friends rather than the official data and figures available in the hospital databases. Chib et al. (2013a) highlight that rural doctors in China utilised both informal and formal networks to address their need for medical information, with informal *guanxi* networks compensating for the limitations of the formal healthcare information system.

Yet another important dimension highlighting the informal versus formal dilemma is the traditional versus biomedical systems of medicine which coexist in a healthcare system. Studies have shown that traditional systems of medicine form an integral part of health-seeking behaviour of the patients, especially in developing countries, and that these systems are central aspects of the logic of care (Miscione 2007; Sujatha 2007). Indeed, many educated patients and intelligent therapists resort to alternative medicine in spite of the limited scientific and statistical support for effectiveness of such therapies (Beyerstein 2001). Nissen and Manderson (2013) map the changing attitude of the healthcare systems in various countries across the world towards CAM (complementary and alternative medicine). CAM is heterogeneous with several societies considering specific CAM as legitimate, for example, Ayurveda in India and Chiropractic in Australia (Nissen and Manderson 2013). The coexistence of these systems affects the healthcare delivery processes. Sachs and Tomson (1992), for example, in their study on drug utilisation among doctors and patients in Sri Lanka illustrate how the sociocultural norms about Ayurveda influenced the doctor–patient interaction and drug usage. Policymakers and healthcare systems in several countries have started recognising the contribution of these systems of medicine in emphasising the logic of care, though the biomedical system proponents raise issues about the "lack of evidence base" in some of these systems. Telemedicine can complicate the already complex relationship between the traditional and biomedical systems of medicine interventions. CAM forms the usual recourse adopted by the patients, especially in the case of primary care in developing country contexts (Sujatha 2007; Shaikh et al. 2006). Telemedicine interventions that are largely restricted to the field of modern medicine emphasise biomedical conceptualisation of health, perpetuate the logic of choice by providing people with an alternative to their usual course of health-seeking in primary care, thus accentuating the conflict between the logic of care and logic of choice (cf. Merrell and Doarn 2012).

Further, the technology itself can be perceived differently by the doctors as well as patients. For example, while doctors are more likely to perceive telemedicine from a formal perspective (e.g. technology as conduit of knowledge), the patients may perceive it as an informal channel for communication (such as appearing on a television screen). The doctor can also invoke formality or informality in a telemedicine encounter by verbal and non-verbal cues. The differential perspectives potentially invoked by the doctors and patient can complicate the interplay between the choice and care logic.

## 7 Presence of Multiple Personnel in the Doctor–Patient Interaction

In this discussion we focus on a specific type of ICT for health intervention, namely, telemedicine. In a doctor–patient interaction, the patient shares personal information related to health, illness and disease and about his or her personal life with the doctor to enable the doctor to reach a particular diagnosis. Maintaining the confidentiality of such personal information and a concern about privacy have been voiced and debated extensively in the healthcare literature (Chalmers and Muir 2003). The discussion on privacy and confidentiality in medical practice involves complex philosophical and conceptual issues (Rothstein 2010) and is out of the scope of this review. Here we highlight an important aspect of telemedicine which can potentially affect the perception of the patient about confidentiality and privacy issues, namely, presence of additional personnel in the telemedicine interaction and the concern about sharing the information with a person via electronic media rather than face-to-face (Miller 2001; Nicolini 2006).

The traditional doctor–patient interaction in a personal visit usually occurs on a one-to-one basis between the doctor and the patient. In tele-visits (telemedicine interactions), however, there are multiple personnel such as technicians, coordinators, and IT assistants who are listening to the interaction, though not directly involved in the medical aspects of the consultation. Involvement of multiple personnel jeopardises the perception of confidentiality and privacy in such consultations (Stanberry 2001). Further, as Labov (1972) highlighted, the phenomenon of the observer's paradox, that is, the difference in behaviour from the usual norms occurring as a result of perception of being observed, can alter the dynamics of doctor–patient interaction. Previous researchers have highlighted other issues arising from one to many medical consultations such as loss of patient centredness in the consultation process (Bristowe and Patrick 2012) and a perception of disempowerment and loss of self-autonomy in the patients (Rees et al. 2007; Maseide 2006). Pilnick et al. (2009) map the studies on conversational analysis of doctor–patient interactions, highlighting the need to look beyond dyadic interactions between doctor and patient to include other health professionals as well and multiparty interactions (see also Rothstein 2010).

Thus it can be conceptualised that the traditional personal visit supports the patient's perception of privacy and confidentiality, thus emphasising the logic of care. Telemedicine consultations, on the other hand, are characterised by the patient's apprehension about sharing his or her personal information with a "remote" consultation over an electronic media and in the presence of multiple personnel, thus jeopardising the logic of care. Scholars have called for further research exploring the dynamics of privacy issues in telemedicine consultations (Fleming et al. 2009). Confidentiality issues have been highlighted in other ICT for health interventions also such as m-health. Chib et al. (2013b) examine the Ugandan HIV/AIDS SMS campaign and highlighted complex gender issues involved in the implementation of the programme. They found that as the primary user of the mobile phone was a male, it jeopardised the freedom and confidence of female patients to share private information over the SMS.

## 8 Discussion and Conclusion

Most of the 2.6 billion people living under USD 2 a day, largely in low- and middle-income countries, have limited access to health services due to limited economic resources, residence in remote or rural areas, and lack of health literacy (Bhattacharyya et al. 2010). This results in significant lacunae in healthcare delivery among a population that, in fact, requires affordable, accessible, and quality healthcare services. In India, for example, 75 % of healthcare facilities (i.e. infrastructure and manpower) is concentrated in urban areas which accounts for only 27 % of the population of the nation. The lack of manpower is mainly at the specialists' level with about half of the posts for surgeons, gynaecologists, paediatricians and physicians lying vacant in rural areas (Bhandari and Dutta 2007). ICT interventions, such as m-health and telemedicine, acting as a conduit for information offer a promise to bridge the knowledge gap between the "haves" and the "have-nots" between the urban and rural areas. However, isolated emphasis on the economic aspects of ICT interventions assumes a linear relationship between knowledge–information transmission and development, which usually underlies such endeavours. ICT interventions considerably alter the institutional dynamics of the existing healthcare system, which is embedded in the sociocultural–political environment. Impact assessment of such interventions would be incomplete without considering the softer dimensions of the "impact", such as the changes in behavioural and cultural aspect of the community, and the effect on the existing institutional systems. Indeed, the longevity of the change process depends upon the softer dimensions such as *depth* (*deep and consequential change* in the processes), *sustainability* (the programme should result in policy implications so that the change is sustained over time), *spread* (involve the diffusion of underlying beliefs, norms and values that form the bases of the programme) and *shift in the reform ownership* (the shift of knowledge, ownership and decision-making from external sources who initiated the project to internal people who are the part of the process) (Coburn 2003: p. 4). Scholars have called for incorporating the cultural and institutional dimensions of the context in the frameworks for assessing the impact of ICT4D intervention (Heeks and Molla 2009). We began with the broad theoretical framework (Fig. 2), emphasising that the ICT for development initiatives, potentially, can affect and are affected by the institutional context.

We attempt to exemplify the importance of institutional context in impact assessment of ICT4D initiatives by examining the interplay of the institutional logic of choice and logic of care in the healthcare system. When an innovative intervention in the form of ICT for health initiative such as telemedicine or m-health is introduced in the system, it juxtaposes the two institutional logics. The evolution of the interaction between the two logics, in turn, determines how the intervention unfolds. The tensions between the key conceptual aspects arise when the logic of choice and logic of care are juxtaposed as a result of ICT4D intervention. Figure 3 details the tensions arising from the interplay of the logic of choice and logic of care in healthcare system when an ICT4D intervention is introduced. We identified four different aspects of the tension between the logics of choice and logic of care, namely, (1) between the proximal and distal variables, (2) an "expert patient" and

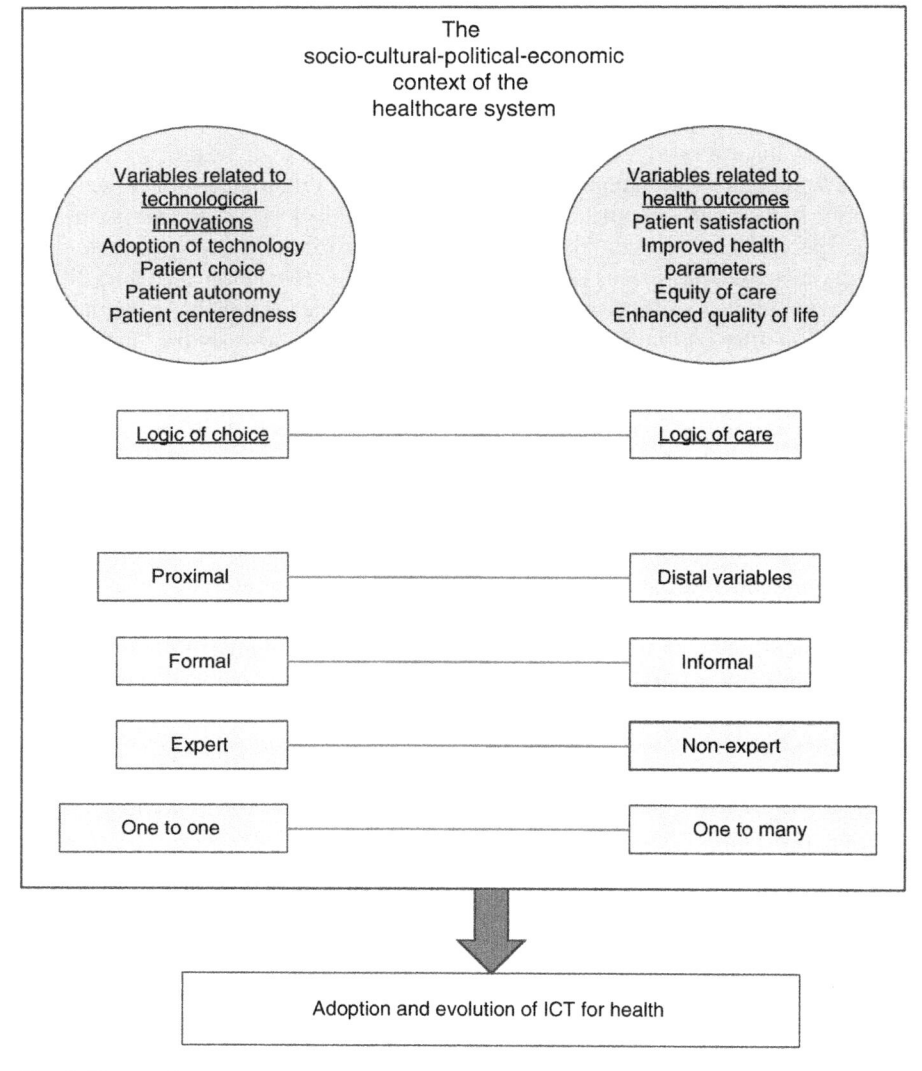

**Fig. 3** The tension between logic of choice and logic of care determining the adoption and evolution of ICT for health initiatives

a "naïve patient", (3) formal and informal systems and (4) presence of multiple personnel in the doctor–patient interaction and one-to-one interaction, respectively. These "tensions" form the two ends of a continuum, and it is the interplay between these factors that determine the adoption and evolution of the ICT for health initiative in the community.

The above analysis reiterates the complex sociocultural context of the healthcare system that determines the care environment. An ICT4D intervention, such as m-health or telemedicine, enables the patient to access the healthcare system through ICT or through the conventional face-to-face encounters. The intervention, thus, enhances the "choice" to the patient to seek access to healthcare. However, the innovative interventions that are driven by the logic of choice and patient centredness are embedded in the context of the logic of care and hence any intervention results in an interaction between the two logics. The adoption of the intervention will be determined by the interplay between the two logics. As "care" forms the core of a healthcare system, we posit that the intervention will be adopted, often in a modified form, so as to facilitate an overlap between the two logics, that of care and of choice. In other words, adoption of the "choice" and the entailing improvements in the system are determined by the alignment of the programme with the broader environment of the "care". The above analysis has an implication for scholars engaged in assessment of the "impact" of ICT4D initiatives as well as the designers and the implementers of the initiative. The programme designers of one of the m-health initiatives in India (http://e-mamta.gujarat.gov.in/), for example, recognised that in rural areas, the community health worker form an important aspect of primary care-seeking. The programme aimed at early detection and treatment of high-risk pregnancies in rural parts of the state of Gujarat, India. The programme designed involved collecting and reporting health information from the patients through the use of SMS in the local vernacular language (in Gujarati). The health workers collected simple information such as vital signs, vaccination status and so on from the patient and sent the information to the State Rural Health Mission, which then set alerts for mother and infants for regular medications and vaccination. The above example highlights that successful implementation of ICT for health initiative emphasises the overlap between the logic of choice (m-health) and logic of care (involvement of community health worker).

The framework highlighting the institutional dualism between logic of choice and logic of care in healthcare reiterates the importance of developing an in-depth understanding of the context and the existing institutional systems. The analysis based on the framework would not only enable a context-sensitive design of the innovation but also outline a roadmap for assessment of an ICT for health intervention for policymakers.

While the above framework pertains to the context of healthcare system, we posit that the core concepts of the framework, that is, the tensions arising from diverse institutional systems being juxtaposed, are generalisable to a broader domain of ICT4D interventions. Scholars have highlighted the issue of "institutional dualism" that ensues when an ICT4D intervention is introduced in an existing system (Heeks and Santos 2007).

## 9 Directions for Future Research

The above analysis reveals several potential areas for future research for examining how the logic of care and logic of choice interplay with each other and the role of ICT in the evolving dynamics. The context of ICT interventions in healthcare such as telemedicine, Web 2.0, HIS systems, electronic medical records, m-health and so on provides fertile areas to investigate the changing logics of the institution of healthcare. For example, studies have shown that Internet can not only enhance clinical and material care which enables managing diseases more effectively but also can act as spaces where people can care for themselves and others (Atkinson and Ayers 2010). As the use of social media becomes ubiquitous and the Internet alters the semantics of "friendship" and "relationships", future research is required to determine the role of the Internet in providing care. Further, as patients and doctors increasingly use Web 2.0 and mobile technologies to gather and share information (see, e.g. Chib 2010), several issues need to be investigated such as the pattern of knowledge sharing and the evolution of the relationships among the healthcare professionals and between healthcare professionals and the patients. Eriksson and Salzmann-Erikson (2013), for example, highlight how the cyber nurses project their expertise on the Web in medical discussion forums. Researchers also need to investigate the gender issues and other sociocultural suspects of the use of ICT in healthcare. Another interesting area of research would be to examine how the doctors and patients make sense of a virtual doctor–patient interaction. For example, tacit clues like non-verbal communications, body language, etc. play an important role in determining the effectiveness of doctor–patient interaction (Henry et al. 2011). To enhance the effectiveness of the virtual tele-consultation, it is essential to examine "how the doctors and patients perceive non-verbal and tacit communication in a virtual interaction" and how this perception affects the quality of the interaction. Further, while we have examined the ICT for health initiatives from the institutional logics perspective, further research from diverse perspectives, such as behavioural sciences and communication studies, will provide a more holistic understanding of the phenomenon.

**Open Access** This chapter is distributed under the terms of the Creative Commons Attribution Noncommercial License, which permits any noncommercial use, distribution, and reproduction in any medium, provided the original author(s) and source are credited.

## References

Arora, P. (2010). Digital gods: The making of a medical fact for rural diagnostic software. *The Information Society, 26*(1), 70–79.
Atkinson, S., & Ayers, A. (2010). The potential of the internet for alternative caring practices for health. *Anthropology and Medicine, 17*(1), 75–86.
Aujoulat, I., Young, B., & Salmon, P. (2012). The psychological processes involved in patient empowerment. *Orphanet Journal of Rare Diseases, 7*(Suppl 2), A31.

Aune, I., & Möller, A. (2010). 'I want a choice, but I don't want to decide'—A qualitative study of pregnant women's experiences regarding early ultrasound risk assessment for chromosomal anomalies. *Midwifery, 28*(1), 14–23.

Barley, S. R., & Kunda, G. (1992). Design and devotion: Surges of rational and normative ideologies of control in managerial discourse. *Administrative Science Quarterly, 37*, 363–399.

Beyerstein, B. L. (2001). Alternative medicine and common errors of reasoning. *Academic Medicine, 76*(3), 230–237.

Bhandari, L., & Dutta, S. (2007). *Health infrastructure in rural India* (India infrastructure report). New Delhi: Oxford University Press.

Bhattacharyya, O., Khor, S., McGahan, A., Dunne, D., Daar, A. S., & Singer, P. A. (2010). Innovative health service delivery models in low and middle income countries – What can we learn from the private sector. *Health Research Policy and Systems, 8*(1), 24.

Birk, H. O., & Henriksen, L. O. (2012). Which factors decided general practitioners' choice of hospital on behalf of their patients in an area with free choice of public hospital? A questionnaire study. *BMC Health Services Research, 12*(1), 126.

Blaya, J. A., Fraser, H. S., & Holt, B. (2010). E-health technologies show promise in developing countries. *Health Affairs, 29*(2), 244–251.

Bristowe, K., & Patrick, P. L. (2012). Do too many cooks spoil the broth? The effect of observers on doctor–patient interaction. *Medical Education, 46*(8), 785–794.

Chalmers, J., & Muir, R. (2003). Patient privacy and confidentiality: The debate goes on; the issues are complex, but a consensus is emerging. *British Medical Journal, 326*(7392), 725.

Chandler, C. I., Hall-Clifford, R., Asaph, T., Pascal, M., Clarke, S., & Mbonye, A. K. (2011). Introducing malaria rapid diagnostic tests at registered drug shops in Uganda: Limitations of diagnostic testing in the reality of diagnosis. *Social Science & Medicine, 72*(6), 937–944.

Chib, A. (2010). The Aceh Besar midwives with mobile phones project: Design and evaluation perspectives using the information and communication technologies for healthcare development model. *Journal of Computer-Mediated Communication, 15*(3), 500–525.

Chib, A., Lwin, M. O., Ang, J., Lin, H., & Santoso, F. (2008). Midwives and mobiles: Improving healthcare communications via mobile phones in Aceh Besar, Indonesia. *Asian Journal of Communication, 18*(4), 348–364.

Chib, A., Phuong, T. K., Si, C. W., & Hway, N. S. (2013a). Enabling informal digital guanxi for rural doctors in Shaanxi, China. *Chinese Journal of Communication, 6*(1), 62–80.

Chib, A., Wilkin, H., & Hoefman, B. (2013b). Vulnerabilities in mHealth implementation: A Ugandan HIV/AIDS SMS campaign. *Global Health Promotion, 20*(1 suppl), 26–32.

Coburn, C. E. (2003). Rethinking scale: Moving beyond numbers to deep and lasting change. *Educational Researcher, 32*(6), 3–12.

Dixon-Woods, M., Williams, S. J., Jackson, C. J., Akkad, A., Kenyon, S., & Habiba, M. (2006). Why do women consent to surgery, even when they do not want to? An interactionist and Bourdieusian analysis. *Social Science & Medicine, 62*(11), 2742–2753.

Dunn, M. B., & Jones, C. (2010). Institutional logics and institutional pluralism: The contestation of care and science logics in medical education, 1967–2005. *Administrative Science Quarterly, 55*(1), 114–149.

Entwistle, V. A., Carter, S. M., Cribb, A., & McCaffery, K. (2010). Supporting patient autonomy: The importance of clinician-patient relationships. *Journal of General Internal Medicine, 25*(7), 741–745.

Eriksson, H., & Salzmann-Erikson, M. (2013). Cyber nursing: Health 'experts' approaches in the post-modern era of virtual encounters. *International Journal of Nursing Studies, 50*(3), 335–344.

Fleming, D. A., Edison, K. E., & Pak, H. (2009). Telehealth ethics. *Journal of Telemedicine and e-Health, 15*(8), 797–803.

Fotaki, M. (2010). Patient choice and equity in the British National Health Service: Towards developing an alternative framework. *Sociology of Health & Illness, 32*(6), 898–913.

Franckel, A., & Lalou, R. (2009). Health-seeking behaviour for childhood malaria: Household dynamics in rural Senegal. *Journal of Biosocial Science, 41*(1), 1.

Fraser, H. S., Biondich, P., Moodley, D., Choi, S., Mamlin, B. W., & Szolovits, P. (2005). Implementing electronic medical record systems in developing countries. *Informatics in Primary Care, 13*(2), 83–96.

Friedland, R., & Alford, R. R. (1991). Bringing society back in: Symbols, practices and institutional contradictions. In W. W. Powell & P. J. DiMaggio (Eds.), *The new institutionalism in organizational analysis* (pp. 232–263). Chicago: University of Chicago Press.

Gomersall, T., Madill, A., & Summers, L. K. (2012). Getting one's thoughts straight: A dialogical analysis of women's accounts of poorly controlled type 2 diabetes. *Psychology & Health, 27*(3), 378–393.

Greaney, A. M., O'Mathúna, D. P., & Scott, P. A. (2012). Patient autonomy and choice in healthcare: Self-testing devices as a case in point. *Medicine, Health Care, and Philosophy, 15*, 383–395.

Greenfield, G., Pliskin, J. S., Feder-Bubis, P., Wientroub, S., & Davidovitch, N. (2012). Patient–physician relationships in second opinion encounters – The physicians' perspective. *Social Science & Medicine, 75*(7), 1202–1212.

Greenhalgh, T. (2009). Chronic illness: Beyond the expert patient. *British Medical Journal, 338*(7695), 629–631.

Hardon, A., Kageha, E., Kinsman, J., Kyaddondo, D., Wanyenze, R., & Obermeyer, C. M. (2011). Dynamics of care, situations of choice: HIV tests in times of ART. *Medical Anthropology, 30*(2), 183–201.

Heeks, R. B., & Molla, A. (2009). *Impact assessment of ICT-for-development projects: A compendium of approaches* (Development Informatics Working Paper No. 36). University of Manchester, Manchester. Retrieved from http://www.sed.manchester.ac.uk/idpm/research/publications/wp/di/di_wp36.htm. Accessed 12 June 2013.

Heeks, R. B., & Santos, R. (2007). *Enforcing adoption of public sector innovations: Principals, agents and institutional dualism in a case of e-Government*. Unpublished manuscript, Development Informatics Group, IDPM, University of Manchester, Manchester.

Henry, S. G., Forman, J. H., & Fetters, M. D. (2011). 'How do you know what Aunt Martha looks like?' A video elicitation study exploring tacit clues in doctor–patient interactions. *Journal of Evaluation in Clinical Practice, 17*(5), 933–939.

Henwood, F., Harris, R., & Spoel, P. (2011). Informing health? Negotiating the logics of choice and care in everyday practices of 'healthy living'. *Social Science & Medicine, 72*(12), 2026–2032.

Hinder, S., & Greenhalgh, T. (2012). 'This does my head in'. Ethnographic study of self-management by people with diabetes. *BMC Health Services Research, 12*(1), 83.

Kahn, J. G., Yang, J. S., & Kahn, J. S. (2010). 'Mobile' health needs and opportunities in developing countries. *Health Affairs, 29*(2), 252–258.

Kaufman, S. R., Mueller, P. S., Ottenberg, A. L., & Koenig, B. A. (2011). Ironic technology: Old age and the implantable cardioverter defibrillator in US health care. *Social Science & Medicine (1982), 72*(1), 6.

Labov, W. (1972). *Sociolinguistic patterns* (Vol. 4). Philadelphia: University of Pennsylvania Press.

Lee, Y. Y., & Lin, J. L. (2010). Do patient autonomy preferences matter? Linking patient-centered care to patient–physician relationships and health outcomes. *Social Science & Medicine, 71*(10), 1811–1818.

Lounsbury, M. (2007). A tale of two cities: Competing logics and practice variation in the professionalizing of mutual funds. *Academy of Management Journal, 50*(2), 289–307.

Maseide, P. (2006). *Body talk: The multiple dimensions of medical discourse*. COMET conference, Cardiff, 29 June–1 July 2006.

Merrell, R. C., & Doarn, C. R. (2012). Real time, old time. *Journal of Telemedicine and e-Health, 18*(4), 251–252.

Miller, E. A. (2001). Telemedicine and doctor-patient communication: An analytical survey of the literature. *Journal of Telemedicine and Telecare, 7*(1), 1–17.

Miscione, G. (2007). Telemedicine in the Upper Amazon: Interplay with local health care practices. *MIS Quarterly, 31*, 403–425.

Mol, A. (2008). *The logic of care: Health and the problem of patient choice.* London/New York: Routledge.
Nicolini, D. (2006). The work to make telemedicine work: A social and articulative view. *Social Science & Medicine, 62*(11), 2754–2767.
Nissen, N., & Manderson, L. (2013). Researching alternative and complementary therapies: Mapping the field. *Medical Anthropology, 32*(1), 1–7.
North, D. C. (1990). *Institutions, institutional change and economic performance.* Cambridge/New York: Cambridge University Press.
Orlikowski, W. J. (1992). The duality of technology: Rethinking the concept of technology in organizations. *Organization Science, 3*(3), 398–427.
Orlikowski, W. J. (1993). CASE tools as organizational change: Investigating incremental and radical changes in systems development. *MIS Quarterly, 17*(3), 309–340.
Pilnick, A., & Dingwall, R. (2011). On the remarkable persistence of asymmetry in doctor/patient interaction: A critical review. *Social Science & Medicine, 72*(8), 1374–1382.
Pilnick, A., Hindmarsh, J., & Gill, V. T. (2009). Beyond 'doctor and patient': Developments in the study of healthcare interactions. *Sociology of Health & Illness, 31*(6), 787–802.
Rees, C. E., Knight, L. V., & Wilkinson, C. E. (2007). "User involvement is a sine qua non, almost, in medical education": Learning with rather than just about health and social care service users. *Advances in Health Sciences Education, 12*(2), 359–390.
Robertson, M., Moir, J., Skelton, J., Dowell, J., & Cowan, S. (2011). When the business of sharing treatment decisions is not the same as shared decision making: A discourse analysis of decision sharing in general practice. *Health, 15*(1), 78–95.
Rothstein, M. A. (2010). The Hippocratic bargain and health information technology. *The Journal of Law, Medicine & Ethics, 38*(1), 7–13.
Sachs, L., & Tomson, G. (1992). Medicines and culture—A double perspective on drug utilization in a developing country. *Social Science & Medicine, 34*(3), 307–315.
Shaikh, B. T., Kadir, M. M., & Hatcher, J. (2006). Health care and public health in South Asia. *Public Health, 120*(2), 142–144.
Sosa-Iudicissa, M., Levett, J., Mandil, S., & Beales, P. F. (1995). *Health, information society and developing countries.* Amsterdam: Ios Press.
Stanberry, B. (2001). Legal ethical and risk issues in telemedicine. *Computer Methods and Programs in Biomedicine, 64*(3), 225–233.
Stenner, K. L., Courtenay, M., & Carey, N. (2011). Consultations between nurse prescribers and patients with diabetes in primary care: A qualitative study of patient views. *International Journal of Nursing Studies, 48*(1), 37–46.
Stoopendaal, A., & Bal, R. (2013). Conferences, tablecloths and cupboards: How to understand the situatedness of quality improvements in long-term care. *Social Science & Medicine, 78*, 78–85.
Sujatha, V. (2007). Pluralism in Indian medicine: Medical lore as a genre of medical knowledge. *Contributions to Indian Sociology, 41*(2), 169–202.
Thornton, P. H. (2004). *Markets from culture: Institutional logics and organizational decisions in higher education publishing.* Stanford: Stanford University Press.
van Schie, T., & Seedhouse, D. (1997). The importance of care. *Health Care Analysis, 5*(4), 283–291.
Walsham, G., Robey, D., & Sahay, S. (2007). Foreword: Special issue on information systems in developing countries. *Management Information Systems Quarterly, 31*(2), 317–326.
World Health Organization 2005 (WHA58/2005/REC/1). (2005, May 16–25). Resolution WHA58.28. eHealth. In *Fifty-eighth world health assembly.* Geneva: WHO. Annex. Resolutions and decisions. Geneva. Available from: apps.who.int/gb/or/e/e_wha58r1.html. Accessed 24 Nov 2013.
Zolfo, M., Bateganya, M. H., Adetifa, I. M., Colebunders, R., & Lynen, L. (2011). A telemedicine service for HIV/AIDS physicians working in developing countries. *Journal of Telemedicine and Telecare, 17*(2), 65–70.

# Cybersex as Affective Labour: Critical Interrogations of the Philippine ICT Framework and the Cybercrime Prevention Act of 2012

Elinor May Cruz and Trina Joyce Sajo

This chapter critically examines the underside of the Philippine Information Society—the cybersex phenomenon. In looking at cybersex, we take stock of the Philippine ICT framework's aim of building "a people-centered, inclusive, development-oriented Information Society" through the state promotion of Filipinos as "world-class" workers and citizens. We also look at how the Cybercrime Prevention Act of 2012 complements this state policy aspiration. We argue that the cybersex phenomenon illustrates how institutional development strategies propelled by ICT could inadvertently exclude already marginalized sectors of society. We use the perspective of affective labour to argue that because ICT-led development failed for these sectors, the response is an illegal service industry that also makes use of, if not feeds off, the same technological infrastructure largely supported by foreign capital. Cybersex is a potent example of how the marginalized learn to transform conditions of exclusion and illegality into creative, practical, and thus, productive strategies of survival. Cybersex is not the solution to achieving a decent quality of life, but the existence and persistence of this phenomenon signifies that the State's vision of ICT for development is not living up to its promise of socioeconomic upliftment. Through the institutionalized uses of technology, a culture of creative ICT use is constrained rather than promoted. Cybersex thus foregrounds some problematic ramifications of the present ICT framework and the institutional mechanisms supporting it. It puts to question the goal of inclusive development in Philippine ICT policymaking and legislation, hinting at the risks and repercussions of creating an "Information Society" under the neoliberal market

E.M. Cruz (✉)
Third World Studies Center, University of the Philippines-Diliman, Quezon City, Philippines
e-mail: elinor.may.cruz@gmail.com

T.J. Sajo
University of Western Ontario, London, ON, Canada
e-mail: trinajoyce.sajo@gmail.com

© The Author(s) 2015
A. Chib et al. (eds.), *Impact of Information Society Research in the Global South*,
DOI 10.1007/978-981-287-381-1_10

economy. Addressing this challenge, we propose, begins with a reconsideration of cybersex as a form of ICT-facilitated affective labour and learning from the multifaceted narratives that constitute informal uses of ICTs.

## 1 Introduction

This chapter looks at the cybersex phenomenon in the Philippines amidst government efforts to promote "ICT for development" (ICTD). Embedded in the purported cybersex capital of the world, the cybersex phenomenon in the Philippines begs for a critical reflection on the implications of ICT-led national growth, especially for those at the margins of techno-social development. The absence of studies in the cybersex phenomenon is glaring, much less in the backdrop of ICTD, and this chapter's work in the Philippine context makes a modest contribution in filling this research gap.

Attempts to grapple with the role of technology in this so-called online version of prostitution have been undertaken through bills filed in the Philippine legislature. In September 2012, the Benigno S. Aquino III administration enacted the omnibus Cybercrime Prevention Act of 2012, which casts cybersex as a cybercrime.[1] Propped up by the Aquino administration's ICT framework also known as the "Philippine Digital Strategy" (PDS), the law promulgates that cybersex, among other cybercrimes, should be prevented to safeguard and promote the integrity of the country's burgeoning ICT industry. Within the purview of the ICT framework and the Cybercrime Law, cybersex serves as a threat to the state project of a "Philippine Information Society." In this chapter, we re-examine the Philippine ICT framework and the Cybercrime Law and how these official discourses impinge on what may be considered as the underside of the Philippine Information Society. In labelling cybersex as a crime, we, the authors, believe that the existing ICT framework and the Cybercrime Law address only the symptoms and not the causes of this phenomenon.

In looking at cybersex, we unpack the Philippine ICT framework's aim towards "a people-centered, inclusive, development-oriented Information Society" (PDS 2010) and how the Cybercrime Law serves this aim. We argue that the cybersex phenomenon in the Philippines illustrates how state developmental strategies propelled by ICT could inadvertently exclude, even alienate, sectors of society for whom development outcomes promise to have an import. Perhaps because ICT-led development failed for these sectors, the response is an illegal service industry that also makes use of, if not feeds off, the same technological infrastructure largely supported by foreign capital. We hinge our arguments on an alternative conceptual

---

[1] The cybersex provision of the law defines cybersex as: "The willful engagement, maintenance, control, or operations, directly or indirectly, of any lascivious exhibition of sexual organs or sexual activity, with the aid of a computer system, for favor or consideration" (Cybercrime Prevention Act of 2012). The law, placed on an indefinite restraining order due to challenges against its constitutionality (Romero 2013), was declared constitutional by the Philippine Supreme Court in February 2014.

understanding of cybersex through the lens of "affective labour," to argue that the excluded members of society learn to transform the conditions of exclusion and illegality into creative, practical, and thus, meaningful and productive strategies of survival. In view of the above, the key question that comes to our mind is this: does casting cybersex as a cybercrime, "(threatening) the moral fibres of our society," lead to a truly inclusive Philippine Information Society?[2]

This chapter begins with a sketch of the Philippine ICT framework motivated by the State's vision of an ICT-led development, particularly through the Philippine branding of "world-class" cyberservice provision. We then zoom in on the underside of the Philippine Information Society, to begin problematizing the cybersex phenomenon in the context of a digital sexual economy. Next, we give a discussion on the relevant conceptualizations of affect in providing the theoretical foundation in our undertaking to view cybersex as affective labour. This is followed by arguments to use affective labour as an alternative perspective to give emphasis on the lived experiences of individuals engaged in cybersex, as well as their agency through the use of ICT. Finally, we offer a critical re-examination of the Philippine ICT framework and the Cybercrime Law and their implications for ICT for development. We wrap up the chapter with some concluding thoughts.

## 2 The Philippine ICT Framework: Exclusions from an Inclusive Growth

In 2006, the Gloria Macapagal-Arroyo administration laid down the Philippine ICT roadmap (CICT 2006), signifying the government's abiding faith in ICTs as an engine of development. The roadmap, created by the Commission on Information and Communications Technology (CICT) under the Office of the President, served as the government's blueprint for creating what it called "a people-centered, inclusive and development-oriented Information Society" (PDS 2010). Strategies in the roadmap included universal access to ICT, developing human capital, efficiency and transparency in governance through ICT, competitiveness in the global ICT economy and crafting an ICT policy framework.

Pushing for ICT-led national growth meant the Philippine government banking on the country's competitiveness as the top provider of cyberservices in the new economy. The Philippines has been branded as a "Cyberservices Corridor"—a 600-mile spectrum of ICT-enabled services, boasting of nothing less than "quality, innovation and world-class sophistication" (CICT 2006). To ensure that world-class service is developed and sustained, the roadmap highlights its "workforce mobilization programme" to enhance English competency and provide industry

---

[2]See Castillo (2010) on Carmelo Lazatin's statement. Lazatin is author of House Bill 1444, also known as the "Anti-Cybersex Act." This bill would go on to constitute, among others, the Cybercrime Prevention Act of 2012.

certification and career advocacy—ensuring a steady supply of "twenty first century knowledge workers" (CICT 2006). Alongside workforce development, the roadmap proposed policy reforms on the safety and reliability of the technological infrastructure to boost confidence in the country's ICT industry. By the end of the Arroyo administration, revenues for the ICT service sector grew sevenfold and employment was five times more than the 2004 baseline figures (PDS 2010)—giving credence to government claims of the benefits of ICT-led national growth: "(G)reater investment flows, more jobs, a better quality of life for the populace" (CICT 2006).

It is not surprising that the Philippine Digital Strategy under the Aquino administration endeavoured to maintain the country's top position in cyberservice provision. The Philippine ICT industry, particularly the Information Technology and Business Process Outsourcing (IT – BPO) sector, contributes almost 5 % of the country's GDP and employs over half a million full-time employees (PDS 2010). Hence, strategies in the PDS also included developing a more coherent national brand,[3] identifying Next Wave Cities[4] for the Philippine Cyberservices Corridor and setting into high gear the "development of world-class knowledge workers and the promotion of a culture of creative ICT use" (PDS 2010).

Further pushing the country's global competitiveness through its "ICT for governance" frame, the PDS implemented security measures for the country's business and technological infrastructure. The Cybercrime Prevention Act of 2012 manifests state intent to combat cybercrime in recognition of the ICT industry's role in the country's socioeconomic development. Cybercrime, as stipulated in the law, constitutes (1) offences against the confidentiality, integrity and availability of computer systems, such as system interference and cybersquatting; (2) computer-related offences, such as computer-related forgery, computer-related fraud and computer-related identity theft and (3) content-related offences, such as cybersex and child pornography (Cybercrime Prevention Act of 2012). While the previous administration aimed to develop "a vibrant, accessible and world-class ICT sector" in sustaining global competitiveness (CICT 2006: p. 2), the current dispensation did this and more—it assured ICT-related businesses the secure environment conducive to their steady expansion (PDS 2010). As a result, the country's ranking on "ease of doing business" in industry and service sectors moved up ten notches since 2010 in the 2012 World Bank International Finance Corporation report (NSCB 2012).

Constituting the collective of cyberservice providers are Filipinos employed as hardware and software developers, web designers and "technopreneurs" (Saloma-Akpedonu 2006: p. 2) and call centre agents, such as technical support specialists and customer service representatives. The PDS (2010), however, in promoting the

---

[3]The PDS (2010) states that the Aquino government plans to streamline its national branding efforts with marketing campaigns of its Southeast Asian neighbours, as in the case of Malaysia— "Malaysia, Truly Asia" for tourism and "Malaysia, Truly Business" for outsourcing.

[4]Defined in the PDS (2010) as "a Philippine-specific term that identifies ICT hubs beyond Manila, based on criteria such as worker supply, telecom infrastructure and other factors necessary to sustain a local BPO industry".

"Philippine brand of cyberservice provision" refers to their communicative, "soft" skills and personable traits rather than their technical competencies:

> English language proficiency, adaptability, educated, a deep-seated value system that prizes serving others, commitment and loyalty, cultural adaptability, and familiarity with Western business culture... (poise Filipinos) to becoming members of the next-generation of highly-valued and fully effective 21st century workers and citizens. (PDS 2010)

The premium placed on these traits led to investments in English proficiency and basic ICT skills training (Diaz de Rivera 2008) that go hand in hand with the Philippine brand of cyberservice provision.[5] This mode of branding Filipino labour is exemplified in call centre work. Labour of this type entails "the simultaneous (overlapping) delivery of physical labour (e.g., use of hardware and software technology); mental labour (e.g., English competency and juggling time zones); and emotional labour (e.g., customer service and developing an American accent)." (Saloma-Akpedonu 2006: p. 50). What binds these forms of "immaterial labour" is "the *affective labour* of human contact and interaction" (Hardt 1999), with the affective qualities of cyberservice provision proving attractive to the international labour market. The Philippines has been consistently the destination of choice of most IT-BPO businesses, producing USD 8.91 billion in revenues in 2010 alone (PDS 2010).[6] In the news, Information Technology and Business Process Association of the Philippines (ITBPAP) Senior Executive Director Gillian Virata lauds the unique traits of the Philippine workforce, that ensure the country remains "the destination of choice for customer-relations management":

> The primary driver (of the industry) is the world-beating performance... unique to the Philippine work force... we are pleasant to work with. We are also easy to train... we are easy to understand; we understand foreigners; and we empathize. (Estopace 2013)

In the 2013 survey of top global outsourcing destinations, Manila outranked New Delhi, placing third before Bangalore and Mumbai (GMA News 2013a).[7] In its race with India as the call centre capital of the world, the Philippines' sunshine industry is considered the country's "new eternal sunshine" or "perennial growth machine", running parallel to revenue contributions of overseas employment (Estopace 2013).[8]

In putting the ICT industry at the helm of economic growth, however, the question remains whether the increase in jobs and growth has actually led to a "better quality of life for the populace" (CICT 2006). Despite the National Statistical

---

[5]It is a common sight in urban communities to display tarpaulin posters of local politicians promoting call centre training as one of the local government's livelihood programmes.

[6]The PDS (2010) boasts of the country's global leadership in voice services in the IT-BPO sector.

[7]The online report may be found here: http://www.cuti.org.uy/documentos/Tholons__Top_100_Ranking_2013.pdf

[8]The ranking report also highlighted the country's efforts not just to maintain its leadership in voice services but also in expanding to other sectors in outsourcing and even going beyond US markets (GMA News 2013a).

Coordination Board (NSCB) reporting a substantial increase in Internet access in the country (Remo 2013)—an increase that positively impacts a country's digital economy ranking—the NSCB also reports that the country's economic situation leaves much to be desired. In the 2013 Philippine Statistics in Brief, the NSCB (2013) reported that first semester figures for the proportion of poor Filipino families including those living under extreme poverty in 2006, 2009 and 2012 have not changed.[9]

## 3 Cybersex Capital of the World: Treading the Underside of the Emerging Philippine Information Society

The achievements of the Philippine ICT industry, especially the IT-BPO sector, according to Budget Secretary Florencio Abad, have "put a face to our aspirations for the economy" (GMA News 2013b). The IT-BPO sector had USD 11 billion revenues in 2011 and USD 13 billion in 2012, even stimulating growth in related industries, from telecommunications to food and beverage services (GMA News 2013b). The IT-BPO sector reportedly projected revenue worth USD 16 billion for 2013 (GMA News 2013b) and the ITBPAP is reportedly targeting revenues worth USD 25 billion for 2016 as well as 1.3 million jobs in the country (GMA News 2013c). Yet, incidental to the country's rip-roaring growth industry is the emergence of a shadow economy as its "sordid side effect" (Cordova 2011). Sprawling all over the country, the cybersex phenomenon adds yet another moniker to the Philippines as cybersex capital of the world.

In November 2013, Terre Des Hommes Netherlands, a Dutch NGO working against "webcam child sex tourism", simulated Sweetie a 10-year-old Filipina, to lure predatory individuals on the hunt for children from poor countries to perform sex acts online (Quigley 2013). While the Aquino government reportedly planned to question the NGO's choice to depict "Sweetie" as Filipina (Fabunan 2013), the group was actually able to snare 20,000 alleged predators exposing the contradiction of competing images the Philippines is beset with (Quigley 2013). Sweetie as the face of cybersex outlines the contours of what may be inferred as the ICT industry's digital sexual economy.

### 3.1 The Digital Sexual Economy

Cybersex, in the case of the Philippines, sprung from the text-heavy online community platforms such as My Internet Relay Chat (mIRC) and Yahoo! Chatrooms, where chat members have sex via sex eyeball (SEBs), sex on phone (SOPs), or with

---

[9] See NSCB press release here: http://www.nscb.gov.ph/poverty/defaultnew.asp/

the use of web cameras (Sabangan 2003).[10] Advocates of the liberatory potential of the Internet, through such prosumer practices may be quick to point to the challenges these practices pose against, if not lead to the effacement of power structures and relations. But this is not often the case, because in a racialized and sexualized digital economy, the Filipino user's body carries an exchange value. Key words most commonly used in these chatrooms, such as *Pinay* pussy, Asian tranny, etc., are examples of commodified content, that reveal the racialized and sexualized nature of "supply and demand" in these online activities. As the Internet serves as an open market for goods and services, these online content and activities have come to generate value for profit for "e-capitalists" (Cote and Pybus 2007)—the most popular example would be the circulation of the commodified, exoticized image of a sexy and subservient Filipina online.

Moreover, multinational companies that provide Internet-related products and services, such as Google, commodify user content and activities or turn them into information goods or services to be sold, for instance, to third-party vendors (Robins and Webster 1988; Carr 2010). Online activities, including sexual content, are thus forms of immaterial labour, quantified and made profitable for Internet businesses (Lazzarato 1996; Cote and Pybus 2007; Dowling et al. 2007). Cybersex, reinvented in multimedia platforms, is now made available in live sex video chat websites, equipped with sophisticated and secure modes of online payment transactions. Such online content and activities now meant creating value for profit, alongside the rapid creation and expansion of business and technological infrastructures to run the digital sexual economy.

In looking at the cybersex phenomenon in the context of a digital sexual economy, we critically rethink the so-called revolutionary nature of ICTs. Examining Filipinas depicted online exclusively as mail-order brides, sex workers, and maids, Gonzalez and Rodriguez argue that ICTs "exacerbate intra-Asian and imperialistic histories of exploitation in the present, creating a specificity of experience that begs for a critical rethinking of the Internet as revolutionary technology" (2003: p. 16). They also contend that pornographic, dating, and marriage websites can be inferred to as "(the circulation of) Filipinas in ways reminiscent of colonial and global histories" (Gonzalez and Rodriguez 2003: p. 217). The simulation of Filipino sex-bot Sweetie is thus indicative of the racialized digital sexual economy and the symbolic violence that accompanies the circulation of Filipino bodies online. This resonates with the observations of ICT scholars examining the digital sexual economy. ICTs appear to facilitate the development

---

[10]In Annie Ruth C. Sabangan's (2003) first of a five-part special report on cybersex titled "In the chatrooms: A look into the minds of cybersex addicts", she defines sex eyeball as "when you get tired of CS [cybersex] and long for the real thing. When your chatmate is ready for SEB, you exchange cell phone numbers, plan where and when you will meet to have the 'big S'". She also defines sex on phone as: "Elevate your verbal CS to auditory CS. Grab the phone, hear each other's moans, while your other hand is busy exploring your body's erogenous zones".

of online sex tourism as a niche market in global tourism (Chow-White 2006), and even function "as an advertising medium for world sex" (Nagel 2003).

While we do acknowledge the exploitative elements of cybersex, we do not want to treat it merely as an issue of gender and sexual exploitation. Such a priori understandings of cybersex could impede productive and meaningful outcomes in research. Cybersex should be seen as mediated by technology, and couched in the networks of digital capitalism which subsume the digital sexual economy. We learned from Ditmore's study of Calcutta sex workers the importance of treating prostitution and trafficking-related abuses as labour issues to advance sex workers' protection. We also recognize a cybersex worker's claim that cybersex resembles call centre work (The Vincenton Post 2009), assimilating other forms of labour—emotional, cognitive, communicative, and technical—opening up the cybersex phenomenon to different alternative readings. Positing cybersex as affective labour, therefore, means that cybersex needs to be understood within these overlapping discourses.

## 4 On Affect: A Prelude to Cybersex as Affective Labour

The utility of affect is significant for its theoretical and practical reconsideration of the neoliberal subject (Paasonen 2011), giving expression to "a new configuration of bodies, technology, and matter" (Clough and Halley 2007: p. 2). Central to theoretical perspectives on affect is "the capacity to affect and to be affected", which stems from Dutch philosopher Baruch Spinoza's philosophy of ethics (Clough and Halley 2007; Clough 2010). This is the underlying logic that compels the moment preceding emotions and psychological states towards movement and action. Affect is not limited to feeling states; it encompasses forms of energies and forces that emanate from the body (Brennan 2004), as well as the body's potentiality, capacities and consequences (Clough et al. 2007; Clough 2010).

Approaches to affect are related but differ in their emphases and concerns (Seigworth and Gregg 2010), and we draw on three of these iterations. The first approach, drawing on thermodynamics, cybernetics and information theory, construes the individual as an organism whose affective capacities enable not only rational thought and action (Clough and Halley 2007) but also irrational, sometimes inconsequential states. Affect is a reservoir of potential matter and energy for determined and unintended actions, and often emerge as intensities (Hayles 1999; Hansen 2006; Massumi 2002; Clough and Halley 2007; Clough 2010; Gregg and Seigworth 2010). It also carries political possibilities or individual capacities for self-organization—what may be referred to as "autopoiesis" (Massumi 2002; Clough and Halley 2007). However, in an era of "intensified financialization", the productivity of affect is a constant struggle between its autonomy and commodification (Clough 2010: p. 219), bound up with forms of neoliberal biopolitics and the regulation of populations. Biopolitics here is defined as "the extension of an economic rationality to all aspects of society, including life-itself" (Clough et al. 2007: p. 72).

The second approach to affect pertains to the reproduction of "affective" responses to risk and uncertainty with the rise of financialization and the role of affective labour to assuage such conditions (Betancourt 2010). Digital capitalism induces and sustains "the illusion of a self-productive domain, infinite, capable of creating value without expenditure, unlike the reality of limited resources, time, expense and so on, that otherwise govern all forms of value and production" (Betancourt 2006). This simulates the "production and maintenance of ignorance" in the realms of consumption and labour (Betancourt 2010). In other words, ICT industries tap on the mental and emotional states of its users' mental and emotional states to keep the Internet economy going, especially its business infrastructure, and to ensure that insidious schemes of capitalist profit extraction remain hidden.

A third view of affect references affective labour as a subset of immaterial labour. The latter denotes forms of linguistic, communicative and knowledge work that have gained prominence in the era of post-industrial, neoliberal capitalism (Hardt and Negri 2000). Affective labour is "immaterial, even if it is corporeal and affective, in the sense that the products are intangible" (Hardt 1999: p. 96). Affect assimilates cognitive elements, and more importantly, modulates other people's affects and one's own:

> Unlike emotions, which are mental phenomena, affects refer equally to body and mind. In fact, affects, such as joy and sadness, reveal the present state of life in the entire organism, expressing a certain state of the body along with a certain mode of thinking. Affective labour, then, is labour that produces or manipulates affects. (Hardt and Negri 2004: p. 108)

Affect is a highly valuable commodity and affective labour's propensity to modulate the former becomes the latter's ultimate advantage. Unlike Patricia Clough's (2010) definition of biopolitics, which suggests a top-bottom or systems-based generation of power, biopolitics comes from the work of labourers producing all sorts of affects (Hardt 1999; Negri and Hardt 1999). The biopolitical power of affective labour is thus double-edged: on the one hand, it is highly valuable for capitalism to generate profits; on the other, it creates an advantage for the labourer because it can be reproduced indefinitely and "beyond measure" (Negri and Hardt 1999)—affective labour as "immaterial" commodity also makes it easy to evade control or surveillance. Ultimately, in this third approach, affective labour is the crucial game-changer for capitalism (Hardt 1999; Negri and Hardt 1999).

## 5 More Than One Side to the Story: Understanding Cybersex as Affective Labour

News headlines banner the proliferation of cybersex dens in the Philippines, where children and women have been trafficked into what mainstream media calls "modern form of slavery" (Lopez 2013). The prevailing narrative which tends to pigeonhole cybersex into one type of experience and excludes other possible experiences, is only one side of the story. Clearly, there are cybersex accounts that speak of the "oppression paradigm" (Weitzer 2010), where children and women fall prey to exploitative operators. The alternative view contends that, more often than not, considering sexual commerce as a form of harm potentially places sex workers in more danger becuase of stigma and discrimination and the lack of protection measures (Ditmore 2007). The very discourse of exploitation can serve as a self-fulfilling prophecy. It adds to the oppression of those who engage in cybersex, instead of recognizing their strategies of survival and self-improvement. The exploitation discourse obscures the enabling aspects and practical benefits of cybersex as a specific form of ICT use in everyday life. The moral binaries with which the State has cast cybersex—legal versus illegal, moral versus immoral—evade the serious ramifications stemming from state efforts to promote ICT for development. How should the state make sense of Internet users engaged in cybersex work if they stand to benefit from it, including their families? What about the instrumental value of cybersex work in situations where quality education and sustainable livelihood are either lacking, ill-suited, or ineffective? In promoting the integrity of the country's business environment and suppressing cybercrimes, whose interest does the Cybercrime Prevention Law protect? What does cybersex tell us about the risks and repercussions in the creation of an Information Society under a neoliberal digital economy?

In viewing cybersex as affective labour, we see the potential of challenging State efforts in building a Philippine Information Society and problematizing the consequences of how the State reproduces modes of affect in the new economy through branding the information technology labor economy as "world class service." Through its ICT framework, the State has inadvertently set the stage where cybersex workers have come to modulate the affects of other individuals and their own and immediately pass judgement on this "sexually explicit" service provision. Cybersex as affective labour thus compels us to critically look at these overlapping spheres institutions and mechanisms that benefit from and allow certain received understandings to be produced, institutionalized and maintained. This pespective makes possible alternative understandings, especially one that brings us back to the cybersex workers, the "bodily bearers" of this phenomenon.[11] This framework emphasizes the experiences of the worker and the work[12] they do through strategies of accommodation, negotiation and resistance in highly precarious conditions.

The affective labour perspective emphasizes agency as much as victimization, without reducing the practice and outcomes of cybersex to one or the other. The cybersex worker may be placed under oppressive conditions, such as exploitative

---

[11] See Tadiar's (2009) argument of Filipina women as "bodily bearers" of sexual economies in the New World Order.

[12] While labelling cybersex workers "labourer" may suggest a structured labour and wage system typical of industrial and manufacturing workshops and factories, we use "labour" and "labourer" to emphasize the mobilization of affects as central to cybersex but use them interchangeably with the terms "work" and "workers".

labor dynamics. At the same time, the worker's affective capacities generate values not just for capitalism but for herself, her family and even her community.[13]

Finally, cybersex differs from other forms of sex work because technology plays a key role in these activities. The role of ICT and the simultaneous material and immaterial elements at work require taking into account the distinctiveness of this type of mediated communication. Simply labelling it as prostitution or pornography or framing it narrowly as a moral issue prevents us from considering the broader contexts to understand why cybersex has developed and proliferated in the first place.

## 6 Re-examining the Philippine ICT Framework and the Cybercrime Prevention Act of 2012: Implications ICT for Development

The Philippine ICT framework programmed the country's ICT industry for the provision of the Filipino brand of "world-class service." It capitalized on the country's "human resource advantage" (CICT 2006) to strategically position the country in the global digital economy. The framework, we argue, inherently lacked enabling requisites for workers and citizens. Instead, it framed the role of the Filipino ICT labourer primarily as service and support, as exemplified by the provision of customer care, customer service or customer-relations management.

In reducing its workforce to front-liners in service provision, the government is pre-empting, rather than channelling, skills and energies toward the creative and meaningful strategies of ICT use. Pushing the country to be a leader in cyberservices destination, the State has exploited foreign capital and technology to rationalize and streamline labor. Affective labour, i.e. the mobilization of affect in cyberservice provision, becomes the product being sold in the global digital economy.

As the State endeavours to leapfrog ICT-led national growth, it excluded sectors it was quick to judge as illegal to sustain the integrity of ICT-related businesses and clear the ground for foreign investors. Thus, the State's limited understanding of cybersex and a priori judgement of cybersex workers has the tendency to be exclusionist. Categorizing cybersex as a crime has been a clear example of this. Understandably, determining which ICT practices are considered acceptable and legitimate, and justifying them based on perceived public threats, is a function of governance. With this approach, the State fails to consider values created from the periphery. ICT governance, preoccupied with creating limits and securitizing the technological infrastructure, unwittingly forces individuals and communities, especially those from the informal sector with little to no income or property, to live in subsistence and to rely on their means of survival. At worst, these people

---

[13]In a separate article, we delve into the lived experiences of cybersex workers, as well as business owners (Cruz and Sajo, forthcoming).

find themselves in the hands of exploitative operators and the cycle of alienation continues. Through institutionalized or formal uses of technology, as articulated and operationalized by public-private partnerships and more so by the State, a culture of creative ICT use is constrained.

The emergence of cybersex could therefore be read as a symptom of the ineffectiveness of ICT-driven development. Cybersex, we argue, is a counter-response to the inherent weaknesses in the State's programmatic approach towards developing the Philippine Information Society. Rather than allowing creative and practical uses of ICT among users, the State chooses to uphold its commitment to ICT-related businesses by giving a higher degree of penalty to crimes found in the Philippine Revised Penal Code when committed through ICT. If it is not working for the marginalized at best, then whose interests does the whole ICT infrastructure serve?

In casting cybersex as cybercrime, the State seems to evade the more fundamental problem of social exclusion that has brought about the informal economy of cybersex in the first place. Creating cybersex as a moral threat to the integrity of the Philippine Information Society obscures the fundamental problem of ICT development: the gap between formalized and informal ICT use to address the practical needs of the excluded for autonomy and self-improvement. Cybersex thus foregrounds the ethical dilemma of the present ICT framework and the institutional mechanisms maintaining it.

## 7 Concluding Remarks

In problematizing the cybersex phenomenon, we outlined the contours of the digital sexual economy in the ICT industry and laid down the theoretical foundation to foreground an alternative understanding of cybersex as affective labour. We offered to broaden the dominant perspective by using an alternative framework and pinpointing the many aspects of cybersex work. We argued that the State's project of ICT-led growth failed to impact on the quality of life of those living in the margins of techno-social development, and as a result, they resorted to cybersex, a creative, entrepreneurial, but morally questionable type of ICT use. Yet these efforts to eke out of poverty are shut down yet again, with the implementation of cybercrime legislation.

We believe that a meaningful ICT framework should aim to uplift the potential of Filipinos by recognizing and enhancing capacities for self-reliance, broadening life choices and chances to reduce socioeconomic inequity, and improving the quality of life especially for the deprived sectors of society. An effective ICT framework and its implementation cannot exclude or simply cast a moral frown on illegal ICT practices and those who engage in them.

A first step towards this goal would be to learn from lived experiences of ICT users in thinking about harnessing ICTs to develop spaces of autonomy, information use, and personal development, while respecting privacy. If the state intends to build "a people-centered, inclusive, development-oriented Information Society", its ICT framework should not exclude cybersex workers and their affective labour, which includes utilizing English language competency, ICT skills, and the (sexualized) mode of "world-class" service. The affective labourer could manipulate consumer fantasies and desires and create strategies of resistance and negotiation in cybersex work. Inadvertently, cybersex creates opportunities for the development of workers' agency and strategies of economic survival strategies. The State should take into account the creative uses of ICT that are practical and meaningful to cybersex workers, and more importantly how they have come to regard these practices as such.

In light of the manifold ways ICT is used, the cybersex worker's capacity to not only be affected but to affect the creation of monetary and non-monetary values could potentially serve as meaningful information for designing more inclusive policies and institutional mechanisms in achieving socioeconomic development. Clearly, cybersex is not the solution to achieving a decent quality of life, but its existence and persistence signifies that the State's vision of ICT for development is not living up to its promise. But if the response, as it were, is to create more legal enclosures, i.e. cybercrime criminalization, we are doubtful whether this is an effective and responsive strategy for improving the quality of life of the marginalized, which can be argued as the root of the State's cybersex problem.

Examining cybersex in terms of affective labour and using this exercise to critique the Philippine ICT discourse have placed us in an uncomfortable position of confronting the complex ethical issues that cybersex engendered. The State's moral approach to cybersex is a means to preserve the integrity and viability of the country's ICT industries or the State's vision of a "Philippine Information Society." As the State takes strides be a leader in the global digital economy, it has marked Filipino ICT workers en masse as cyberservice providers, along with the proprietary "world class service" seal.

We argued that in rendering cybersex work illegal and immoral, the State has chosen to institutionalize the provision of service instead of recognizing the informal yet productive means by which cybersex workers attempt to make a living. Cybersex might just be the necessary threat to the pursuit of ICT-driven development to compel the State to recognize that Filipino ICT workers are much more than pawns for capital.

In the absence of a grounded understanding of the lived experiences of those engaged in this rampant and illicit phenomenon, we wonder whether cybersex even needs to be legislated. Cybersex demands complicating and superseding the dominant view that construes it as moral transgression, as preying on victims, as a criminal activity. How do we make sense of cybersex that eventually turns into a meaningful sexual relationship? How do we respectfully respond to cybersex workers who are using cybersex to provide for their economic needs and those of their families? Would it even be fair and ethical to point cybersex workers to more socially accepted ICT work such as call center work, knowing that this form of cheap, outsourced labor is also subject to new forms of surveillance and servitude (Mirchandani 2012)? It remains a challenge for Philippine ICT policy to address fundamental issues regarding genuine inclusive development. Responding to this challenge requires developing an ICT framework that starts from an empathic consideration of the grounded, multifaceted narratives and strategies on technology use and adaptation coming from marginalized individuals and communities.

**Open Access** This chapter is distributed under the terms of the Creative Commons Attribution Noncommercial License, which permits any noncommercial use, distribution, and reproduction in any medium, provided the original author(s) and source are credited.

# References

Arviddson, A. (2007). Netporn: The work of fantasy in the information society. In K. Jacobs, M. Janssen, & M. Pasquinelli (Eds.), *C'lick me: A netporn studies reader* (pp. 69–78). Amsterdam: Institute of Network Cultures. http://www.networkcultures.org/_uploads/24.pdf. Accessed 30 Nov 2013.
Betancourt, M. (2006). The aura of the digital. *CTheory.net*. http://www.ctheory.net/articles.aspx?id=519. Accessed 30 Nov 2013.
Betancourt, M. (2010). Immaterial value and scarcity in digital capitalism. *CTheory.net*. http://www.ctheory.net/articles.aspx?id=652. Accessed 30 Nov 2013.
Brennan, T. (2004). *The transmission of affect*. Ithaca: Cornell University Press.
Carr, N. (2010). *The shallows: What the internet is doing to our brains*. New York: W.W. Norton & Company.
Castillo, L. (2010). *Solon seeks to prohibit cybersex operations in the country*. http://www.congress.gov.ph/press/details.php?pressid=4703. Accessed 30 Nov 2013.
Chow-White, P. A. (2006). Race, gender and sex on the net: Semantic networks of selling and storytelling sex tourism. *Media, Culture and Society, 28*(6), 883–905. doi:10.1177/0163443706068922.
Clough, P., & Halley, J. (2007). *The affective turn: Theorizing the social*. Durham: Duke University Press, 2007.
Clough, P. T. (2010). The affective turn: Political economy, biomedia, and bodies. In M. Gregg & G. J. Seigworth (Eds.), *The affect theory reader* (pp. 206–225). Durham/London: Duke University Press.
Clough, P. T., Goldberg, G., Schiff, R., Weeks, A., & Willse, C. (2007). Notes towards a theory of affect-itself. *Ephemera: Theory and Politics in Organization, 7*(1), 60–77.
Commission on ICT. (2006). *Philippine ICT road map*. http://www.unapcict.org/ecohub/resources/philippine-ict-roadmap. Accessed 30 Nov 2013.
Cordova, J. (2011). *Cybersex dens: A sordid side effect of the call center boom*. http://asiancorrespondent.com/45961/cybersex-a-sordid-side-effect-of-the-call-center-boom-in-philippines/. Accessed 30 Nov 2013.
Coté, M., & Pybus, J. (2007). Learning to immaterial labour 2.0: MySpace and social networks. *Ephemera: Theory and Politics in Organization, 7*(1), 88–106.
Cybercrime Prevention Act of 2012. (2012). Retrieved from Official Gazette. http://www.gov.ph/2012/09/12/republic-act-no-10175. Accessed 30 Nov 2013.
Diaz de Rivera, A. T. M. (2008). Digital libraries and the Philippines' strategic roadmap for the ICT sector. *Journal of Philippine Librarianship, 28*(1), 55–77.

Ditmore, M. (2007). In Calcutta, sex workers organize. In P. T. Clough & J. Halley (Eds.), *The affective turn: Theorizing the social* (pp. 170–186). Chapel Hill: Duke University Press.

Dowling, E., Nunes, R., & Trott, B. (2007). Immaterial and affective labour: Explored. *Ephemera: Theory and Politics in Organization, 7*(1), 1–7.

Estopace, D. (2013). *From sunrise to 'eternal sunshine' industry.* http://www.businessmirror.com.ph/index.php/en/features/perspective/15226-from-sunrise-to-eternal-sunshine-industry. Accessed 30 Nov 2013.

Fabunan, S. S. D. (2013). *Palace warns vs cyber sex.* http://manilastandardtoday.com/2013/11/08/palace-warns-vs-cyber-sex/. Accessed 30 Nov 2013.

GMA News. (2013a). *Manila and Cebu in top 10 global outsourcing destinations.* http://www.gmanetwork.com/news/story/292785/economy/business/manila-and-cebu-in-top-10-global-outsourcing-destinations. Accessed 30 Nov 2013.

GMA News. (2013b). *Abad: Gov't will help booming BPO industry.* http://www.gmanetwork.com/news/story/300947/economy/business/abad-gov-t-will-help-booming-bpo-industry. Accessed 30 Nov 2013.

GMA News. (2013c). *BPO revenues rose 18% in 2012 – IBPAP.* http://www.gmanetwork.com/news/story/298940/economy/business/bpo-revenues-rose-18-in-2012-ibpap. Accessed 30 Nov 2013.

Gonzalez, V., & Rodriguez, R. M. (2003). Filipina.com: Wives, workers and whores on the cyberfrontier. In R. C. Lee & S. L. C. Wong (Eds.), *Asian America.Net: Ethnicity, nationalism and cyberspace* (pp. 215–234). New York: Routledge.

Hansen, M. (2006). *Bodies in code: Interfaces with new media.* New York/London: Routledge.

Hardt, M. (1999) Affective labor. *Boundary 2, 26*(2), 89–100. http://www.jequ.org/files/affective-labor.pdf. Accessed 30 Nov 2013.

Hardt, M., & Negri, A. (2000). *Empire.* Cambridge, MA: Harvard University Press.

Hardt, M., & Negri, A. (2004). *Multitude.* New York: Penguin Press.

Hayles, N. K. (1999). *How we became posthuman: Virtual bodies in cybernetics, literature, and informatics.* Chicago: The University of Chicago Press.

Jacobs, K. (2007). *Netporn: DIY web culture and sexual politics.* Lanham: Rowman & Littlefield.

Lazzarato, M. (1996). Immaterial labour. In P. Virno & M. Hardt (Eds.), *Radical thought in Italy: A potential politics* (pp. 132–146). Minneapolis: University of Minnesota Press.

Lopez, R. (2013). *Virtual Filipina kid lures 1,000 'sexual predators' worldwide.* http://www.mb.com.ph/virtual-filipina-kid-lures-1000-sexual-predators-worldwide/. Accessed 30 Nov 2013.

Massumi, B. (2002). *Parables for the virtual: Movement, affect, sensation.* Durham/London: Duke University Press.

Mirchandani, K. (2012). *Phone Clones: Authenticity Work in the Transnational Service Economy.* Ithaca, NY: Cornell University Press.

Nagel, J. (2003). *Race, ethnicity and sexuality: Intimate intersections, forbidden frontiers.* New York: Oxford University Press.

National Economic and Development Authority. (2011). *Philippine development plan (2011–2016).* http://www.neda.gov.ph/PDP/2011-2016/frontmatter.pdf. Accessed 30 Nov 2013.

National Statistical Coordination Board. (2012). *Statistical indicators on Philippine development* (Chapter/sector: Competitive industry and services sectors). http://www.nscb.gov.ph/stats/statdev/2012/ch2_industry.asp. Accessed 30 Nov 2013.

National Statistical Coordination Board. (2013). *Philippine statistics in brief.* http://www.nscb.gov.ph/download/NSCB_PhilippinesInBrief_Apr2013.pdf. Accessed 30 Nov 2013.

Negri, A., & Hardt, M. (1999). Value and affect. *Boundary 2, 26*(2), 77–88.

Paasonen, S. (2011). *Carnal resonance: Affect and online pornography.* Cambridge, MA: MIT Press.

Philippine Digital Strategy. (2010). http://icto.dost.gov.ph/index.php/philippine-digital-strategy. Accessed 30 Nov 2013.

Quigley, J. T. (2013). *Virtual 10-year-old Filipina reveals thousands of cybersex predators.* http://thediplomat.com/2013/11/virtual-10-year-old-filipina-reveals-thousands-of-cyber-predators/. Accessed 30 Nov 2013.

Remo, M. V. (2013). *NSCB says PH Internet penetration rate at 36 %*. http://technology.inquirer.net/30571/nscb-says-ph-internet-penetration-rate-up-36. Accessed 30 Nov 2013.

Robins, K., & Webster, F. (1988). Cybernetic capitalism: Information, technology, everyday life. In V. Mosco & J. Wasko (Eds.), *The political economy of information* (pp. 44–75). Madison: University of Wisconsin Press.

Romero, P. (2013). *SC extends TRO on cybercrime law*. http://www.rappler.com/nation/21089-sc-extends-tro-on-cybercrime-law. Accessed 30 Nov 2013.

Sabangan, A. R. C. (2003). In the chatrooms: A look into the minds of cybersex addicts. *Manila Times*. http://trafficking.org.ph/v5/index.php?option=com_content&task=view&id=407&Itemid=56. Accessed 30 Nov 2013.

Saloma-Akpedonu, C. (2006). *Possible worlds in impossible spaces: Knowledge, globality, gender, and information technology in the Philippines*. Quezon City: Ateneo de Manila University Press.

Seigworth, G. J., & Gregg, M. (2010). An inventory of shimmers. In *The affect theory reader* (pp. 1–25). *Durham y Londres: Duke University Press*.

Tadiar, N. X. M. (2004). *Fantasy-production: Sexual economies and other Philippine consequences for the new world order*. Hong Kong: Hong Kong University Press.

Tadiar, N.X.M. (2009). If not mere metaphor, sexual economies reconsidered. *The scholar and feminist online*. http://sfonline.barnard.edu/sexecon/tadiar_01.htm. Accessed 21 Nov 2014.

The Vincenton Post. (2009). *Special report: Cybersex trade thrives in a time of crisis*. http://fvdb.wordpress.com/2009/11/10/cybersex-trade-thrives-in-time-of-crisis/. Accessed 30 Nov 2013.

Weitzer, R. (2010). The mythology of prostitution: Advocacy research and public policy. *Sexuality Research and Social Policy, 7*(1), 15–29.

# The Internet and Indonesian Women Entrepreneurs: Examining the Impact of Social Media on Women Empowerment

Ezmieralda Melissa, Anis Hamidati, Muninggar Sri Saraswati, and Alexander Flor

## 1 Introduction

### 1.1 Background

Unemployment is one of the major challenges that Indonesia faces. Home to over 230 million, Indonesia has an unemployment rate of 2.6 % among a labour force of 107.7 million (Badan Pusat Statistik 2010). Of this number, the working population is pegged at 104.9 million, comprising 66.8 million men and 38.1 million women (ibid.). The unemployment rate appears to be moderate. However, the Indonesian Statistics Bureau estimates that only 67.72 % of the working force is regularly employed. Therefore, the actual unemployment rate at any given time is higher than the official figure.

There are many reasons behind unemployment in Indonesia. According to Danani (2004: p. 2), unemployment is 'the result of a combination of a multitude of circumstances and distinct factors, some acting on the supply side and others on the demand side'. Examples of supply factors are the age structure of population, fertility, education and the sustained rise in the economic well-being of the population. Meanwhile, the demand factors include the sectoral and status structure of employment as well as the pace of economic growth (ibid.).

E. Melissa (✉) • A. Hamidati
Department of Communication and Public Relations, Swiss German University, Banten, Indonesia
e-mail: ezmieralda.melissa@sgu.ac.id

M.S. Saraswati
Asian Research Centre, Murdoch University, Murdoch, WA, Australia

A. Flor
Faculty of Information and Communication Studies, University of the Philippines (Open University), Los Baños, Laguna, Philippines

The Indonesian government has implemented several strategies to counter unemployment. In addition to opening job opportunities in the market through customary measures such as opening factories or other labour-intensive projects, it has encouraged its citizens to create work opportunities through entrepreneurship. A number of policies have been formulated to provide loans for new entrepreneurs to start or expand their businesses (Ardieansyah et al. 2011: p. 40). The government also provides training to improve citizens' business skills, organize exhibitions to showcase their products and so on (ibid.). All of these activities are conducted to increase the number of entrepreneurs from 0.18 to 2 % of the total population, which is considered as the ideal ratio by the government (Antara 2012).

Many entrepreneurial projects in Indonesia target women mainly because the unemployment rate among women is higher than that of men. According to BPS (2010), the unemployment rate among women is 3.6 % compared to 2.0 % among men. Another equally important reason is the fact that Indonesian women are not particularly ignorant about entrepreneurship. Instead, Indonesian women have been involved in business despite limited public acknowledgement. In 2011, the Ministry of Women Empowerment and Child Protection reported that there are 55,206,444 micro-, small and medium enterprises[1] in Indonesia. It is estimated that 60 % of these enterprises are run by women (Kementrian Koperasi dan Usaha Kecil Menengah 2008). As a result, women are seen not only to have potential but as capable and resilient entrepreneurs as well.

Another significant fact about entrepreneurship in Indonesia is revealed by the Indonesian Women's Business Association (IWAPI). This association reported that around 85 % of businesses run by women are micro- and small businesses, while 13 % are medium enterprises and only 2 % are big businesses (IWAPI in Widyadari et al. 2006: p. 4). Most of these micro- and small businesses are operated from home and manufacture products such as handicrafts, traditional food, clothing items and so on (ibid.) Unfortunately, there is no existing data that explains how many of these businesses are ICT based.

Entrepreneurship provides both benefits and challenges to women. On the one hand, it enables them to generate additional income for themselves and their families, create jobs and, eventually, actively contribute to the country's economy (see Sarri and Trihopoulou 2005). On the other hand, women also have to face many challenges involving financing, business skills, networks and support groups, among other things, to be able to set up their own business (Coleman 2000; Cromie and Birley 1992).

---

[1] In Indonesia, microenterprises refer to businesses with assets that are up to Rp 50 million and turnovers that are up to Rp 300 million. Meanwhile, small enterprises refer to business with assets that are in between Rp 50 million and Rp 500 million and turnovers that are in between Rp 300 million and Rp. 2.5 billion. Finally, medium enterprises refer to business with assets that are in between Rp 500 million and Rp 10 billion and turnover that are in between Rp 2.5 billion and Rp 50 billion (Regulation No. 20/2008 on Micro, Small and Medium Enterprises).

One of the main issues faced when opening a business is funding. Since many women quit their jobs after getting married, they do not have much savings to start their own company. In addition, Indonesians traditionally give more property ownership rights to men in conjunction to their role as head of the family.[2] As a result, women normally have fewer assets that may be used as collateral for bank loans (Widyadari et al. 2006). Otherwise, they should get consent from their husbands – the ones holding the rights to the property – stating their legal responsibility to complete payments of the loans. Hence, if their husbands disapprove of their intention to work or if they are unmarried, it becomes difficult for women to fund their business.

The second challenge faced by women in becoming entrepreneurs is related to education. In many developing countries, women tend to have lower education than men (Coleman 2000; Cromie and Birley 1992). In Indonesia, merely 11.62 % of Indonesian female entrepreneurs engaged in micro- and small businesses graduated from high school (Kementrian Pemberdayaan Perempuan dan Perlindungan Anak 2011). The majority (75.5 %) of them only finished elementary education. This is far lower than male entrepreneurs in the same business scale with 19.9 % of them graduating from high school and 63.4 % graduating from elementary schools (Kementrian Pemberdayaan Perempuan dan Perlindungan Anak 2011). As a consequence, women tend to have limited technical, technological and managerial skills to implement their business ventures.

Moreover, women have to face challenges in starting, grooming and expanding their business due to their limited social circles. It is widely known that social networks are important factors in entrepreneurship, particularly within the context of Asian countries (Gold and Guthrie 2002; Siong-Choy 2007; Carney et al. 2008). Without enough social capital gathered from these networks, entrepreneurs might not be able to survive in today's competitive market.

In addition to those factors, Indonesian women have to deal with balancing their career and family. This is actually a typical issue faced by many women workers in other countries (see Segal et al. 2005: p. 3; Lombard 2001: p. 216), but it is more prominent in developing countries, such as Indonesia and Vietnam (Van der Merwe and Lebakeng 2010; Nguyen 2005). The majority of Indonesians position men as breadwinners and women as homemakers. Therefore, although women, particularly educated women in urban areas, can enter the workforce, they cannot escape from their traditional role as homemakers.

---

[2]Many ethnicities in Indonesia practise patriarchal culture. In this culture, properties such as houses and land are inherited only to the males in the family. In addition, many Indonesians are also Muslims; according to the Islamic law, males receive twice as much inheritance as those received by the females. Both of these practices are implemented with reasoning that because men have a role as the head of the family, then they have to be responsible for the other family members.

## 1.2 Rationale

The introduction of new information and communication technologies (ICTs) has brought new hopes to women entrepreneurship (Ndubisi and Kahraman 2006; Duncombe et al. 2005). One prominent feature of the Internet is the introduction of Web 2.0 enabling users to form and expand their networks online (O'Reilly 2009). This feature has been proven capable to support business entities, particularly in marketing-related activities (Jones 2010: p. 150). For existing and future women entrepreneurs, an innovation that is capable of forming and expanding networks can support them in overcoming challenges related to social capital. Considering the increase in the number of Indonesian Internet users from 500,000 in 1998 to 55 million users in 2011 (Markplus 2011), generating social capital through expanding networks can become a concrete possibility.

In Indonesia, the majority of Internet[3] users access it mostly for social networking purposes. Indonesian Internet users have been widely acknowledged as ardent social media users.[4] It is reported that 47.5 million Indonesians subscribed to *Facebook* in 2012 (Internet World Stats 2012), putting Indonesia fourth after the USA, Brazil and India in terms of total number of users (Socialbakers 2012). In addition to *Facebook*, there are around 19.5 million Twitter accounts in Indonesia, making it the fifth largest user population in the world in 2012 (Socialbakers 2012). Conventional blogging, which is also a form of social media, also remains popular among Indonesian Internet users. The number of blogs in Indonesia rose from less than 1 million in 2009 to 5.3 million in 2012 although only 32.7 % were regularly updated (SalingSilang 2011). It is estimated that currently there are 2.7 million bloggers and more than 25 blogger communities in the country (Nugroho and Syarief 2012: p. 37). It would only be logical to assume that social media would have a significant social impact on Indonesians, particularly on those engaged in networking such as women entrepreneurs.

---

[3] Internet penetration level in Indonesia is actually low. It is reported in 2010 that only 9.1 % of the population had access to the Internet. Meanwhile the neighbouring countries reached higher rate, for instance, 55 % in Malaysia and 27.6 % in Vietnam (International Telecommunication Union 2011). Access to the Internet is also mostly centralized in urban areas in the country. Despite the not-so-promising performance, the International Telecommunication Union (2011) reported that there were 220 million mobile phone users in Indonesia and many of them subscribed to mobile Internet services. Hence, this new service is able to increase Internet penetration in the country.

[4] Social media platforms such as *Facebook*, *Twitter* and *YouTube* have become standard features of affordable mobile phones. A Chinese brand mobile phone with SNS platforms is available in the Indonesian market for as low as Rp 100,000 (approximately USD 10). Branded mobile phones like *Nokia* can also be bought for less than Rp 500,000 (approximately USD 50). Moreover, Indonesian telecommunication providers offer reasonably reliable Internet connection for social networking activities for as low as Rp 25,000 (approximately USD 2.50) a month. Hence, mobile Internet users increased from 16.2 million in 2010 to 29 million in 2011 (Markplus 2011).

## 1.3 Research Objectives

This study attempts to answer the following research questions: How does social media empower women? How does Internet technology impact women entrepreneurship?

Therefore, the main objective of this study is to contribute to the body of knowledge on the role of social media in empowering women through entrepreneurship. Additionally, the study aims:

- To generate insights on the potentials of social media in empowering women
- To determine how social media entrepreneurship can provide a balance between career and family life

This study is particularly significant since emerging studies on business and the Internet, more specifically social media, in Indonesia have been focusing only on marketing activities and rarely discuss the element of women entrepreneurship (see Handayani and Lisdianingrum 2011).

## 2 Literature Review

### 2.1 The Internet, Social Media and Social Capital

The Internet, considered as one of the greatest inventions of the twentieth century (Gates 2000), is considered a window for information democracy. The arrival of the Internet arguably opens many possibilities for many people. This is due to the fact that this new technology allows a more efficient way to communicate, so that not only communication is made faster but also relatively cheaper and accessible to many people with minimum skills. Internet power as the platform for the World Wide Web is manifested in Metcalfe's law, which states, 'the power of the Web is enhanced through the network effect produced as resource links by network members' (Esplen and Brody 2007: p. 14). As the number of people in a network grows, the connectivity between members also increases (ibid.). This characteristic is said to increase social capital among network users.

The power of the Internet in enhancing users' social capital can be seen more clearly in a new kind of service which has gained tremendous popularity, social media. With social media, people no longer form groups and connections exclusively within their physical locale. Instead, relationships are developed around geographically dispersed social networks. Social media themselves can be described as applications that 'allow the user to articulate an egocentric network, anchored by a profile' (Gross and Acquisti in Ellison et al. 2011: p. 876). One advantage of social media is their ability to help users manage both their weak and strong ties. Especially on sites such as *Facebook*, 'features to search people by name, region and school allows them to find and keep in touch with friends with whom they might have otherwise lost touch' (Ellison et al. 2011: p. 874). At the same time, users can also connect to casual acquaintances that might have some similarities with them, such as similar hobbies, occupations, educational backgrounds and so on.

Bourdieu (in Ellison et al. 2011: p. 874) defines social capital as 'the aggregate of actual or potential resources linked to a durable network of more or less institutionalized relationships of mutual acquaintance and recognition'. By this definition, social capital is the value or benefit that one gains from his or her social relationships and interactions (Burt in Ellison et al. 2011: p. 874). Similar to other types of capital, these benefits can come in many forms, such as access to information, companionship, financial gain and so on. Before the arrival of communication technologies, social capital was usually gained from face-to-face interaction and direct relationship; however the arrival of the Internet and other new media technologies changed this (Boase et al. 2006: p. 5). It is now argued that social capital is an integrated part of social media structures (Burt in Ellison et al. 2011: p. 874). It is because the significant number of users on social media with a diversity of backgrounds has resulted in increased social utility of these sites. Hence, social media can help users to be readily available and visible to a lot of different people (Brandtzæg et al. 2010: p. 1007).

Following its development, social media has been used more than just to find friends but to promote a social cause, to introduce a political candidate and more recently to conduct business. In Indonesia, particularly, there is a growing trend to open business ventures through social media. For instance, *Facebook* alone records that there are 549,740 users in Indonesia who are owners of small and medium enterprises (SMEs) and 176,300 of them are women (Facebook 2013). Flexibilities offered by social media become prominent contributing factors to the emergence of this trend. First of all, since social media accounts can be accessed through mobile technologies, such as smart phones or tablet computers, business can be done whenever and wherever they like. Furthermore, social media also allows products to be showcased in the virtual stores, thus eliminating the need to have a physical store – which is costly. In addition, the applications in social media are very user-friendly; users can tag pictures and provide information of products to potential customers by a single click.

The flexibilities described above also encourage women to become entrepreneurs. This arrangement is seen to be an ideal option for women for several reasons, among them being society's perception that favours women to stay at home, the dilemma faced by women between having a career and taking care of the children, the relatively low cost needed to open the business and so on. Among these reasons, the second one particularly attracts the interest of some scholars studying women online entrepreneurship (see Segal et al. 2005: p. 3; Lombard 2001: p. 216).

## 2.2 *Social Media Entrepreneurship*

The growing interest of women in developing an online business may also be supported by the nature of women who like to make contact with others. Furthermore, due to the limited time available, today's customers like to shop online as they can do it through the comfort of their computers or mobile phones. In terms of potential consumers for this kind of business, Indonesia is very promising. It is reported that Indonesia has one of the highest number of social media users with more than 35 million people (FI Management 2010). This resulted in the proliferation of businesses using social media as their communication platform, whether through sales offers or advertising.

Social media entrepreneurship offers many advantages to women. Not only does this business potentially support women to be financially independent, but in a more substantial level, social media business encourages women to be more self-actualized. According to Abraham Maslow, which places self-actualization in his last level of hierarchy of needs, self-actualization is the one factor that makes a human being 'fully human' (in Goble 1970: p. 25). The importance of self-actualization helps women not only to cultivate themselves but also to increase their self-esteem. All of these lead to women having a stronger position in the family and society, thus contributing to their empowerment.

## 2.3 Women Empowerment

According to Mosedale (2005: p. 246), the concept of women's empowerment began in the 1970s, initiated by third-world feminists and women organizations. In the 1990s, the concept evolved, and many agencies began to associate it with strategies to provide support to increase women productivity in relation to the increasing withdrawal of state responsibility for broad-based economic and social assistance. However, despite the fact that the strategies provide women with more access and control over household finances, there are those who argue that this claim may be overrated. According to Mayoux (in Mosedale 2005: p. 246), without the proper support networks and empowerment strategies, this condition actually gives more disadvantages to women since they will have to assume the burden for household debt and subsistence.

Empowerment itself is defined differently by different scholars. Constructing his definition on Narayan, Bennett (in Malhotra et al. 2002: p. 4) describes empowerment as 'the enhancement of assets and capabilities of diverse individuals and groups to engage, influence and hold accountable the institutions which affect them'. Meanwhile, Malhotra et al. (2002: p. 4) define empowerment as a term that is embedded by two important elements, which differentiate it from the general concept of 'power'. The first element is 'process' or 'change', which signifies empowerment as an improved condition from having less power to having more power and which enables individuals to gain social inclusion. The second element of empowerment is 'human agency', which refers to the freedom and willingness to exert one's choice without severe consequences. More detailed explanations of the terms will be discussed in later paragraphs.

As mentioned above, in the first element of power, there is an important note on social inclusion that can be defined as 'the removal of institutional barriers and the enhancement of incentives to increase the access of diverse individuals and groups to assets and development opportunities' (Williams 2005: p. 3). This view of empowerment is supported by Kabeer (1999: p. 436) who argues that being empowered is not only about having choices but also more importantly about having the freedom to exercise these choices. Based on these definitions, it can then be concluded that empowerment incorporates three interrelated dimensions, which are resources (preconditions), agency (process) and achievements (outcomes). Moreover, they highlight that empowerment is not only referring to access to resources, usually associated with the poor. Those who are not considered poor may also be involved in the process of change to be empowered, such as having more choices, control, freedom and so on. In addition, resources should not only be understood as material resources in the more conventional economic sense (such as money, land, etc.) but also the various human and social resources which serve to enhance the ability to exercise one's choice (skills, education, supports, etc.). Resources in this broader sense are acquired through different social relationships, such as family, market and community (Kabeer 1999: p. 437).

On the other hand, the second dimension of power is related to agency or the ability to define one's goals and act upon them (Kabeer 1999: p. 437). Agency itself does not simply mean an action to exercise one's choice; instead it also 'encompasses the meaning, motivation and purpose which individuals bring to their activity, their sense of agency, or "the power within"' (ibid.). Apart from being commonly understood as a decision-making process, agency can actually take a number of other forms, such as bargaining and negotiation, deception and manipulation, subversion and resistance as well as more intangible, cognitive processes of reflection and analysis (ibid.). This dimension can also be exercised either by an individual or by groups. In relation to power, agency has both positive and negative implications. In the positive sense, agency refers to people's capacity to define their own life choices and to pursue their own goals, even in the face of opposition from others. On the other hand, agency can also mean a 'power over' or 'the capacity of an actor or category of actors to override the agency of others, for instance, through the use of violence, coercion and threat' (ibid.). The two dimensions of power – resources and agency – combined reflect what Sen (1985) refers to as capabilities or 'the potential that people have for living the lives they want'.

There are several aspects of women empowerment as proposed by different theorists such as Acharya and Bennett (1983), Kabeer (1997) or Frankenberg and Thomas (2001). However, for purposes of this study, only the economic, socio-cultural, familial/interpersonal and psychological dimensions will be examined. In addition, this study will use several indicators, which are proposed by Malhotra et al. (2002: p. 26) to measure women empowerment. These include domestic decision-making, access to or control over resources, freedom of movement, economic contribution to household, appreciation in household, sense of self-worth, time use/division of domestic labour and knowledge.

Reflecting back from the above explanations, it can be concluded that empowerment is an ongoing process and should not be seen as a result since the environment within which empowerment operates continuously changes. Hence, the empowered condition cannot be maintained unless the agents continuously challenge the power system that suppresses their rights. Entrepreneurship is seen as one potential tool to fulfil this objective. It is due to the fact that the increasing role of women as entrepreneurs or as financial contributors in the family leads to increasing power over the decision-making in the family. Researches have concluded that the increase of women's access to activity within the market correlates with the increase of their authority within the household and community (Coughlin and Thomas 2002). Positive results also arise with the increasing empowerment of women as entrepreneurs which improves the education, health and productivity of the household members, particularly the children (Coughlin and Thomas 2002: p. 557). Nonetheless, it should be noted that although these literature and research on entrepreneurship emphasize its ability as an economic engine with potential social impacts, Calás et al. (2009: p. 560) suggest a different point of view of entrepreneurship that entrepreneurship should be seen more as a social change activity with a variety of possible outcomes that may or may not be beneficial for the entrepreneur.

## 2.4 Women Entrepreneurs

Sicat (2007) identified that there are generally two types of women entrepreneurs. The first is the subsistence entrepreneur where women, particularly in low-income households, are driven to generate supplementary income due to their concern about the basic needs of the family. The second type is defined as the growth entrepreneur, where women tend to be the modern, career-minded and working for the purpose of self-actualization. Many women state that their primary motivation in starting up a business is to meet personal and business goals, as well as to go against stereotypes imposed by society (Roffey et al. 1996).

The reality today is that women play a much bigger role in the economic empowerment within the family as well as within the society. In Indonesia, the State Ministry of Cooperatives and Small and Medium Enterprises estimates in 2006 that 60 % of the micro-, small and medium enterprises are owned by women. In the same year, about 85 % of Indonesian Women's Business Association members were owners of micro- and small businesses (Sarinastiti 2006: p. 10). Other parts of Asia also experience the growing phenomenon with about 35 % of small and medium enterprises (SMEs) headed by women (Chiam 2001). Taking into account that SMEs make up 95 % of all enterprises in the Asia-Pacific region and contributing between 30 and 60 % of GDP in each APEC economy (Chiam 2001), it is not hard to see the significant role women play within the economy. This is also evident in China where about 25 % of business start-ups are conducted by women (Chiam 2001) as well as in Japan which holds the staggering statistical domination by women in the SMEs, where four out of five small business owners are women (Chiam 2001). Given these trends, the study intends to contribute to the existing body of knowledge on women, social media, entrepreneurship and empowerment.

## 3 Research Methods

Five focus group discussions with women entrepreneurs in the five largest Indonesian cities (Jakarta, Surabaya, Bandung, Medan and Makassar) were conducted. Fifty-two women participated in these FGDs. The participants were snowball sampled using the following selection criteria: married women in their productive age, lived in one of the five largest cities in Indonesia and owned an online business that made use of social media, either individually or with a partner(s), as a full-time job for at least 2 years.

The FGDs solicited their stories on their use of social media in their businesses and its impact on their life. Following this, in-depth interviews were conducted with one participant in each city. The interviews were conducted in the participant's house with the purpose of experiencing first-hand how the participant manages her business in between her daily activities. Both the FGDs and interviews were video and audio recorded.

At the beginning of the FGDs, questionnaires were distributed to solicit respondents' socio-demographic information and online business details including type of business, clientele, assets, liabilities, income and so on (Fig. 1).

**Fig. 1** Focus group discussion in Makassar, South Sulawesi

## 4 Findings and Discussion

### *4.1 Respondents' Profiles*

From the questionnaires, the following data can be derived:

First of all, with regard to demographic characteristics, most of the women are between the ages of 31 and 40 years old (54 %), 22 % are between the age of 20 and 30 years old and 20 % are between the age of 41 and 50 years old, while the oldest are between the ages of 51 and 60 (4 %).

In addition, most respondents have high education level with 25 % claiming to hold a master's degree, 73 % holding a bachelor's degree and only one person (2 %) stating that she only graduated from high school. Many of these women once pursued a professional career (70 %), while 22 % stated that they had never worked, and 8 % started the business straight away after finishing their studies.

Furthermore, some information was also gathered with regard to their business. The questionnaire revealed that around 70 % of these women sell home-made cakes or other food products, while the rest mostly sell fashion products targeted mainly for women. In terms of income made from these businesses, 40 % reported that they earn approximately 2.5 to 5 million a month (approximately USD 250–USD 500), while 26 % mentioned that they earn between 5 and 7.5 million, 24 % earn below 2.5 million, and only 10 % earn above 7.5 million.

Although there were not any specific questions in the questionnaire asking about their computer literacy and usage, all respondents mentioned in the FGDs that they learned some basic computer skills during their studies. Those who worked acquired more advanced skills at the office, while others only developed their skills when they decided to establish the online business.

### *4.2 Networking and Social Capital*

Social capital has been one of the main drivers of women's entrepreneurship. Online businesses are started, developed and maintained with the support of online communities and friends. Such was validated in this study.

Among the respondents, leaving work to take care of families served as a catalyst for opening an online business. Initially, the loss of daily interaction with peers at work or within their social circles prompted them to find other means to interact. The Internet offered a solution to be virtually present to their peers by interacting through technology-mediated conversations while at the same time being available for their families and their daily domestic activities. In time, these online interactions grew and turned into a potential source of income.

**Fig. 2** A respondent in Surabaya controlling the production process in her home

Aside from being a medium of communication between the respondents and their peers, the platform also attracted other bloggers who were interested in similar topics. Conversations were initiated through comments or requests. The networks grew and some groups became established communities.

> This (business) community is very supportive. We can ask each other information on what products are trendy right now. Then, I can ask others if I run out of stock. Simply, we help each other. (A respondent from Bandung)

Online community members take friendship in the virtual world to be as serious as in the physical world. They have devoted as much time and effort or perhaps even more to these friendships. Their activities may be the means to achieve an end of 'what would I post on my social media today?' Baking cookies and trying out certain recipes become the activities that were not purely motivated to provide for the family but also to provide the experience and photos to be posted up and shared among peer groups in social media (Fig. 2).

These online interactions grew. To accommodate the communities' need to interact in person, gatherings are often conducted. Depending on the common interest of the group, these gatherings are also done for a certain purpose. In the culinary online business, for example, events would include workshops in making currently popular food products or about food photography. There are also communities that conduct events for networking among members in support of their online businesses.

### 4.2.1 Knowledge Sharing and Reuse

The majority of the respondents opened their social media accounts for the sole purpose of interacting with their peers rather than to set up online businesses.

However, the medium has proved to be valuable where their interactions enabled them to identify peers who have similar interests, including in establishing online businesses. In such cases, these friendships were developed and strengthened even further by the sharing of experiences of their entrepreneurial activities. Some even started out their business inspired by peers who had been online entrepreneurs. Reciprocally, the entrepreneur's increasing involvement in the business and the more success they received from their business encourage them to give support to inspired peers who want to follow on their footsteps in online entrepreneurship.

However, it was found that a significant number of respondents started from zero online capital and were advised by their offline environment to use social media to market their products. Their network increased gradually by adding more people based on the personalities that matched the products or services they offered. Their closest peers' networks serve as their initial points in searching for personalities in the category of their target market. Their peers' network is likely to be more easily approached, and request to enter their target consumer's network is more easily approved due to the existence of mutual friends.

These mutual friends are considered very important not only in expanding one's networking but also as points of references in promoting their online business. The women entrepreneurs often used testimonials from friends to convince others about the quality of their products and services. Photos posted up on the Internet would not be convincing enough because they only appeal to the visual sense. Other senses such as tasting and smelling of a product may have to depend on others' testimonials. Testimonials such as the fast response of the shops in answering customer's inquiry as well as fast delivery are also considered important.

> In NCC (a food and cooking community) I can pass orders to other members. For instance if I'm busy and can't take orders, I just call my friend to take that order. My regular customers trust me as I am the one recommending this friend. (A respondent from Bandung)

## 4.3 Self-Actualization and Economic Gain

The biggest benefits of doing online business from homes for the women are self-actualization and economic gain. Being merely a housewife makes most women feel uncomfortable and unproductive. Online business also enables women to make money and contribute to their household finance.

### 4.3.1 Sense of Self-Worth

The change of pace from working outside to becoming a housewife took its toll on some. Many expressed their frustration in the initial stage of changing from a busy office worker to a housewife. This is particularly true for those whose babies have grown and demand less of their time. Being 'idle' after completing daily housework has led them to seek other forms of self-actualization. The Internet then became a primary source where they can start exploring activities that can be done while raising the family. Those who are interested in culinary arts would seek recipes from the Internet to make at home and would share their experience on blogs and the likes.

Testimonies from the respondents reveal how social media increased their sense of self-worth. Most of them used blogs and other personal accounts from social networking services to present themselves virtually to their peers. These platforms are used to post their daily achievements such as personal cooking recipes and photos of their finished products.

> I want others to see that I do not only take care of the children, that I actually work although I stay at home. (A respondent from Jakarta)

### 4.3.2 Appreciation by the Family

Upon leaving their employment and focusing on their families, the respondents may feel that they are not contributing much. This has to do with the pressure from the society who sees that being a stay-at-home mother is somewhat an unproductive choice. In many cases, those who have acquired a university degree may receive some criticism from family members who see it as a waste of time to study after a degree if not to be used for work for some companies. In facing such dilemma, the respondents would turn to others who are experiencing similar situations within their offline and online communities for support. These supports may be in the form of motivation and encouragement from the communities for them to continue with their decision to work while staying at home and be successful with their choice.

> At the beginning, my husband was in doubt. He didn't think that I could create something that people would buy. After my business started to grow he began to believe in me. With this business, I feel more appreciated, that we are actually not inferior to the husband. (A respondent from Makassar)

Mastering the Internet becomes a challenge for the respondents who may even have to keep up with their digitally native children. It becomes a need to know the Internet and not being considered as 'web illiterate', both at home and outside. Self-actualization is recognized by the respondents as an individual need that should be satisfied whether from working outside or from home. But just as their decision to leave work was motivated by family, so too is the need for self-actualization. Through succeeding in their online business, they aim to make their family proud of them as businesswomen and to improve the family's welfare through gaining financial freedom and contributing to the family's expenses.

Many of the women used the knowledge that they gained while under employment to start their business. Those who may have been using the Internet extensively prior to quitting work and focusing on family may find it easier to start their online business since they are somewhat familiar with virtual environment. Some respondents have already opened up the online business while they are under employment but had to choose to focus on the latter as the business grew and proved to have more potential than their current work. For these women, focusing on their online business serves as a favourable solution where they can enjoy the flexibility of working at home while taking care of family.

### 4.3.3 Economic Contribution to the Household

The main motivation for the respondents to set up their online business is for financial gain while at the same time being able to stay home. For most of the respondents who were under employment, quitting work and focusing on the family means cutting the family's source of income to solely rely on their husbands. With no fixed individual income, they no longer have the freedom to purchase things to their likings and have to focus more on the basic family needs. Their desire is to have financial freedom without having to leave home to work.

> Although, I don't earn so much but at least I can supplement my husband's income. I'm grateful about that. (A respondent from Medan)

> Now, if I want to buy something for my children or take her for leisure activities, I don't need to ask my husband's permission. (A respondent from Makassar)

Encouragement and support from husbands are most important in helping the women to set up their online business. Many of the husbands feel that the respondents need self-actualization, to be productive and to continue to socialize by expanding their knowledge and capacity as individuals. The family feels the benefit from the respondents as they started their online business as it gives good examples for the children to learn from as well as giving the financial contribution.

> I don't want my daughter to think that I am only a mere housewife. I want to set good examples for her. (A respondent from Jakarta)

## 4.4 Domestic Issues

The main motivation for the women to quit working is due to the demand to take care of the growing family. In some cases, the respondents' husbands specifically requested them to put their career on hold and to focus more on the children, as having both parents working outside is not seen to be favourable for the children. Leaving children in the hands of others is considered quite risky for their development. Moreover, getting hired to help raise children is considered increasingly difficult, and many would have to resort to relinquishing work to focus more on the family.

### 4.4.1 Freedom of Movement

The respondents felt that the online business enabled them to be self-actualized without the restraint of working for others. Some may argue that not having a superior to watch over them or to take orders from and being able to do the business to their liking and to suit their needs would be much more preferable.

> Now we can decide when we want to work. When I worked (as a teacher), my daughter always complained that I didn't have time to play with her. Now, we are with her every day. She is so happy when she tells the neighbour "My mommy doesn't work anymore". (A respondent from Medan)

Moreover, many of the respondents feel that by working outside home, they miss out on their children growing up. They feel that working in offices takes up too much of their time and energy, and by doing so, oftentimes they feel that they are abandoning their children. This attitude is also applied to their online business where for some respondents, the business comes second after their children, and they only respond to the customers' queries when they are free from their domestic duties. Although they realized that this is not a good way to run their business, they continuously return to why they quit work in the first place (Fig. 3).

However, the above attitude is not followed by others. There are cases where their thriving business comes first and they would spend more time on it than they do on their families. Some admit of not taking much care of their children because they are too heavily involved on the online interaction, starting their online interaction once they are awake and go right through the late night hours. This usually occurs in the initial stage of their business where they have to spend more effort in setting up their business before reaching their learning curve. Balancing between their business and family is what most would strive for as most admit that it is not easy. Their neglect of the family has more to do with their need to generate income from the online business for the family.

**Fig. 3** A respondent from Bandung had to bring her child to the FGD session as she could not find anyone to look after him

### 4.4.2 Domestic Decision-Making

Having their own income also improves these women's position in the family. Respondents reported that they can now contribute more actively to the family decision-making and other family members give more values towards their opinions.

> When my husband saw the potentials of this business, he was impressed. Now he even decides to resign from his work so that he can manage this business together to make it grow. (A respondent from Surabaya)

In addition, these women also feel that they can now negotiate more equality in the household chore division. Before, their husbands seemed to think that it is their role as housewives to take care of the house and the children. Now, by seeing that these women have many things in between their roles as housewives and as entrepreneurs, husbands tend to be more willing to help at home, such as by feeding the children, washing the dishes and so on.

## 5 Conclusions and Implications

As the narrative of social media and women entrepreneurs in Indonesian society is still evolving, this study contributes to the conversation. The findings confirm that online businesses have great potential in empowering women by assisting them to become entrepreneurs. In addition, social media entrepreneurship can be seen as a solution to the dilemma faced by women in managing and balancing between their career and family life.

### *5.1 How Does Social Media Empower Women?*

The study showed that social media expanded the social networks of the respondents and subsequently increased their social capital as argued by the current literature on networking. Social media contributed to their sense of self-worth and increased their knowledge.

### *5.2 How Does Internet Technology Impact Women Entrepreneurship?*

The narratives of the women entrepreneurs who participated in the study submit that Internet technology added to their self-actualization and economic gain. They were more appreciated by their family including their husbands who initially objected to their self-employment. They were able to contribute substantively to the household economically. They had freedom of movement and more active participation in domestic decision-making.

## 5.3 Implications

In spite of the positive testimonials in the foregoing discussion, Indonesian society still views conventional employment as more favourable because of its stability when compared with high-risk self-employment. The low number of entrepreneurial activities in Indonesia may be an indicator of this trend. The introduction of online entrepreneurship may change this perception.

The potential of social media for businesses has yet to reach its full potential. While some respondents agree that the offline market may be too saturated, the online market is still far from being explored. The low start-up cost of the social media enabled those who are lacking in capital to still open up their businesses. For businesses that produce the products such as cake shops, they can buy the raw materials when they received orders, reducing the risk of sales. Businesses that only act as distributors can distribute photos of their products and only buy the products when they received the orders. This way, the capital needed to run the business is at its lowest.

Ultimately, women social media entrepreneurship would progress in Indonesia due to its unique features that include: mobility and flexibility promoted by social media, social capital gained through social media interactions, unequal distribution of products in Indonesian cities and lack of time among customers to shop and confidence and satisfaction women gained from doing this business.

**Open Access** This chapter is distributed under the terms of the Creative Commons Attribution Noncommercial License, which permits any noncommercial use, distribution, and reproduction in any medium, provided the original author(s) and source are credited.

## References

Acharya, M., & Bennett, L. (1983). *Women and the subsistence sector: Economic participation and household decision-making in Nepal* (Working Paper No. 526). Washington, DC: World Bank.

Antara. (2012). *Indonesia butuh 4,76 juta wirausahawan*. Viewed on October 21, 2012, from http://www.antaranews.com/berita/338483/indonesia-butuh-476-juta-wirausahawan

Ardieansyah, A. H., Rozman, A. M. Y., & Buang, A. (2011). The imperative of training for women economic empowerment – Statistical evidence from Indonesia. *World Applied Sciences Journal, 13*, 39–45.

Badan Pusat Statistik. (2010). *Keadaan ketenagakerjaan Agustus 2010* (Official Report No. 77/12/Th. XIII). Jakarta: Badan Pusat Statistik.

Boase, J., Horrigan, J. B., Wellman, B., & Rainie, L. (2006). *The strength of internet ties: The internet and email aid users in maintaining their social networks provide pathways to help when people face big decisions*. Washington, DC: Pew Internet and American Life Project.

Brandtzæg, P. B., Lüders, M., & Skjetne, J. H. (2010). Too many Facebook "friends"? Content sharing and sociability versus the need for privacy in social network sites. *International Journal of Human-Computer Interaction, 26*(11–12), 1006–1030.

Calás, M., Smircich, L., & Bourne, K. (2009). Extending the boundaries: Reframing entrepreneurship as social change through feminist perspectives. *Academy of Management Review, 34*(3), 552–569.

Carney, M., Dieleman, M., & Sachs, W. (2008). The value of social capital to family enterprises in Indonesia. In P. H. Phan, S. Venkataraman, & R. Velamuri (Eds.), *Entrepreneurship in emerging regions around the world: Theory, evidence and implications*. Cheltenham: Edward Elgar.

Chiam, V. (2001). *E-commerce technologies and networking strategies for Asian women entrepreneurs*. In Proceedings of the 2nd conference on Women Entrepreneurs in SMEs, OECD. Bercy: France.

Coleman, S. (2000). Access to capital and terms of credit: A comparison of men- and women-owned small business. *Journal of Small Business Management, 38*(6), 523–541.

Coughlin, J. H., & Thomas, A. R. (2002). *The rise of women entrepreneurs: People, processes, and global trends*. Westport: Quorum Books.

Cromie, S., & Birley, S. (1992). Networking by female business owners in Northern Ireland. *Journal of Business Venturing, 7*(3), 237–251.

Danani, S. (2004). *Unemployment and underemployment in Indonesia, 1976–2000: Paradoxes and issues*. Geneva: International Labour Organization.

Duncombe, R., Heeks, R., Morgan, S., & Arun, S. (2005). *Supporting women's ICT-based enterprises: A handbook for agencies in development 2005*. Manchester: Institute for Development Policy and Management (IDPM). Viewed October 1, 2012, from http://www.sed.manchester.ac.uk/idpm

Ellison, N. B., Steinfield, C., & Lampe, C. (2011). Connection strategies: Social capital implications of Facebook-enabled communication practices. *New Media & Society, 13*(6), 873–892.

Esplen, E., & Brody, A. (2007). *Putting gender back in the picture: Rethinking women's economic empowerment*. Sussex: BRIDGE (Development – Gender), Institute of Development Studies, University of Sussex.

Facebook. Viewed April 25, 2013, from http://www.facebook.com/ads/create/

FI Management. (2010). Indonesia: A social media nation. *FI Management*, 12 April 2011. http://www.fimanagement.com

Frankenberg, E., & Thomas, D. (2001). *Measuring power: Food consumption and nutrition division* (Discussion Paper No. 113). Washington, DC: International Food Policy Research Institute.

Gates, B. (2000). *Shaping the Internet age*. Washington, DC: Internet Policy Institute.

Goble, F. (1970). *The third force: The psychology of Abraham Maslow*. Richmond: Maurice Bassett Publishing. http://en.wikipedia.org/wiki/Maslow%27s_hierarchy_of_needs-cite_ref-6

Gold, T., & Guthrie, D. (2002). *Social connections in China: Institutions, culture, and the changing nature of Guanxi* (Vol. 21). Cambridge: Cambridge University Press.

Handayani, P. W., & Lisdianingrum, W. (2011). Impact analysis on free online marketing using social network Facebook: Case study SMEs in Indonesia. In Paper presented at the Advanced Computer Science and Information System (ICACSIS), 2011 international conference. Jakarta.

International Telecommunication Union. (2011). *ICT statistics, Geneva*. Viewed on October 1, 2012, from http://www.itu.int/ITU-D/ict

Internet World Stats. (2012). *Internet World Stats*. Viewed October 12, 2012, from http://www.internetworldstats.com/stats.htm

Jones, B. (2010). Entrepreneurial marketing and the Web 2.0 interface. *Journal of Research in Marketing and Entrepreneurship, 12*(2), 143–152.

Kabeer, N. (1997). Women, wages and intra-household power relations in urban Bangladesh. *Development and Change, 28*, 261–302.

Kabeer, N. (1999). Resources agency, achievements: Reflections on the measurement of women's empowerment. *Development and Change, 30*, 435–464.
Kementrian Koperasi dan Usaha Kecil Menengah. (2008). Undang Undang No. 20 Tahun. Republik Indonesia. Jakarta.
Kementrian Pemberdayaan Perempuan dan Perlindungan Anak. (2011). *Perkembangan data usaha mikro, kecil, dan menengah (UMKM) dan usaha besar (UB) tahun 2010–2011*. Jakarta: Kementerian Pemberdayaan Perempuan dan Perlindungan Anak.
Lombard, K. V. (2001). Female self-employment and demand for flexible, non-standard work. *Economic Inquiry, 39*(2), 214–237.
Malhotra, A., Schuler, S., & Boender, C. (2002). *Measuring women's empowerment as a variable in international development*. International Center for Research on Women and the Gender/Development Group of the World Bank. Washington D.C.
MarkPlus Insight. (2011). *Pengguna Internet di Indonesia*. MarkPlus Insight: Jakarta.
Mosedale, S. (2005). Assessing women's empowerment: Towards a conceptual framework. *Journal of International Development, 17*, 243–257.
Ndubisi, N. O., & Kahraman, C. (2006). Malaysian women entrepreneurs: Understanding the ICT usage behaviour and drivers. *Journal of Enterprise Information Management, 18*(6), 731–739.
Nguyen, M. (2005). *Women entrepreneurs: Turning disadvantages into advantages*. Pre-Flight Ventures. Viewed October 1, 2012, from http://www.preflightventures.com/resources/WomenEntrepreneursLitSurvey012005.htm
Nugroho, Y., & Syarief, S. S. (2012). *Beyond click-activism? New media and political processes in contemporary Indonesia*. Jakarta: Friedrich-Ebert-Stiftung.
O'Reilly, T. (2009). *What's next for web 2.0, O'Reilly*. Viewed August 12, 2011, from http://oreilly.com/web2/archive/what-is-web-20.html
Roffey, B., Stanger, A., Forsaith, D., McInnes, E., Petrone, F., & Symes, C. (1996). *Women in small business: A review of research*. Canberra: Australian Government Publishing Service.
*SalingSilang*. (2011, February). Indonesia social media landscape. Viewed April 12, 2011, from http://www.salingsilang.com
Sarinastiti, I. (2006). *IFC highlights women's entrepreneurship in Indonesia: IFC-PENSA and IWAPI launch "voices of women in the private sector"*. Jakarta: International Finance Corporation.
Sarri, K., & Trihopoulou, A. (2005). Female entrepreneurs' personal characteristics and motivation: A review of the Greek situation. *Women in Management Review, 20*(1), 24–36.
Segal, G., Borgia, D., & Schoenfeld, J. (2005). The motivation to become an entrepreneur. *International Journal of Entrepreneurial Behaviour & Research, 11*(1), 42–57.
Sen, A. (1985). *Commodities and capabilities*. Amsterdam: North-Holland.
Sicat, M. (2007). Impact of improved communication on women's transport needs and empowerment in Bangladesh. In *Gender and transport* (Transport and communications bulletin for Asia and the Pacific, Vol. 76, pp. 72–88). New York: United Nations Economic and Social Commission for Asia and the Pacific.
Siong-Choy, C. (2007). Theorising a framework of factors influencing performance of women entrepreneurs in Malaysia. *Journal of Asia Entrepreneurship and Sustainability, 3*(2), 42–59.
Socialbakers. (2012). *Indonesia Facebook statistics*. Viewed December 12, 2012, from http://www.socialbakers.com/facebook-statistics/indonesia
Van der Merwe, S. P., & Lebakeng, M. (2010, August 25–26). *An empirical investigation of women entrepreneurship in Lesotho*. Paper presented at the African International Business and Management conference. Nairobi.
Widyadari, S., Shrader, H., & Pranoto, H. (2006). *Suara-suara perempuan pengusaha*. Jakarta: IFC-PENSA and IWAPI.
Williams, J. (2005). *Measuring gender and women's empowerment using confirmatory factor analysis*. Boulder: Population Program, Institute of Behavioural Science, University of Colorado.

# The Use of Mobile Communication in the Marketing of Foodstuffs in Côte d'Ivoire

Kabran Aristide Djane and Richard Ling

This paper examines the use and non-use of mobile communication in the food distribution system of Côte d'Ivoire. We examine how this mediation forms in the existing systems for producing, transporting and marketing of food by women in that country. We base the analysis on five focus groups and 51 interviews carried out between May 2012 and February 2013. We find that mobile communication allows for more responsive and flexible planning on the part of large- and small-scale wholesalers. They are able to use it to manage and adjust the delivery of products to different retailers as surpluses and shortages develop. We also find that the mobile phone is not used by some producers since there are well-established systems for the marketing of their products that preclude the need for mobile communication.

## 1 Introduction

A food market in Côte d'Ivoire is a seeming welter of sounds, sights and odours. To an uninformed visitor, the whole can seem overwhelming. The market consists of many small stalls where women are hawking different foodstuffs. Around the perimeter, there may be buildings with cloth stores, hardware stores and the like. Large trucks move about, and bags of food are loaded and unloaded while shoppers

---

K.A. Djane (✉)
Université de Korhogo, Korhogo, Côte d'Ivoire
e-mail: aristide.djane@upgc.edu.ci

R. Ling
Division of Communication Research, Wee Kim Wee School for Communication and Information College of Humanities, Arts, & Social Sciences, Nanyang Technological University, Singapore, Singapore
e-mail: rili@ntu.edu.sg

move about looking at the different offerings and making their purchases. In one stall there might be tomatoes and in another live snails. On the edge of the market, there may be a small group of elderly women who are keeping a careful watch over the situation. Indeed, these women are often the central actors in the organization of these large-scale markets.

While it may seem chaotic, there are well-entrenched routines and ways of organizing the production, transport and marketing of the wares. There are flows of wares, information and monies that have been developed over time. In some cases, these are being changed by the adoption of mobile communication, but our work suggests that in other cases, this type of mediated interaction is not widely accepted and utilized for this purpose.

A central idea motivating this work is that improving information flow in the value chain supports development. If the information contained in this value chain is easily accessible by the players, it will facilitate market logistics. Abraham (2006), also echoed in the work by Rashid and Elder (2009), suggests a strong link between mobile phone access and increased economic opportunities for farmers. Among other things, these authors note the increased ability on the part of the producers to follow price information, hence allowing them to better judge when to sell their crops.

In this chapter, we will first examine the role of mobile communication for marketing in developing countries. We will then review this in terms of marketing foodstuffs. Following a discussion of the methods, we will look at the role of mobile communication in the case of large-scale wholesaling, smaller "petite" wholesalers and producers who do not find the mobile phone useful. In conclusion, we will offer some comments on the broad effect of mobile communication in the marketing of food in Côte d'Ivoire.

## 2 Literature Review

### 2.1 Distribution of Food Products in African Countries: A Food Security Problem

Turning to the question of food production in Africa, for nearly 20 years (1970–1989), the dominant issue in discussions on food consumption in African cities has been the supply to the city by the countryside. The 1950s saw the push to develop cash crops that disrupted the domestic production of foodstuffs. From the mid-1990s, however, there has been a renewed interest by, for example, the Food and Agriculture Organization (FAO) of the United Nations to supply the cities in Africa with domestic foodstuffs (FAO 1996). The change in approach coincides with the urban explosion. The push has been in response to the growing demand for food in urban areas by developing domestic production. This implies reduced dependence on imports and a focus on least cost routing of commodities. This approach also has the potential to create jobs, but it becomes more complex when one considers the relationship between urbanization, poverty and food insecurity (Molony 2006).

Earlier studies have been devoted to the food supply of large cities in Africa. These include Goossens et al. (1994) who examined the system in Kinshasa and Chaléard (1996) who studied Abidjan and Bouake in Côte d'Ivoire. According to Argenti (1997), there is the need to give "priority to improving the efficiency of marketing systems and links between areas of production and consumption, to facilitate access to food and thus improve food security". When thinking of these production and marketing systems, the food supply and distribution involve several spatial scales: first, there is the suburban or rural areas in which there is the production and collection of food products; second, there is the intervening logistical space of exchange and redistribution from the rural collection to the urban markets; and finally, there is the city as a location of final consumption. If it is suggested that supplying urban centres with food is an opportunity for rural producers, this depends on efficient connections between the production, processing and storage areas via the transport system into the markets of the city. According to Courade (1985), urban-based demand will be an engine of change provided that the agricultural supply chains and transport are efficient and transaction costs are as low as possible.

## 2.2 Mobile Communication and the Logistics of Food Marketing

The mobile phone has been widely adopted in developing countries in the past decade (Donner 2008). A major effect of the mobile phone is that it has been used by impoverished people to change their life situation (Souter et al. 2005), and it has affected the way that food production has taken place (Jensen 2007). Flor (2009) sees the institutionalization of the so-called e-agriculture as an important instrument with which to address agricultural problems. Following the work of Flor, an analysis by the World Bank (2012) examines how mobile telephony can strengthen access to agricultural information and provide access to new markets. Porter (2012) who examines sub-Saharan Africa and Beuermann et al. (2012) who study Peru also confirm this economically based approach examining the effect of mobile telephony on community wellbeing. They found that mobile phone coverage increases the income, assets and expenditures of rural consumers.

Mobile communication gives people access to time-sensitive information that can be used in commercial settings to facilitate the functioning of markets. This finding is not new. For example, Fischer (1992) discusses how farmers in the USA used the landline telephone to follow the ebbs and flows of crop pricing. This allowed them to consider when to hold and when to sell their crops by allowing for the diffusion of pricing information and the relative excess or shortage of crops at different points in time. The landline has also been used in developing countries for these purposes (Clark 1995) albeit the lack of access meant it was not a commonly used tool. The mobile phone has changed people's lifestyle and habits in the developing world.

Indeed, there are more mobile phones in developing countries than in the developed world. While our research shows that the functionality is not universally adopted, we are able to see the contours of when and how it has been seen as useful.

The analysis in Côte d'Ivoire enters into a series of studies that look at the use of mobile communication in marketing in developing countries. In general, research points to the way in which mobile communication reduces the costs and time required for commercial activities (Jagun et al. 2008; Overå 2006). The work of Jagun et al. (2008) documents changes in the logistics associated with cloth production in Nigeria. The mobile phone allowed the substitution of phone calls for journeys. The ability to make this substitution, however, rested upon the mutual trust between the interlocutors and necessary training farmers and the use of mobile phones for accessing innovations on agriculture.

Time saved per call was typically several hours, and, overall, this had meant that the turnaround time between the first order and final fulfilment was reduced. Money saved was typically understood by comparing call costs with transport costs: for example, interviewees talked about a call rate of N50 (ca. USD 0.40) per minute being cheaper than a taxi cost for an average journey of, say, N1,000 (ca. USD 8.00), given that calls were normally completed in less than 5 min (Jagun et al. 2008: p. 57).

There was still the need for travel and transport in this system, but the mobile phone streamlined this need. People who had a legacy of interaction and who were embedded in a network of both trust and mutual obligation used the mobile phone to facilitate the commercial process. They summarized by saying that, "in conceptual terms, this study confirms the need to understand mobile phones as devices for communication of information" (Jagun et al. 2008: p. 60). The communication of information as a way to facilitate the flux of commercial interactions is a theme that runs through the literature on mobile communication in these contexts.

The same theme is seen in the work of Robert Jensen in what is perhaps the most exhaustive studies of mobile communication's role in foodstuff production in developing countries (Jensen 2007). Jensen was interested in the effect of market information on the pricing of fish in Kerala, India. He was able to gather the spot price of fish every day for approximately 5 years for 15 different "beach markets" where boats sold their catch. Jensen's period of data collection covered the time in which mobile phones were introduced into this area. Thus, his work studied the effect of access to more timely and ubiquitous market information. Prior to the adoption of mobile phones, the boats would return to their homeport with their daily catch. Following the notion of supply and demand, if there was a good catch by the fleet for a particular port, the price of fish would be low and vice versa. In scenarios where late-coming boats could not sell their fish, the catch would be dumped into the harbour. Yet, it is likely that at a neighbouring port, there would not be enough fish to meet demand. Thus, the lack of information meant that there was not an optimal distribution of the fish.

The analysis of Jensen dramatically illustrates how the advent of mobile communication changed this situation. After buying mobile phones, the boats could call up the ports and compare the prices of fish. If they could realize a better price at another market instead of their homeport, they were able to act on this information. Jensen's material shows that on any given day, between 30 and 40 % of all boats delivered their fish to markets outside their homeport. Over time, the price of fish stabilized and dumping of fish totally ceased. According to Jensen:

> We find that the addition of mobile phones reduced price dispersion and waste and increased fishermen's profits and consumer welfare. These results demonstrate the importance of information for the functioning of markets and the value of well functioning markets; information makes markets work, and markets improve welfare (Jensen 2007: p. 919).

Jensen also makes the point that facilitating the flow of information via the use of mobile communication was not a traditional development project. Rather, it was the diffusion of technology in a situation where there was a convergence among various commercial actors. Both Abraham (2006) who also examines the same production chain and Aker (2008) who studies the effect of mobile phones in Nigerian grain production note that mobile communication increases the market efficiency and reduces risk and uncertainty.

While the work of Jensen points to the potential for enhanced information flow, there are limitations. For one, there is not a system for storing the fish for more than a few hours after being caught. This means that the flow of information is particularly useful in this case since the producers and the consumers both benefit from expediting the process. There are likely similar possibilities with products such as milk, eggs, fruit, vegetables and spot-labour markets (Jensen 2007: p. 920). In addition, the market studied by Jensen has a large number of actors. There are no cartels or major actors who exert disproportionate influence.

The work of Barrantes Cáceres and Fernandez-Ardevol (2012), in their study of market traders in Peru, corroborates that of Jensen on several points. Their work takes the perspective of the individual entrepreneur having wares to sell at the weekly markets in rural Peru. Echoing Jensen, they describe how these actors poll other sellers in order to seek out the best weekly markets to attend, given that they have several alternatives to choose from. They find that the traders who deal in perishable commodities and have more to lose by not actively participating in the markets were more likely to use mobile phones for such purpose. It is interesting to note that there are some traders in this system that did not particularly benefit from using the mobile phone, namely, those who did not deal in perishable commodities and those who were more buffered from the immediate need to participate.

Rahman (2007) examines how pricing of foodstuffs that can be stored (such as rice) do not necessarily follow the same dynamics as noted by Jensen. In this case, poor farmers who need to repay loans can be forced to sell even when the price is low. Rahman notes that in contrast to "ideal-type" markets, the daily situation faced by farmers can be characterized by diverse cross-cutting formal and informal institutions, where the access to information noted by Jensen does not necessarily have the same effect.

There is a legacy of work showing that mobile communication facilitates the trading and commerce in developing countries. As noted by Overå in her analysis of the mobile phone use in Ghana, the device is a tool that facilitates information exchange and can be used to cultivate reliability in various marketing chains. In the same spirit, Prahalad and Hammond (2002) notes that mobile telephony gives economic actors access to information regardless of their position in the income pyramid. As such, it helps in the distribution of wares, particularly when there are gluts or shortages of perishable commodities. In summary, the mobile communication facilitates the flow of information allowing actors in some situations to better orientate themselves to the ebbs and flows of the market.

## 2.3 Barriers to Mobile Phone Use

While there are many possibilities associated with the adoption of mobile communication, it is important to point out that there are also barriers. One point is that ownership of a mobile phone can disturb gendered ideas regarding power (Chib and Hsueh-Hua Chen 2011) and notions of propriety (Cohen et al. 2007). Another is that there is a long tradition of oral interaction (Barrantes Cáceres and Fernandez-Ardevol 2012). The mobile phone contravenes traditional ways of interacting. The oral tradition is a well-established mode of knowledge transmission between various castes in Africa. While this is evolving in African matriarchal societies, as is the case in most traditional societies, the family's secret knowledge is owned by women. The woman is, in the eyes of African society, the foundation of the family. The balance of the family rests on this outlook.

Moreover, the mobile phone is sometimes treated with suspicion. In some situations, the mobile phone carries with it certain taboo. The issue of myths and fears associated with the use of mobile phones is a recurring theme in the research. In their work in South America, Fernández-Ardèvol et al. (2011) note that the question of mobile-based myth and taboo is linked to the perception of in-group versus out-group and even can be seen as an invention of the devil. Similarly, in some sub-Saharan countries, the mobile phone is seen as the tool of sorcerers and other "marabouts" who use it to locate their enemies and cast spells on them from a distance. The spell can be received just by picking up the phone. These perceptions are common in rural areas but less so in the cities (Gakuru et al. 2009).[1] These negative understandings of the mobile phone in rural areas can hinder its use in the agricultural value chain.

## 2.4 Women, Food Production and Mobile Telephony

In many West African countries, women play a central role in the production and marketing of foodstuffs. In Ghana, for example, there is the *ohemma*

---

[1] This difference indicates that these unfavourable perceptions of the mobile phone can eventually be changed through the use of education (Sauvé and Machabée 2000).

(or the Queen Mothers) (Overå 2006), and as we will see below, Côte d'Ivoire has the tradition of the "Chain of the Grandmother". Each case study describes a system whereby women are the key actors in the production, transportation and sale of foodstuffs. This does not mean that they do the lower-level work and men are in charge of wholesale functions. Rather, it is women who are active at all levels. According to Overå (2006), these women are the negotiators when it comes to conflicts; they interact with authorities, organize the marketing chain and administer the marketing areas. To be sure, there are men who, for example, drive the trucks, help with the loading and unloading and help to organize the marketplace. However, it is women who sit in central positions of authority.[2]

Work by Hafkin and Odame (2007) outlines how women have a central role in the preservation of food security. As such, the women are an essential element in the social structure. This means that women's use of mobile commutation can be a major linchpin in the resolution of social issues. Their mastery of the technology affords social benefits (Johnson 2004). Especially in the development of new practices, intended crops reduce the level of food insecurity. Rural women who engage in agricultural become a hub through which many partners communicate via the mobile phone. The assertion of Souter et al. (2005) is echoed in the work of Fernández-Ardèvol et al. (2011), namely, that there is an important role for the mobile phone in the social relationships between the actors in a female-based agricultural system.

In this context, the mobile phone has helped to change the situation of women (Bayes 2001). Research has shown that the mobile phone is an important tool for women farmers. In Uganda, for example, the use of mobile phones enables rural women to increase their income by better following market developments. Dimitra found that there was a link between poverty reduction and ICT use among women (Asenso-Okyere and Mekonnen 2012).[3]

In the case of Côte d'Ivoire, an important element in the logistics of food distribution between producer and retailer is the so-called Chain of the Grandmother. This draws on use of well-entrenched traditional symbols to facilitate the distribution process. The existence of this augurs against use of electronic mediation.

In addition to celebrating the legacy between generations, it is also useful since it allows the women, who are largely illiterate, to organize a complex distribution system. When a buyer in the countryside secures produce for a retailer in the city, they negotiate a price and then mark the bags of produce with a piece fabric that has the individual pattern of the retailer. This piece of cloth, with a unique colour and pattern, allows the truck driver to route the produce to the correct women in the city. The truck driver becomes the instrument of communication in this case between the buyer and the producer. The system relies heavily on interpersonal "mouth-to-ear" interactions. Even though the majority of respondents consider that this approach is

---

[2] http://www.irinnews.org/report/92213/cote-d-ivoire-women-bring-food-to-market-against-all-odds

[3] See also http://www.fao.org/dimitra/home/en/

quite difficult when determining the scheduling and pricing of wares, the physical contact has the advantage of maintaining trust between stakeholders in the food distribution business.

The marketing of food products Côte d'Ivoire has some elements of a cartel. There is one ethnic, the "Gouro", who dominates in this system. Because of the structural adjustment programme of 1983, many Gouro men lost their jobs. This meant that in order to take care of their family, women were obliged to work. They started working with producers in different Ivorian villages and organized cooperatives and provided them with seeds and fertilizer. They were able to secure locations within the cities, interact with drivers and set up wholesale operations with drivers, wholesalers and retailers within the different market spaces.

Coming back to our general research theme, the research question that drives the analysis is to study if "rural women will use the mobile phone in food production and distribution if they perceive mobile phone to be a tool that improves the distribution of their food products and it also improves their revenue stream". In the sections below, we will examine various stakeholders' understandings of the mobile phone. We will particularly focus on the role of the phone vis-à-vis food production, transportation and marketing.

## 3 Methods and Data

We drew on qualitative data to examine the role of mobile communication in food production, distribution and marketing based on the situation in Côte d'Ivoire. Our research draws on a series of field investigations conducted from May 2012 to February 2013, in the department of Soubré and Korhogo in Côte d'Ivoire. Each site was visited three times.

Soubré is a town in southwestern Côte d'Ivoire. The inhabitants of the department of Soubré are from the Kru group.[4] Soubré is an important agricultural area that produces some of the best quality cocoa and coffee. Our reason for choosing this area was that it would give us insight into an area that has an active and lucrative agricultural sector. The sites or villages that were visited during the fieldwork in this region included the town of Soubré as well as Koreyo I, Logboayo, Sokozoua and Oureyo.

Korhogo is a town in northern Côte d'Ivoire. It is the capital of the savannah region of the country. It is also the fifth largest city in Côte d'Ivoire. Because of its geographical location situated in an arid area, the department of Korhogo is not as verdant as Soubré. Korhogo is the location of many agricultural projects focused

---

[4]The Bete constitute the vast majority of the indigenous population, followed by Bakoué and some Kouzié. Soubré is a prefecture and the capital of the new region of Nawa, born from the split of the former Bas-Sassandra. The department includes the subprefectures of Soubré, Liliyo, Oupoyo, Okrouyo, Buyo, Méagui and Grand Zattry.

on increasing food security. In addition to visiting Korhogo, interviews were also carried out in Gbalogo, Gbaméleguekaha, Takalé and Natchokobadara.

During the visits to the two sites, the first author and the data collection team conducted a series of focus group discussions and individual interviews with farmers, labourers, truck drivers, traders, commission agents and commercial organizations (including food cooperatives) involved in the agricultural sector. The team conducted five focus group discussions with people involved in the food cooperatives that are central in the food production system. In addition, there were 51 individual interviews in two subprefectures and eight villages. In total, approximately 17 farmers and eight *jardin* were interviewed. The informants who were interviewed included both those with and without mobile phones. This allowed us to compare the situation of people in these two situations. The interview materials were transcribed and analysed.

With the exception of the investigation in Korhogo and Soubré markets, all the locations were rural village with populations ranging from 250 to 600. All interviewees were over the age of 18. None of the interviewees had more than a primary level of education. This was especially the case for women.

The farmers, truck drivers, wholesalers, retailers and resellers who were interviewed grew, transported and sold a wide variety of foodstuffs including rice, corn, bananas, cassava, yams, eggplants, okra, cabbage and pepper. Almost all farmers practised multiple cropping with cabbage being the most common crop. The household income of the farmers ranged from CFA 5,000 to CFA 850,000 (approximately USD 10 to USD 1,700) per month and the average household size was fewer than six people. The people associated with the cooperative have the highest income in this system ranging between CFA 180,000 and CFA 1,000,000 (approximately USD 365 to USD 2,000) per month. The interviewees in these two regions also had access to irrigation, but they noted that storage facilities and credit institutions were a problem. When considering the transportation workers and their equipment, the average age of trucks for food transportation is 13 years. Wholesalers and retailers make up the largest number of workers in this sector and are largely women (approximately 92 %). We found that 68 % of interviewees working in the urban areas (Korhogo and Soubré) have at least one mobile phone with the capacity for two SIM cards.

Some interviewees in the region of Korhogo in rural areas refused to be recorded because they believed that some evil spirits could use their recorded voice for spell casting. In these cases, we transcribed the interviews on a notepad.

The following sections turn to the findings from the fieldwork, beginning with an overview of the food system before the advent of mobile phone in Côte d'Ivoire. We then report on how our interviewees perceived the effect of the mobile phone on their supply chain and economic activities. In conclusion, the factors that limit adoption of mobile phone by women in food supply system are discussed.

## 4 Findings and Discussion

We will analyse the results by examining mobile phone use in the case of managing over and undersupply of goods by large-scale wholesalers, its use by what we call *petit* wholesalers and different food production systems where the use of mobile phone was not adopted for a variety of reasons. Our analysis indicates that the mobile phone has a different value depending on the position of the stakeholder. Below, we examine the production and marketing aspects of this perspective.

### 4.1 Management of Logistics by Large-Scale Wholesalers

As noted above, a central actor in Ivorian food production/marketing is the large-scale wholesalers or cooperatives that can function as a cartel. These organizations often have an integrated position that includes contact with food producers (with whom they provide seeds and fertilizer), transportation workers who truck the products from the producers into the cities and the provision of a market space in the cities where retailers can order foodstuffs from the cooperative and then sell them to the consumers. These cooperatives have accounting services and a leader who is commonly chosen by the old women who have originally established the cooperative. As such, these cooperatives have a pivotal position in the Ivorian food production system.

While the "Chain of the Grandmother" system is firmly in place, it has been enhanced with the use of the mobile phone. This is reflected in the comments of *Ngnan*[5] Catherine, one of the elderly women who manage the cooperative:

> My son, I'm glad you came to ask me that ... because it is we who have fed the Côte d'Ivoire. Even when there's been a crisis, we are dribbled the ball to send food to Abidjan ... It's not us that were going to Abidjan or in villages Soubré but these are our daughters. You see my daughter Zita, she is brave, she is like me and it is not the only one to do that way, to sell food products, and the other women of the cooperative also their daughters who are very involved in the production of food. In fact, they are over. You see it is not as a string. We moms, it was for us to feed Côte d'Ivoire and now they are our girls, yes it is the chain of Grandmother because we began, our daughters follow and their daughters will follow. But we are here to give advice. But now it has become even easier, it's true that the roads are not good but there's phone to call quickly, otherwise you had to pay before transporting someone to send message others producers. *Ngnan* Catherine (Market Soubré)

Informants such as *Ngnan* Catherine, a woman who managed the cooperative in Soubré, said that the mobile phone has revolutionized their work. They noted that they use the mobile phone to interact with truck drivers (sometimes as often as

---

[5] An honorific version of "Grandmother" in the local language.

twice a day to each truck driver) to maintain an overview and manage the collection of food from producers. All the truck drivers interviewed in this study possessed and used mobile phones in their work. This was seen as an essential item. In the words of Konan, a driver:

> My brother, if you do not have mobile today how you can do business. Today it has become necessary ... otherwise; women cooperatives cannot call you to pick up their goods. Without it, you lose money. Konan (Soubré Market)

Another driver, Vincent, also responded:

> (He laughs)... I can say that with mobile phone, it is *Money Horse Course*, the money comes quickly ... Well I do not know how women were before but I think the phone mobile is important in the food trade and the phone companies have an incentive to cover all areas where there's a strong production that is otherwise difficult. Vincent (carrier) (Korhogo Market)

The women who run the cooperatives visit the producers to gauge whether there will be gluts or shortages of different crops. At the time of harvest, the actual surpluses and shortage become obvious, the women are able to call other cooperatives to either buy or sell particular crops as required. In some cases, this can be done before they are being transported into the cities.

In other cases, just as with Jensen's fishers (2007), there is the need to manage the transportation in real time. The women can divert trucks that are en route:

> The phone today is important because you can call the driver while he is on his way to tell him to complete the inventory of goods... before, if you had to start over ... it lost much of money. Communication is really important. *"non fô lèhè"* Kouyaté (Korhogo)

According to the women who manage the cooperatives, the mobile phone allows real-time ability to balance the distribution of crops.

There is another dimension to the marketing of foodstuffs in Côte d'Ivoire. There are one or more main markets managed by the cooperatives in the city. In these markets, there can be several hundred retailers who are active. The women who run these cooperatives ensure that there is enough but not too much of the different products in this location. In addition, there are also smaller street-side and local satellite markets that rely on the women running the larger markets. Where oversupply or undersupply of goods can happen in the main market, the same applies to these smaller markets. Again, the mobile phone allows a real-time ability to deal with these distribution problems.

The use of the mobile telephone is similar to that described by Jensen in India and the Peruvian sellers depicted by Barrantes Cáceres and Fernandez-Ardevol (2012). That is, the women running the cooperatives use the device to gather information and to address logistical issues more quickly than prior to the adoption of the mobile phone. Notably, however, the situation described here is from a somewhat different angle. With Jensen, Barrantes Cáceres and Fernandez-Ardevol, their analysis was from the perspective of the retailer, whereas this particular study focused more on the perspective of those managing the wholesale side of the equation.

## 4.2 Petit *Wholesalers/Retailers*

Other groups that use the mobile phone to arrange logistics are the small (*petit*) wholesalers and the retailers. In the larger markets, there are some women who have a dual role. These *petit* wholesalers are retailers who have small sales outlets in one of the larger food cooperatives. Each retail outlet is about a few square metres in the market area (the land is often owned by the cooperative and rented to individuals). However, they are also small-scale entrepreneurs who, on a small scale, seek out alternative marketplace and supply systems. Rather than relying totally on the foodstuffs that are available through the cooperative, they seek out other types of foods from alternative producers that are transported to the market using noncooperative trucks. Similar to the managers of the cooperatives, these women described the mobile phone as an innovation in their job.

In their role as a *petit* wholesaler, they might use the phone to organize purchase of wares. They might negotiate with the producers about prices and organize the wares to be picked up and transported to the market with noncooperative trucks. Since their consignments might be smaller, these *petit* wholesalers might have to cooperate with other such wholesalers to share the cost of a truck. They may also have to arrange for the delivery of particular type of their foodstuff to the retailers in the satellite markets.

In their role as a retailer in the market, their interactions with the final users are face-to-face with no need for mediation. However, there are occasions where the retailers may need to quickly sell leftover produce. If the produce is highly perishable and there are not any storage possibilities, it is important for them to find other small retailers who will be able to quickly sell the surplus. The women must act quickly so as to not be caught with extra wares. The mobile phone becomes an important tool in this situation.

In this role, they become the buyer for another small satellite market that is not organized like the cooperative system. This position can lead to intense mobile phone use. This was perhaps more the case in Soubré than in Korhogo. There can be a series of calls early in the morning about the amount of produce transacted.

Based on these interactions, the retailers at the satellite markets will come to the main market to retrieve the bags of produce. There are also financial negotiations during such interactions:

> In fact, I'm a relay for other markets. So, when I have many goods and products (that may) come to expire, I called the other traders, other markets so that they come to buy from me. It solves the problems of waste quickly... Monique

Like the large wholesalers running the big collectives, these *petit* wholesalers use the mobile phone when they have an excess or shortfall of foodstuffs. The mobile phone enables them to contact other producers and places quickly to either dump or buy a particular product. The material from the interviews indicates that there is a seasonal aspect to this issue. In the period after the rainy season, when there is an abundance of crops, the ability to buy or more often dispose of excess foodstuffs is important:

> After rainy season, we have many products on the market so you need to call other women in other markets they are taking it on credit or it may rot in our hand here. Yoboue Suzanne

In the case of managing the large cooperatives, the situation of the *petit* wholesalers and the case of the retailers, the mobile phone facilitates the logistics of food distribution by opening up the flow of information. The need for this varies according to the position of the stakeholder and in relation to the supply of various foodstuffs. It was also clear that such occurrence was more often evident in Korhogo than in Soubré.

## 4.3 Non-use of Mobile Communication

While the research shows that there were groups using mobile communication to facilitate the marketing of foodstuffs, it also presents groups which, for one reason or another, did not find the adoption useful.

### 4.3.1 *Jardin* vs. Larger-Scale Production

One group who made only limited use of mobile communication was women who had small "kitchen gardens" or *jardin* of approximately 10–20 m$^2$ that were used to both provide food for the family and, when there was a surplus, to sell in the local market. This practice was more common in the areas near Korhogo. In the words of one informant:

> ...When you have a kitchen garden, it prevents you from hunger in the family. You plant what you want around your home and can prepare in your kitchen, you will cut it in the garden ... generally, it's not great but it helps a lot ... but when the rainy season it is great, we will sell what we could not eat in the market. Mariam

The concept of *jardin* appears often in the discourse of respondents. Informants note that they grow the tomato, pepper, carrot, eggplant, cabbage and others foodstuffs. However, when this production exceeds the needs of the home, the woman sells the surplus in the nearest market. Thus, the *jardin* becomes an economic opportunity for women. The revenues obtained from the selling of the *jardin* products range approximately USD 10–25 per day of sales. This cash becomes a fund with which these women can purchase small things for the home and for their children. This income may also become a point of contest between the husband and wife. According to the women interviewed, if the husband is aware of the income, he may withhold other monies from the wife and family. Conversely, nearly all the women noted that they sometimes help their husband monetarily with these revenues:

> ...When my husband needs money to pay for the children education or farm products of maintenance, and he has no money, I lend him money (that) comes from sales of products of my Jardin. Mariam

As one might expect, the interaction between the woman and the market is very informal. Mobile phone is not used extensively in this context nor were there many of these informants who actually owed a phone. Of the 17 producers interviewed in this analysis, eight fell into this category. In general, these producers grew less than 250 kg of produce per year.[6] They were also the least likely to have a mobile phone. Only three of the eight interviewees owned one. Furthermore, as opposed to the reported use of larger producers, these women did not use their phones for business purposes.[7] Familial interaction was often the main motivation for owning a phone. The informants noted that mobile telephony was used to facilitate logistics and to save time in organizing activities.

Our finding indicates that it was only the larger producers who used the mobile phone in the line of work to interact with the truck drivers, retailers and wholesalers. The smaller producers noted that they found it easiest to accept bids for produce face-to-face. The informants felt that it was too complicated to carry out these negotiations via the mobile phone. There are well-established meeting points when the producers and wholesalers can meet to negotiate the prices and quantities of goods. In these situations, there is a legacy of trust and also well-established routines that augur against use of the mobile phone:

> The prices here are the result of markets. But usually it is the amount that determines it. But when the distance is too great and the producer knows you, you can negotiate by phone if not in this environment benefit goes more to meetings as in the old time.... *non fô lèhè*
> Kouyaté (Korhogo)

That said, when the producer interacted with a single wholesaler, the use of the mobile phone was more obvious. The mobile phone was used to agree on the logistics of delivering the food from the farm to the market. This type of interaction was more common among the larger producers when they are selling to the large cooperatives.

### 4.3.2 Effect of the Distance to Market

Another element that limits the usefulness of the mobile phone is when the market is located near the farm. If the distance is great, then the farmer will need to work out a plan to transport the goods by, for example, contacting a truck driver. However, where the proximity is close, the farmer can carry the wares to the market, either on her head or in a cart, without a need for a mobile phone. In the rural areas where the

---

[6] The categorization of the producers is based on the quantity of crops produced and not the area of production. The quantity of products produced is a more reliable measure since some producers have a large area but only have a small output. Thus, there is often not a high correlation between the area of a farm and its production.

[7] In the case of medium-scale producers who grew between 250 and 1,000 kg of produce, five out of six had a mobile phone, and among the three large-scale producers, two of them owned a mobile phone.

roads are in poor conditions, it is difficult to contract a truck to come to retrieve a small consignment of, for example, 5–100 kg.

While proximity to the market might suggest that an individual can forego use of the mobile phone, there is also an issue of safety. Our informants shared that there is the question of highway robbers who prey on women moving their wares to market. In this case, the mobile phone is seen as a security link. In the words of Akissi:

> Here, There are too many highway robbers... in fact it is the demobilized (military) men... become armed robbers, they know we poor women, we are no force, they attack to rob us. I remember there's my cousin who went to meet Adjoua producing credits to pay for his goods 630,000 CFA. (She) was stripped. She even had the opportunity otherwise, they would rape her. Fortunately, the area was covered by the MTN (mobile phone) network, so she was able to call his brother in town who came to fetch her. Mobile, it helps us ... Akissi.

The ability to transport crops to nearby markets means that the individual does not need to arrange other forms of transportation and thus does not need to use a mobile phone. However, there is safety and security associated with having a device.

### 4.3.3 African Taboo and Mistrust of the Technology

The concept of *"African taboo"* plays on the idea that ownership and use of a mobile phone can lead to jealousy and mistrust. Since a mobile phone is a sign of wealth, in some situations, it can arouse jealousy from the other members of the community. Therefore, it is best not to "flash" a mobile phone. In the words of one informant Soro Kolo: "It is better to live hidden if you want to live long time".

Another issue is that the use of the mobile phone engenders a fear of technological systems that in turn leads to a reluctance to adopt. In the words of Silué:

> Portable (a mobile phone) is good ... but when you know what is behind, you're not even going to buy (one). Wizards move quickly inside, and then they can quickly know where you are ... so around here I tell the people of my family not to pay it.

A number of the informants who had not used the mobile phone reported having a relative that went mad after receiving a call on their mobile phone starting with the number +223. When asked how this madness manifests, one informant, Konate Djafolon, said of her husband, "he speaks alone all the night long". Another informant, Miriam, noted that it was the light function of the mobile phone that disturbed the spiritual world:

> Since Lacina got mobile phone, he is always getting sick... he says that someone that he did not know just called his number and then hung up; when we went to look for his illness there, the healer said that the spirits are angry against him because every night when they come to see him... he turns the light of his cell phone and it bothers them, so that's why they hit him.

The discussion of these taboos is more common among those in the rural areas. As noted above, it is also in these areas that there is the least functional value for the mobile phone. There are often well-established alternative methods that allow people to avoid use of the device.

## 5 Conclusion

We have examined how the mobile phone effects (or alternatively, does not affect) food distribution systems in Côte d'Ivoire. In the case of large-scale wholesalers who manage the food cooperatives, the mobile phone plays role in allowing the women who run these organizations to "fine-tune" the transport of wares when there are surpluses and shortages. The managers of the cooperatives can buy or sell products while they are being transported. We find that the mobile phone was used by *petit* wholesalers in a similar way. However, they also use the device to organize their somewhat more scattered distribution chain and to interact with other small-scale wholesalers to collaborate on transportation. The informants clearly see that the device helps to facilitate logistics. Lastly, it allows the retailers to avoid holding on to too much perishable foodstuffs.

On the other side of the equation, there are those who are less attuned to mobile communication in the production of food. We find that some smaller producers in the more rural areas do not use the mobile phone since there are alternative systems of organizing the distribution chain that, to some degree, preclude the need for its use. In addition, there are those who live so near the market that there is no need to use the mobile phone to arrange for logistics. There is also a sense among some of the informants that the mobile phone is a source of bad spirits.

Overall, these findings fit into other work that has examined the role of information flow in rural areas. Looking at the people who have adopted the use of mobile phone, their usage provides efficiency and robustness to the distribution system that was not possible before (Clark 1995). As noted by Argenti (1997), the mobile phone is an important element in the development of food security.

Mobile communication improves the flow of real-time information that allow the food sector, or for that matter, any entrepreneurial activity, to function better (Jagun et al. 2008). Because of this, they supplement existing systems of interaction and improve the efficiency of these systems. In addition, they reduce various types of embezzlement (Porter 2012).

It is interesting to note that the mobile phone system was not used to any degree for the transfer of monies. In other countries, mobile phones have gained a central role with which people transfer money for purchases (Hughes and Lonie 2007). This can be partially explained by the existence of well-established banking flows in the case of large-scale wholesalers and the selective use of mobile phones among the smaller producers. Notably, such money transfer system was not utilized by the *petit* wholesalers. It is possible that this particularly entrepreneurial group could lead the way in the use of this type of application. However, if there is no critical mass of users, it may be difficult to institutionalize use (Ling and Canright 2013).

Another element that may retard the use of mobile communication in food distribution is the cartel-like structure of the large food cooperatives (Harre 2001). The markets are run like cartels that control the access to market space and deter other actors from penetrating the market. This can both facilitate food production and limit innovation. To the degree that alternative supply chains that are perhaps based on the use of ICTs are frustrated by the existing cartel-like structure of the marketing chains, there will not be innovation. The tradition-bound "Chain of the Grand Mother" is a well-functioning system that facilitates distribution for retailers who are illiterate and want to conduct their daily business. That said, it also hinders the motivation to adopt new technologies such as mobile phone and mobile payment.

Another issue here is the taboos and mistrusts associated with mobile communication. These were largely prevalent in the rural areas. It is interesting to note that their social position resulted in a type of anti-mobile ideology (Ling 2012). This helped them to justify their situation vis-à-vis the mobile phone. An ideology of use, or in this case non-use, is often central to our relationship to technology. In this case, it seems that the informants had constructed a series of rationales regarding this anti-mobile ideology.

It is clear that there are limitations to this work. We have only been able to interview people in two sections of Côte d'Ivoire. There are likely other dimensions to the use of the mobile phone that were not uncovered in this analysis. Nonetheless, these insights help us to understand the ways in which the use of mobile phone influences pre-existing systems of distribution.

Finally, this study recommends that training in local languages should be conducted by private companies, especially mobile telephone service providers and interested development agencies. This is to enable essential improvement in accessing mobile telephone services by rural farmers, especially the women in Côte d'Ivoire. Consequently, the mobile phone would be a boon, especially for them to send e-moneys because it seems secured. Due to the bad condition of roads which impedes transportation via trucks, they are forced to walk short distances with their produce to speak directly to wholesalers and cooperatives to sell their goods. They admit that this approach is exhausting, but it avoids heavy financial burdens, e.g. transportation charges and physical efforts to find coverage area of the mobile network.

**Open Access** This chapter is distributed under the terms of the Creative Commons Attribution Noncommercial License, which permits any noncommercial use, distribution, and reproduction in any medium, provided the original author(s) and source are credited.

# References

Abraham, R. (2006). *Mobile phones and economic development: Evidence from the fishing industry in India*. International conference on Information and Communication Technologies and Development, 2006 (ICTD'06), Barkeley, CA (pp. 48–56).
Aker, J. (2008). *Does digital divide or provide? The impact of cell phones on grain markets in Niger* (Center for Global Development Working Paper No. 154). Retrieved from http://papers.ssrn.com/sol3/papers.cfm?abstract_id=1093374
Argenti, O. (1997). Approvisionnement et distribution alimentaire des villes de l'Afrique Francophone. In *Seminaire Sous regional FAO-ISRA*. Presented at the Seminaire Sous regional FAO-ISRA. Dakar: FAO.

Asenso-Okyere, K., & Ayalew Mekonnen, D. (2012). *The importance of ICTs in the provision of information for improving agricultural productivity and rural incomes in Africa* (UNDP Working Paper). http://web.undp.org/africa/knowledge/WP-2012-015-okyere-mekonnen-ict-productivity.pdf

Barrantes Cáceres, R., & Fernandez-Ardevol, M. (2012). Mobile phone use among market traders at fairs in rural Peru. *Information Technologies and International Development, 8*(3), 35–52.

Bayes, A. (2001). Infrastructure and rural development: Insights from a Grameen Bank village phone initiative in Bangladesh★. *Agricultural Economics, 25*(2–3), 261–272.

Beuermann, D. W., McKelvey, C., & Sotelo, C. (2012). *The effects of mobile phone infrastructure: Evidence from rural Peru.* Retrieved from http://www.aae.wisc.edu/mwiedc/papers/2011/Beuermann_Deither.pdf

Chaléard, J.-L. (1996). *Temps des villes, temps des vivres: l'essor du vivrier marchand en Côte d'Ivoire.* KARTHALA Editions. Retrieved from http://books.google.com/books?hl=en&lr=&id=bFMPBWQJq3cC&oi=fnd&pg=PA5&dq=Temps+des+villes,+temps+des+vivres:+l%27essor+du+vivrier+marchand&ots=JRRhyBEi4D&sig=PmMG9tf1xENDzXE0GRZPApgqT0s

Chib, A., & Hsueh-Hua Chen, V. (2011). Midwives with mobiles: A dialectical perspective on gender arising from technology introduction in rural Indonesia. *New Media and Society, 13*.

Clark, G. (1995). *Onions are my husband: Survival and accumulation by West African market women* (1st ed.). Chicago: University of Chicago Press.

Cohen, A., Lemish, D., & Schejter, A. M. (2007). *The wonder phone in the land of miracles: Mobile telephony in Israel.* Cresskill: Hampton Press.

Courade, G. (1985). Villes/campagnes: Les liaisons dangereuses. In N. Bricas et al. (Eds.), *Nourrir la ville en Afrique subsaharienne* (pp. 67–81). Paris: L'Harmattan.

Donner, J. (2008). Research approaches to mobile use in the developing world: A review of the literature. *The Information Society, 24*(3), 140–159.

FAO. (1996). *Déclaration de Rome sur la sécurité alimentaire mondiale et Plan d'action du sommet mondial de l'alimentation, rapport du sommet mondial de l'alimentation 13–17.* Rome: FAO.

Fernandez-Ardevol, M., Barrantes Cáceres, R., & García, A. A. (2011). Mobile telephony in a rural areas:a case. In *Todos tienen celular* (WP, 161., Serie Economía, pp. 50–60). Lima: Instituto de Estudios Peruanos.

Fischer, C. (1992). *America calling: A social history of the telephone to 1940.* Berkeley: University of California.

Flor, A. G. (2009). *Developing societies in the information age: A critical perspective.* Quezon City: UP Open University.

Gakuru, M., Winters, K., & Stepman, F. (2009). *An inventory of innovative farmer advisory services.* Accra: Agricultural Research in Africa (FARA).

Goosens, F., Minten, B., & Tollens, E. (1994). *Nourrir Kinshasa: L'approvisionnement local d'une métropole africaine.* Paris: L'Harmattan.

Hafkin, N. J., & Odame, H. H. (2007). *Gender, ICTs and agriculture.* A situation analysis for the 5th consultative expert meeting of CTA's ICT observatory meeting on Gender and Agriculture in the Information Society, Sageningen.

Harre, D. M. (2001). Formes et innovations organisationnelles du grand commerce alimentaire à Abidjan, Côte-d'Ivoire. *Autrepart, 3*, 115–132.

Hughes, N., & Lonie, S. (2007). M-PESA: Mobile money for the "unbanked" turning cellphones into 24-hour tellers in Kenya. *Innovations Technology Governance Globalization, 2*(1–2), 63–81.

Jagun, A., Heeks, R., & Whalley, J. (2008). The impact of mobile telephony on developing country micro-enterprise: A nigerian case study. *Information Technologies and International Development, 4*(4), 47–65.

Jensen, R. (2007). The digital provide: Information (technology), market performance and welfare in the South Indian fisheries sector. *The Quarterly Journal of Economics, 122*(3), 879–924.

Johnson, D. G. (2004). *Ethical, psychological and societal problems of the application of ICTs in education*. Moscou: UNESCO Institute for Information Technologies in Education (IITE).

Ling, R. (2012). *Taken for grantedness: The embedding of mobile communication into society*. Cambridge, MA: MIT Press.

Ling, R., & Canright, G. (2013). *Perceived critical adoption transitions and technologies of social mediation*. Cell and Self conference, Ann Arbor

Molony, T. (2006). "I don't trust the phone; It always lies": Trust and information and communication technologies in Tanzanian micro-and small enterprises. *Information Technologies and International Development, 3*(4), 67–83.

Overå, R. (2006). Networks, distance, and trust: Telecommunications development and changing trading practices in Ghana. *World Development, 34*(7), 1301–1315.

Porter, G. (2012). Mobile phones, livelihoods and the poor in Sub-Saharan Africa: Review and prospect. *Geography Compass, 6*(5), 241–259. doi:10.1111/j.1749-8198.2012.00484.x.

Prahalad, C. K., & Hammond, A. (2002). Serving the world's poor, profitably. *Harvard Business Review, 80*(9), 48–59.

Rahman, T. (2007). *"Real markets" in rural Bangladesh: Institutions, market interactions and the reproduction of inequality*. Manchester: IPPG, University of Manchester.

Rashid, A. T., & Elder, L. (2009). Mobile phones and development: An analysis of IDRC-supported projects. *The Electronic Journal of Information Systems in Developing Countries, 36*(2), 1–16.

Sauvé, L., & Machabée, L. (2000). La représentation: Point focal de l'apprentissage. *Éducation relative à l'environnement-Regards Recherches Réflexions, 2*, 175–185.

Souter, D., Scott, D., Garforth, C., Jain, R., Mascarenhas, O., & McKemey, K. (2005). *The economic impact of telecommunications on rural livelihoods and poverty reduction*. Reading: Gamos. Retrieved from http://gamos.org.uk/KaR8347Summary.pdf

World Bank. (2012). *Information and communications for development 2012: Maximizing mobile*. Washington, DC: International Bank for Reconstruction and Development/The World Bank.

# Designing Web 2.0 Tools for Online Public Consultation

**Fabro Steibel and Elsa Estevez**

Public consultation is a formal mechanism of social participation where government invites citizens to participate in policymaking. Increasingly, public consultations are being held online, where Web 2.0 tools and other information and communication technology (ICT) tools become central to understand the design of virtual spaces for government-citizen interaction. Through the analysis of two case studies from Brazil—the "Gabinete Digital" and the "Marco Civil Regulatório" initiatives—this chapter discusses how online public consultation spaces are designed, using a combination of ICTs. Based on three frameworks of deliberative theory and characteristics of Web 2.0 tools, the aim of our paper is studying what aspects of Web 2.0 tools are useful for online consultation and what democratic environments they might generate when combined. The main contribution of this work is raising awareness on how the usage of certain Web 2.0 tools can reinforce or diminish some attributes of political communication and therefore, as a result, produce different models of online democratic communication.

## 1 Introduction

Public consultation is a formal mechanism of social participation where government invites the general public to participate in policymaking (Shane 2004). Increasingly, public consultations are being held online, where Web 2.0 and other ICT tools

F. Steibel (✉)
Institute for Technology & Society of Rio de Janeiro, Praia do Flamengo, 100, Rio de Janeiro, RJ 20010-030, Brazil
e-mail: ofabro@ofabro.com

E. Estevez
United Nations University, Operating Unit on Policy-driven Electronic Governance (UNU-EGOV), Rua Vila Flor 166, 4810-445 Guimarães, Portugal
e-mail: estevez@unu.edu

become central to understand the design of virtual spaces for government-citizen interaction (Coleman and Shane 2012). Given the relevance of public consultation for democratic events, and the potential of applying ICT tools in such processes, investigations on this field have recently raised interest within the research community. Application areas of public consultations include drafting legislation, urban planning, election monitoring and participatory budgeting preparation, among others. This research focuses on online public consultations applied for drafting legislation.

The medium of law gives legitimacy to the political order and provides it with its binding force. Legitimate lawmaking itself is generated through a procedure of public opinion and will-formation (Habermas 1996). However, drafting legislation is a hermetic process mostly driven and managed by governments. Such process can be opened to the citizenry through public consultation, which usually follows two types of approaches—top-down and bottom-up. Experiences following a top-down approach are started by government and aim at engaging society. Experiences following a bottom-up approach are started in society and aim at mobilising government for listening to public concerns. In particular, top-down initiatives of e-rulemaking—"a deliberate agenda-setting process designed to elicit, sort, and clarify fact and opinion from a wide variety of interested parties" (Fountain 2003)—are known for being too technical for non-specialised audiences, bureaucratically complex and polarised (Åström and Grönlund 2012). At the same time, such initiatives are known for producing policymaking outputs tied in an accountable way to actual governmental policymaking (Coleman and Price 2012). Consequently, challenges remain high for those willing to enrol in lawmaking experiences even when opportunities to participate are real.

To overcome some of the challenges described above, Web 2.0 tools appear useful and innovative. Web 2.0 tools are increasingly popular across different segments of society, and digital barriers, although still present, are gradually being reduced (Norris 2001). Although a complex term to be set, Web 2.0 can be understood in at least three different ways: (1) as a set of social relationships and practices, (2) as a mode of production and (3) as a set of values (Postigo 2011). In this work, Web 2.0 refers to the term as coined by O'Reilly (2005), focusing on the use of collaborative tools that require no specialised knowledge and allow content generated by users to be reused by others in their own social networks.

From this point of view, it seems that Web 2.0 tools have the potential to revolutionise how legislation is made, but it is not clear how politics and technology should interact (Wright and Street 2007). Based on that, the aim of our research is to study how Web 2.0 tools are used to enhance collaborative practices in policymaking, particularly in drafting legislation, and how they affect political communication during the first two stages of the policy life cycle—agenda setting and policy formulation. To achieve the aim, we address the research questions of (1) what aspects of Web 2.0 tools are used by online consultations and (2) what democratic environments are generated by combining different Web 2.0 tools. Our methodology to address these questions is based on an evaluation of two case studies of online public consultation that took place in Brazil. For each case, we analyse what Web 2.0 tools were used and group them according to their features. We then assess to what extent they are able to deliver a set of indicative attributes based on three models of democratic communication (as suggested by Freelon 2010).

The rest of this chapter is structured as follows. Section 2 presents our literature review, focusing on defining online consultation as a form of collaborative policymaking, analysing how web tools relate to policymaking life cycle stages and how a framework of democratic theory can explain the uses of web tools for opening up government. Section 3 introduces our methodology, justifying the rationales for selecting case studies and conducting data analysis. Section 4 presents and analyses the two case studies, while Sect. 5 presents our contribution to forecast future uses of online tools for policymaking, as well as introduces policy recommendations to aid policymakers, ICT developers and civil society activists to improve the use of Web 2.0 tools in opening up government for collaborative policymaking. It also draws some conclusions and outlines future work.

## 2 Literature Review

This section presents related work that enables to frame our research work within the state of the art. First, we discuss the relationship between democratic experiences and online consultations. By defining online consultations as object of study, we explain their relevance within current issues of political theory and public policy formulation (Sect. 2.1). Second, we introduce main concepts related to policy design and policy cycle stages, illustrating them with practical cases of online consultations (Sect. 2.2). Third, we introduce models of democratic communication and explain the relevance of using Web 2.0 tools as part of such models (Sect. 2.3).

### 2.1 Democratic Experience and Online Consultations

Relations between members of the public and holders of political authority are in a period of transformative flux. Claims regarding the Internet's potential to reshape democratic life have proceeded through several waves, from early enthusiasm to pessimistic reactions, from theory-driven speculations to empirically driven approaches (Chadwick 2006; Coleman and Blumler 2009; Norris 2000). It is not an objective of this chapter to align the role of online consultations to one wave or another, but rather to support the argument that the Internet "possesses a vulnerable potential to revitalise our flagging political communication arrangements by injecting some new and different elements into the relationship between representatives and represented and governments and governed" (Coleman and Blumler 2009: p. 13). In other words, we assume that Internet has the potential to reshape democracy, although the realisation of this potential relies on a case-by-case analysis.

Online public consultations, and e-rulemaking in general, are far from a new phenomenon (Benjamin 2006). The interest of using the Internet to revive participatory democracy emerged in the mid-1990s, but the overall participatory movement in governance is situated in the broader context that took place during the 1960s and 1970s (Chadwick 2006). "Listening to the public" increasingly emerged as a necessity of democratic government when citizens became more educated, politically volatile and less differential, what lead politicians to increasingly rely on mechanisms of consultation to plan and evaluate policy decisions (Bimber 2003).

The debate on online public consultations influences the process of formulation of public policies as much as democracy and what it means (Coleman and Shane 2012). Some argue that online consultations increase the level of participatory democracy by lowering the cost of participation and opening up the administrative process to individual citizens (Angeles et al. 2004). For example, at the most basic level, integrating the Internet into rulemaking makes it easier for citizens to comment on government proposals (Åström and Grönlund 2012) or to participate in a shared project and collaboratively draft a comment (Noveck 2010). We should also consider that online consultations have the potential to legitimise the rulemaking process, making it easier to form and maintain virtual deliberative communities (Coleman and Gøtze 2001). Online consultations also push government agencies to open up. At the most basic level, each agency has to redesign its website and alter its existing rulemaking procedures to some degree (Jaeger and Bertot 2010); however, in the broad level, citizens' electronic participation can require agencies to reconsider their entire rulemaking process:

> E-rulemaking does not fundamentally change the need for rules. What makes e-rulemaking potentially revolutionary is that it necessitates mapping the processes of rulemaking onto interactive software, embedding the desired practice into the design of the virtual spaces for rulemaking. By virtue of having to 'translate' rulemaking into a set of software specification, agency officials have to focus on the practice and precise how-to's of rulemaking. Technology design opens the political imagination to better way of organizing, not simply documents, but the interpersonal relationship of the rulemaking processes. (Noveck 2010: p. 435)

Nowadays, every democratic system has some mechanisms for obtaining public inputs in policymaking processes. Examples of such tools include letters to elected politicians, public hearings, notice requirements, focus groups, surveys, citizen juries, community advisory boards and consensus conferences, among others (Dryzek 1993; Goodin 2006; Innes and Booher 2005; Parkinson 2004). Such tools invite, or at least welcome, comments by the public on important policy initiatives. However, seriously taking public feedback is another step. For example, it is out of question that governments are well positioned to control the shape of whatever discursive spaces are being opened; and we should also wonder if when public speaks, is anybody listening to it (Coleman et al. 2008).

A challenge to study online public consultations is to identify serious and systematic democratic deficits and then to devise appropriate institutional remedies. Some have argued in favour of using online public consultations as a mechanism to address such deficits while contributing to provide such remedies. Four arguments follow.

First, online deliberative interactions offer excellent opportunities for knowledge sharing and long-term community building (Cindio and Schuler 2007; Plant 2004). Second, public consultations often devise innovative policy designs that experts alone would not have developed (Innes and Booher 2003). Third, we can argue that open discussion of matters affecting the public exposes people to other views, what enables them to evaluate the best arguments (Sunstein 2007). Fourth, they go beyond affecting outcomes of individual policymaking episodes and provide large flow of political communication within a society (Shane 2012: p. 14).

Looking from a different perspective, we may argue that online public consultation tools, instead of reducing democratic gaps, are reinforcing them. Some have argued that unequal access to the Internet skew online consultations towards more favoured groups in terms of race, geography, gender and class (Norris 2001). Others have argued that online political debate reinforces extremist beliefs and shrinks the interested public (Sunstein 2007), while others have witnessed little impact of online consultations in policymakers' decisions (Blumler and Coleman 2001). However, the most challenging argument is certainly one of scale: the basic factual point that "most exercises in online deliberation attract relatively small numbers of participants" (Shane 2012: p. 3), referring to the limited number of contributors who participate.

## 2.2 Policy Design and the Policy Cycle

For some time, policy sciences are trying to design methods to help groups articulate and structure their preferences in ways that will enable them to reach decisions that are consistent with their values and intentions. The argument supporting such quests is that to understand and explain why a policy has a particular design, we need to examine the process and the discursive practices leading to the policy's processes of design and implementation (Fischer and Hajer 1999; Schneider and Ingram 1997). Institutional arenas have rules, norms and procedures that affect choices, strategies, decisions, perceptions and preferences (Wright and Street 2007). Thus, the design of particular tools for institutional arenas has political consequences and may lead to different policy decisions (Coleman and Shane 2012).

The policymaking process can be modelled in terms of a cycle of sequential and discrete stages. Although the policy cycle perspective allows us to systematise and compare the diverse approaches in the field of policy sciences, it has been criticised in terms of its theoretical construction as well as in terms of its empirical validity (Jann and Wegrich 2007). The two main criticisms against the model refer to (1) descriptive inaccuracy, "because empirical reality does not fit with the classification of the policy process into the sequential and discrete stages", and (2) its conceptual value, because "stages model does not offer causal explanations for the transition between different stages" (p. 56). At the same time, if we read the stages model not as a perfect model to describe policymaking, rather using it to group research questions (more or less explicitly) derived from the policy cycle framework, we can evaluate actual impacts of particular interventions or in terms of democratic governance understand which actors are dominant and which are not:

> Therefore, the policy cycle framework... offers a perspective against which the democratic quality of these processes could be assessed (without following the assumption of a simple, discrete sequence and clear separation of stages). Additionally, the cycle framework allows the use of different analytical perspectives and corresponding research questions that will stay among the most important ones in policy research, although the stages heuristic of the policy cycle does not offer a comprehensive causal explanation for the whole policy process and even if the fundamental theoretical assumptions, on which initial versions of the framework were based, have long been left behind. (p. 58)

In general terms, the policy cycle perspective considers "the policy process as evolving through a sequence of discrete stages or phases" (Jann and Wegrich 2007: p. 44). Although a number of different variations of the stage typology have been put forward, it is largely accepted that the differentiation between agenda setting, policy formulation, decision-making, implementation and evaluation is the conventional way to describe the chronology of a policy process. Therefore, in this work, we adopt the 5-stage policy cycle model, as depicted in Fig. 1.

Among the five policy cycle stages, it has been argued that the most successful cases of citizens' participation in policymaking refer to cases where the consultation took place particularly during the agenda-setting and policy formulation stages (what we refer to as the *early stages of policymaking*) (Albrecht 2006; Jensen 2003; Macintosh 2003). Based on such argument and since our research focuses on these two stages, below we present more insight about agenda setting and policy formulation.

As the first policy cycle stage, agenda setting is a process of structuring the policy issue regarding potential strategies and instruments that shape the development of a policy in the subsequent stages of a policy cycle (Jann and Wegrich 2007). The agenda-setting stage focuses on problem recognition and issue selection:

> An agenda is a collection of problems, understandings of causes, symbols, solutions, and other elements of public problems that come to the attention of members of the public and their governmental officials. An agenda may be as concrete as a list of bills that are before a legislature, but also includes a series of beliefs about the existence and magnitude of problems and how they should be addressed... (Birkland 2007: p. 63)

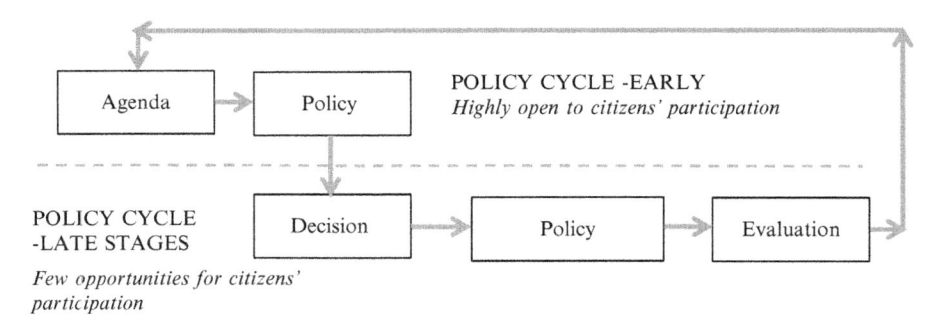

**Fig. 1** Policy cycle stages

Applying online consultations at the stage of agenda setting presupposes that government has recognised by itself a social problem and defined the necessity of state intervention. Citizen's contributions refer to suggesting a list of subjects or problems, articulating preferences, reframing problems and connecting them to relevant aspects of policymaking that should be addressed by the government, the private sector and non-profit organisations or through joint action by some or all of these institutions (Fischer 2000; Goodin 2006; Kingdon 2002).

The policy formulation stage refers to the process where expressed problems, proposals and demands are transformed into government programmes. It includes definition of objectives and the consideration of different action alternatives, drawing goals, priorities, cost-benefit trade-offs and expected externalities of available alternatives:

> Formulating the set of alternatives... involves identifying a range of broad approaches to a problem, and then identifying and designing the specific sets of policy tools that constitute each approach. It involves drafting the legislative or regulatory language for each alternative... and articulating to whom or to what they will apply, and when they will take effect. Selecting from among these a smaller set of possible solutions from which decision-makers actually will choose involves applying some set of criteria to the alternatives, for example judging their feasibility, political acceptability, costs, benefits, and such. (Sidney 2007: p. 79)

The policy formulation stage expects fewer participants to be involved when compared to the stage of agenda setting and expects more of the work to take place out of the public eye. Most of policy formulation requires expert knowledge, which includes government bureaucrats, legislative committee rooms and think tanks (Fischer 2000; Kingdon 2002). Nonetheless, consultations related to formulating policy take up a variety of issues and can redefine the interests of those involved and the balance of power held by participants. It is expected for citizens to participate in policy formulation as well. However, because few voices are able to engage simultaneously in policy formulation, any large-scale argumentation forum is a challenging set-up at this state of policymaking.

## 2.3 Web 2.0 Tools and Models of Democratic Communication

Web 2.0 tools refer to the new generation of ICT tools and functionality enabling the design of websites beyond the static pages used in the web. The concept refers to various changes applied to the design and usage of web pages. Examples of Web 2.0 tools, as referred in this work, include social media—like Facebook, Twitter, blogs, online forums as well as functionality enabling the use of avatars (nicknames), the inclusion of tags to content of web pages and later the retrieval of content based on such tags, among others.

The basic observation that technological design of ICT tools can produce political consequences is well documented (Street 1992; Winner 1980), and a number of scholars have applied this perspective to the study of online spaces (Beierle 2004; Coleman and Gøtze 2001; Fishkin 2009; Freelon 2010; Sack 2005; Taylor and Preece 2010; van der Heijden 2003; Wright and Street 2007; Wright 2006). Related to this area of study, another number of scholars have proposed several policy designs intended to increase the capacity of ordinary citizens to participate in policy development and implementation (Fischer 2003; Hajer and Wagenaar 2003; Macintosh 2003), and the same can be said about governmental programmes oriented to foster citizen's participation online (Noveck 2010; Tambouris and Tarabanis 2007).

If we consider that online discussion forums can influence politics, we can argue that online space design choices can powerfully influence the nature of citizens' engagement and have a real effect on political participation and deliberation (Wright and Street 2007). To evaluate the role of Web 2.0 tools in online policy consultation, we concentrate on indicative attributes of democratic forums. The framework adopted in this work comes from Freelon (2010), who based on the work of Jürgen Habermas (1996) and Lincoln Dahlberg (2001) identifies three distinct, overlapping models of democracy that online political forums may manifest: (1) the liberal individualist, (2) the communitarian and (3) the deliberative democracy. The three models may overlap, and this is likely to happen. For each model, the framework also defines a set of indicative attributes. As Freelon explains, "what the framework contributes is a suite of normative standards for conversation quality that advances the literatures to which it is applied by introducing new conceptual distinctions between divergent notions of democracy" (p. 12). The framework is presented in Table 1.

In search of a model to understand how different kinds of public spheres exist and to measure conversation in itself, Freelon's three-model typology describes communication characteristics of online political communities, specified in terms of coherent clusters of indicative attributes. The liberal individualist model encompasses characteristics of online conversation involving personal expression and the pursuit of self-interest. The communitarian model upholds the cultivation of social cohesion and group identity; and the deliberative democracy model is marked by rational-critical arguments, by focusing on public issues and putative equality. The full list of indicative attributes for each model is shown in Table 1 and discussed in detail via our research question that states that Web 2.0 tools, when combined, generate different models of democratic environments which in turn addresses different indicative attributes of political communication.

## 3 Methodology

We address in this paper two research questions: (1) what aspects of Web 2.0 tools are used by online consultations and (2) what democratic environments are generated by combining different Web 2.0 tools. These questions are addressed based on a combination of models of democratic communication and their attributes (see Table 1), and as we have argued, we do not intend to understand what models provide higher impact on democracy or if they can exist alone or only in combined

# Designing Web 2.0 Tools for Online Public Consultation

**Table 1** Three models of online democratic communication and their indicative attributes

| Model of democratic communication | Indicative attribute | What the attribute stands for |
|---|---|---|
| *Liberal individualist* Encompasses characteristics of online conversation involving personal expression and the pursuit of self-interest. It refers to the "democratic traditions, which posit the individual as a rational, autonomous subject who knows and can express their own best interests. This knowing subject is assumed by a diversity of liberal democratic theories, from classic liberalism to libertarianism" (Dahlberg 2001: p. 160) | Monologue | The individual pursuit of his or her own interest (in opinion expression) at the expense of dialogue |
| | Personal revelation | The disclosure of information about oneself in a public forum |
| | Personal showcase | Advertising platforms for content that individuals have created apart from the forum itself (i.e. other than standard text posts), be it visual, aural or textual |
| | Flaming | Engaging in flaming to derive personal satisfaction by harassing political opponents, releasing tension associated with suppressing their unpopular opinions in offline life or simply antagonising others for its own sake |
| *Communitarian* Upholds the cultivation of social cohesion and group identity. It "argues that sustainable democracy must be based upon the shared values and conceptions of 'the good' that bind people into community... the communitarian self is understood to be constituted within relationships structured by social roles and shared subjectivity. Democratic dialogue serves the common life of the group... it enables members of a community to discover their shared identity and purpose" (Dahlberg 2001: p. 163) | Ideological homophyly | The proposition that citizens tend to assemble themselves into politically homogeneous collectives that rarely engage with outsiders |
| | Mobilisation | The promotion of conducive atmosphere for the furtherance of collective political objectives |
| | Community identification | The extent to which participants view themselves as members of a community, such as the use of community language and collective pronouns |
| | Intra-ideological response | Conversational communication primarily established with ideological similar persons (i.e. in-group members) |
| | Intra-ideological questioning | Questioning, as users must respond to each other in order to ask direct questions, addressed to ideological in-group members |

(continued)

**Table 1** (continued)

| Model of democratic communication | Indicative attribute | What the attribute stands for |
|---|---|---|
| *Deliberative democracy* Pursues rational-critical arguments, by focusing on public issues and putative equality. "For both liberal individualism and communitarianism, the source of democratic legitimacy (and rationality), is based upon the will of all, which is derived from the expression of already formed wills... In contrast, deliberative democracy relies upon inter-subjectivity. All pre-discursive interests and values are up for grabs. A legitimate (and rational) decision rests not upon the expression of pre-given wills but upon the deliberative process by which everyone's will is formed" (Dahlberg 2001: p. 168) | Rational-critical argument | The willingness (to say nothing of the ability) to use logical, methodical appeals to the common good in arguing for one's position |
| | Public issue focus | Discussions focusing primarily on issues traditionally considered political |
| | Equality | Equality is operationalised as the extent to which forum contributions are spread evenly among participants (the opposite case of a small number of users contributing the vast majority of posts) |
| | Discussion topic focus | As distinct from public issue focus, discussion topic focus assesses the extent to which posts within discussion threads address the initial thread topic |
| | Inter-ideological response | Participants actually communicate across lines of difference to fully realise the ideal of debate |
| | Inter-ideological questioning | The asking of questions between members of mutual out-groups |

Based on Freelon (2010)

forms. Nonetheless, our aim is to evaluate how these models can aid us to understand the quality of democratic communication in online forums based on the types of Web 2.0 used.

The two case studies analysed in this chapter refer to online policy consultations that took place in Brazil, and they refer each to a different policy stage. The "Marco Civil Regulatório da Internet" (MC, *Civil rights regulatory framework for Internet*) ran a public portal to consult citizens on a drafted bill of Internet rights, hosting therefore a "policy formulation" online consultation (Segurado 2011; Steibel and Beltramelli 2012). Running as an "agenda-setting policy consultation", the case of "Gabinete Digital" (GD, *Digital Cabinet*) invited citizens to present issues, suggestions and challenges on the theme of urban traffic in the southern city of Porto Alegre (Brisola and Leal 2012; Resende and Mata 2012). The portals were analysed based on print screens and downloaded versions of their content, collected during their own consultation periods.

# Designing Web 2.0 Tools for Online Public Consultation 253

The reasons for selecting the two case studies are because of their pioneering nature and political relevance. The MC was the first major online policy consultation held in Brazil. The initiative was led by the Ministry of Justice, working with the Ministry of Culture and the civil society think tank Centre for Technology and Society (FGV-RJ, *Centro de Tecnologia e Sociedade*). The general objective of the MC project was to draft a bill of law for the Internet in Brazil, and it was based on collaborative practices. The project ran from October 2009 to May 2010 (in two phases of 45-day each, with a break period in between) and resulted in an online forum where stakeholders interested in the topic could post, debate and comment on the possible design of future Internet legislation, attracting as a whole over 2,000 contributions and considerable amount of media coverage (Steibel and Beltramelli 2012).

The MC consultation made use of several Web 2.0 tools (mainly a WordPress platform, Twitter, RSS feeds and blogs). Divided into two rounds of discussion, people were first invited to comment on a *white paper* with a set of general ideas to broadly orient the draft legislation, and later citizens were invited to comment on the draft legislation as formatted to be sent to Congress. As the policymakers involved in the process describe the initiative, the first round of consultation tested a set of normative standards, predefined by those sponsoring the initiative, that were considered important to include in future legislation, while the second round focused on receiving feedback on the draft itself (Steibel and Beltramelli 2012: p. 96).

The second case study analysed, the GD, is the first online consultation portal run by a state-level government to achieve significant number of participants (Brisola and Leal 2012). Released in May 2011, the portal had previously made use of a wide range of public consultations (some of them include answering questions posted by citizens through YouTube and others like the one under analysis designed as proper agenda-setting consultations). In this chapter, we analyse the last agenda-setting consultation run in the GD portal on road traffic. The consultation ran in two phases: in the first phase, citizens could submit policy propositions related to five themes of policy agendas, where 2,100 policy suggestions where made; in the second phase, policymakers grouped the previous submissions in 600 policy propositions that were voted by citizens online (more than 240,000 votes, given by 100,000 registered citizens, were received during this phase). As a result of the consultation, the top 10 policies voted by citizens were prioritised by government and selected for policy formulation and implementation.

The GD consultation also made use of several Web 2.0 tools (mainly HTML5, YouTube, Twitter, Facebook, RSS feeds and blog). Divided into two rounds of consultation, the portal was inspired by the "All our ideas" project, the *wiki survey* tool which is an academic initiative from the Department of Sociology at Princeton University, and sponsored by Google (Salganik and Levy 2012). Using such a platform, the portal allowed citizens and users to write and publish online short policy programmes that were at the end of the phase analysed and merged by policymakers and grouped in five thematic groups of policy propositions. During the second phase, users were presented with a pair of random ideas selected out of the existing proposal and belonging to the same theme and asked to choose between them. After making this vote, participants were immediately presented with another randomly selected pair of ideas. All phases were supported by strong use of other network sites sponsored by the project leaders.

In the next section, we first analyse the Web 2.0 tools used in each portal, grouping them according to their features. We then match the possible uses of Web 2.0 tools selected with the models of communication and attributes suggested by Freelon (2010).

## 4 Case Study Analyses

This section compares the Web 2.0 tools used in each case study, aligning them with the three democratic models presented in Sect. 2.3.

### 4.1 Web 2.0 Tools Used

The first note we should describe refers to the overall objective of each portal. The MC portal was designed as a document-centred website: a bill of law. The portal is based on a WordPress content management system[1] and is hosted in a multi-user WordPress server sponsored by the Ministry of Culture, the CulturaDigital.br server (Savazoni and Cohn 2009). Due to these characteristics, the portal is actually a blogging tool, and it had to be adapted to work as a document-centred website in which users could comment in a set of policies online. The initiative made use of a blog section as well, but the consultation itself had to adapt from the original use of the portal to work. Due to similarities shared between a static page and the overall structure of a legal document (i.e. a document written in short textual parts: chapters, articles and paragraphs), the blogging interface could easily accommodate a static page where its content (i.e. the drafted bill of law) would be able to receive comments online.

In contrast with the MC portal, the GD portal was designed from the start to host online public consultations. Written in HTML5, the portal has different templates to fit different types of consultations. The portal section "Governor replies", for example, received questions from general citizens that after a period of time were analysed by a team of policymakers, shortlisted and replied by the governor itself using YouTube videos. All these tasks are designed from scratch to be performed through the website. For example, the road traffic online public consultation analysed in this chapter was tailor-made for its purpose. Inspired by the project "All our ideas", the GD portal hosted during its first phase a web page where citizens could leave their contributions, and during the second consultation

---

[1] See more at http://developer.wordpress.com/

phase, citizens were displayed with two proposal, one against each other as in battle of ideas and had to vote which one of the two they liked the most.

In terms of access and openness to contributions, the two portals used a two-stage approach to design the consultation. In both portals, citizen's contributions and policies under review were visible to anyone visiting the portal's URL. However, in both cases, only registered users could post contributions. In the case of MC, comments were accepted only from users with a CulturaDigital.br's login (which was free for all and required no proof of identity). In the case of the GD, comments were accepted by users with portal's login (which was free for all and could also be completed by associating a Facebook or a Twitter account). An important difference between the profiles in the case studies analysed is that in the MC, each user received, associated to its login, a public profile that listed all contributions made by the user as well as displayed blogs, posts, pages and comments related to the user; in the GD, no user had a personal page, and therefore, citizens could not see the profile of others participating in the consultation.

In relation to tools used to integrate the consultation portal with other social networks, in both cases the portals made use of Twitter, but only GD used Facebook. In the case of MC, the portal created a public profile "@marcocivil" that was regularly updated and promoted the hashtag "#marcocivil", which was promptly adopted by other twitters. Twitter was a very important social media for the MC initiative to use due to the social movement "#MegaNão" which was one of the key reasons why the MC consultation started (Steibel and Beltramelli 2012). In the case of GD, the initiative also set up a Twitter account profile "@gabinetedigital", regularly updated it and started a set of hashtags to promote it, e.g. "#gabinetedigital" and "#GovRS". They also created a Facebook profile and its page "/gabinetedigital" was also regularly updated. They also generated a Google+ profile, which was eventually updated.

Table 2 presents a detailed comparison of Web 2.0 used by each portal. The categories and tools coded in the table were defined after a case study analysis of each portal. The starting point to categorise the Internet-based tools refers to those topics referred as benefits and challenges of e-rulemaking process (see Sect. 2). Once the list of tools was completed, these broad issues were used to collate the findings and compare both initiatives side by side.

An important note to make about both projects is that while the MC was mainly promoted as a policy consultation event (although eventually referred to as an initiative sponsored by the Ministry of Justice), the GD was regularly associated (and promoted as) an activity of the incumbent government and even as an event promoted by the governor itself. It is also important to note that in relation to other Web 2.0 tools used by the portals, both made use of RSS feeds for spreading news, and while the GD created a specific email account to receive comments from the general public, the MC did not set up such tool to consider email as a "private" technology (a one-to-one flow of communication, instead of many-to-many technologies such as Twitter and WordPress architectures) (Steibel 2012).

**Table 2** Comparison of Web 2.0 tools adopted in each case study

| Web 2.0 tool by functionality type | GD | MC |
|---|---|---|
| *Personal identification* | | |
| Login authentication | Yes | Yes |
| Personal and public profile page | No | Yes |
| User avatar | No | Yes |
| *Publishing contributions* | | |
| Comment identified by tag word | No | Yes |
| Possibility to leave written contribution | No | Yes |
| Text-length limitation | Yes | No |
| Possibility to leave multimedia contribution | No | No |
| *Sharing contributions* | | |
| User's contributions could be shared by other users | No | No |
| Portal's contributions could be shared by others | Yes | Yes |
| *Interaction with social media* | | |
| Portal's profile on Facebook | Yes | No |
| Portal's profile on YouTube | Yes | No |
| Portal's profile on Twitter | Yes | Yes |
| Portal's profile on Google+ | Yes | No |
| Dedicated email account | Yes | No |
| RSS feed link | Yes | Yes |
| *Promoting an informed debate* | | |
| Hosting a blog with experts' contribution | Yes | Yes |
| Hosting a forum/section for citizens to exchange ideas | No | Yes |
| Promotion of user's in-depth contributions in the blog | No | Yes |

Table 2 aids us to consider some aspects of Internet-based tool on democracy. Three arguments can be made. First, both initiatives present some tools to interact with social media; however, none of the portals have tools to hosts policy debates outside of their own environment. Social media tools are present to connect people, to promote the overall initiative and to facilitate login processes. However, links and content shared in other networks are oriented to bring citizens back to participate in the main URL. Second, the MC case uses tools to promote an informed debate more heavily than the DG case does. This is likely to be related to the difference in terms of policy cycle. It is reasonable to expect text-based comments to be more important during a policy formulation stage than during an agenda-setting stage. Third, both cases present some tools to promote personal identification. Although such tools are not based on official records, they make a clear statement that anonymous contributions are not welcomed and refer to a need for online consultations to record individual, person-based, contributions by design.

## 4.2 Web 2.0 and Democratic Models

This section applies the Web 2.0 tools presented in Table 2 with the three models of democracy and their indicative attributes presented at the literature review. The explanation follows and the summary is presented in Table 3.

The Liberal Individualist framework considers imperative to make possible for individuals to express themselves. This framework requires Web 2.0 tools to allow citizens to create personal spaces for self-expression and to advertise their own preferences to others. In the MC portal, for example, Web 2.0 tools allow users to make their own voice heard ("monologue" attribute), while in the GD approach, citizens can express their personal opinion only in a limited way (writing a short, limited size, policy proposal). The MC portal also allows citizens to present "personal revelation"—to disclose information about oneself in a public forum—which can be done either in the comment section and in the personal profile page and can also promote "personal showcase" because contents created by users are directly linked to them as their contributions (which in both cases cannot be done in the GD portal). We should also consider that the MC is open to "flaming" while the GD (due to the limitation in terms of tools for individual expression) is closed to it. Since "flaming" can be both democracy-enhancing and democracy-detracting conversation, it was a main concern of policymakers promoting the MC portal; however, it never occurred (Steibel and Beltramelli 2012).

The communitarian framework considers as a priority strengthening ties between citizens involved in policymaking and creating user groups by similarity of ideo-

**Table 3** Analysis of case studies based on the framework of online democratic communication

| Model of democratic communication | Indicative attributes | Cases GD | MC |
|---|---|---|---|
| Liberal individualist | Monologue | No | Yes |
|  | Personal revelation | No | Yes |
|  | Personal showcase | No | Yes |
|  | Flaming | No | Yes |
| Communitarian | Ideological homophyly | No | Yes |
|  | Mobilisation | No | Yes |
|  | Community identification | Partly | Yes |
|  | Intra-ideological response | No | Partly |
|  | Intra-ideological questioning | No | Partly |
| Deliberative | Rational-critical argument | Yes | Yes |
|  | Public issue focus | Yes | Yes |
|  | Equality | Partly | Partly |
|  | Discussion topic focus | Partly | Yes |
|  | Inter-ideological response | No | Yes |
|  | Inter-ideological questioning | No | Yes |

logical alignment or social ties or both. In terms of "ideological homophyly", we should consider it as a possibility in the MC portal. However, it cannot happen in the GD portal due to the absence of interaction between users. In the same direction, "mobilization" is related to the offer of a conductive atmosphere for the furtherance of collective political objectives, an important feature to other online and also offline set-up of communitarian spaces, which can happen in the MC case but not in the GD case. Regarding "community identification", although this can be partially achieved in the GD portal (because users can view contributions of others), this cannot be fully accomplished there because people cannot align themselves to others' policy proposals (except voting on them, considering however that proposals are randomly drawn), and they cannot make use of collective pronouns such as "we", "us" and "our". In terms of "intra-ideological response" and "intra-ideological questioning", these two attributes can only happen in the MC portal, and even so, partially. In the MC portal, there is only one communicative arena where the consultation happens, a space shared by all participants, which does not allow discussion organised by groups to take place inside the portal.

The deliberative model is marked by Habermas' conceptual trio of (1) rational-critical argument, (2) public issue focus and (3) putative equality. In this regard, both portals are able to promote "rational-critical argument" because tools used in each case allow citizens to be willing to use logical methodical appeals to the common good. The same is true about the "public issue focus" characteristic because it is possible in the MC and the GD portal for citizens to focus on strong public issues traditionally considered political (referring here, in the case of the MC, to civil rights related to Internet use and, in the case of the GD, to road traffic–related issues). The "equality" attribute is possible to be achieved in both portals, but cannot be measured in neither of them. Although Web 2.0 tools allow users to participate in equal terms in both cases, inequalities related to access, interest or knowledge may provide inequality as a result of tools' use in the MC and the GD. Regarding the "discussion topic focus" attribute, we notice that in the case of the GD, there is no thread of discussion allowed by tools in use (unless we consider voting in a "battle layout" as a form of thread discussion). In the case of the MC, however, the topic focus can be achieved and it's actually enforced by the use of document-centred discussion. When we analyse the "inter-ideological response" and "inter-ideological questioning", it is clear that this can only be achieved in the MC case (due to tools allowing citizens to express themselves and ask questions to others).

## 5 Conclusion and Future Research

The two research questions presented in the Introduction section were addressed respectively in Sects. 4.1 and 4.2. As regards the first research question, we mapped what Web 2.0 tools and functionalities were used in the two case studies, clustering them according to what political action they enabled citizens to achieve (i.e. enacting personal identification, publishing contributions online, sharing comments with others, linking different social media networks and promoting an informed debate). Pertaining to the second question, we noted that Web 2.0 tools, when combined, generate different models of democratic environments which in turn addresses different indicative attributes of political communication. Read together, our two research questions validate the perspective of Wright and Street who claim that design choices of the use of ICT in online space influence the nature of citizens' engagement and have a real effect on political participation and deliberation (Wright and Street 2007). Although our current research cannot assess real impact related to design choices, we expect future research to fill this gap by analysing the comments and interactions established in the two case studies here addressed.

We do however expect to find differential impact based on what Web 2.0 tools are in use. The presence or absence of text boxes to receive comments from users considerably restrings the use of deliberative and communitarian models. Having in mind that the policy stage of policy formulation is closely linked to attributes of these models, we can predict that this Web 2.0 tool is mandatory at this stage. Another possible prediction is that the policy stage of agenda setting requires fewer number of Web 2.0 tools than the policy formulation stage does (which is true when we consider that the GB initiative employed fewer tools than the MC did).

As we evaluate the case studies, it becomes clear that when combining Web 2.0 for online public consultation, different models of democratic communication can overlap. The overlapping of different models is predicted by Freelon, who supports that a common virtual space has little chances to adhere to a singular set of standards or to encompass all possible modes of political expression. "Thus, rather than unilaterally declaring a forum more or less 'deliberative' after analysing its contents, the new framework permits more precise conclusions such as 'communitarian with some deliberative aspects' or 'solidly liberal individualist'" (Freelon 2010: p. 6). The Web 2.0 tools used in the GD portal, for example, have the potential to deliver aspects of the deliberative model of communication, but can only partly deliver attributes of the communitarian model and are unable to deliver any aspect of the libertarian model. The MC case, however, can deliver aspects of the three models, what is particularly true regarding the libertarian and the deliberative models.

We should take into consideration that there is no normative standard arguing that one model of democracy is better than the others. Answering the question of which model is preferred to others requires defining what is the purpose of the undergoing process of policymaking. Policy study is a problem-oriented field of research (Lasswell 2003), and unless we define beforehand what is the problem to address, we cannot evaluate the impact or the success of one model over the others. We also need to consider that the ability to deliver models of communications is not the same as delivering it. To do so, we still need to evaluate the content of online public consultations, particularly the posts and contributions of general users, and evaluate if the usage of Web 2.0 tools was indeed used for the purpose they were designed for.

In future research, we plan to apply our framework to other stages of policy cycle and to evaluate, for example, if online public consultation is taken into consideration by policymakers during the decision-making stage. We also plan to increase the sample of observed case studies, including, particularly, websites and online platforms built with technologies other than WordPress. Finally, we realise a need to evaluate the impact of experiences among users, understanding the effect of Web 2.0 tools in strengthening citizens' engagement.

**Open Access** This chapter is distributed under the terms of the Creative Commons Attribution Noncommercial License, which permits any noncommercial use, distribution, and reproduction in any medium, provided the original author(s) and source are credited.

# References

Albrecht, S. (2006). Whose voice is heard in online deliberation?: A study of participation and representation in political debates on the internet. *Information Community and Society, 9*(1), 62–82.

Angeles, L., Sevilla, G., Shulman, S., & Tiller, E. (2004). E-rulemaking: Bringing data to theory at the federal communications commission. *Duke Law Journal*, http://scholarship.law.duke.edu/dlj/vol55/iss5/3/.

Åström, J., & Grönlund, Å. (2012). Online consultations in local government: What works, when, and why? In S. Coleman & P. M. Shane (Eds.), *Connecting democracy: Online consultation and the flow of political communication*. Cambridge, MA/London: MIT Press.

Beierle, T. C. (2004). Digital deliberation: Engaging the public through online policy dialogues. In *Democracy online: The prospects for political renewal through the Internet* (Vol. 155, p. 155). Publisher: Routledge; 1 edition (August 4, 2004), ISBN-10: 0415948657, http://www.amazon.com/Democracy-Online-Prospects-Political-Internet/dp/0415948657/ref=sr_1_1?ie=UTF8&qid=1423535343&sr=8-1&keywords=Democracy+online%3A+The+prospects+for+political+renewal+through+the+Internet.

Benjamin, S. M. (2006). Evaluating E-rulemaking: Public participation and political institutions. *Duke Law School Faculty Scholarship Series*. Paper 73.

Bimber, B. (2003). *Information and American democracy: Technology in the evolution of political power*. Cambridge: Cambridge University Press.

Birkland, T. A. (2007). Agenda setting in public policy. In F. Fischer, G. J. Miller, & M. S. Sidney (Eds.), *Handbook of Public Policy Analysis: Theory, politics and methods*. Boca Raton: CRC Press.

Blumler, J. G., & Coleman, S. (2001). Realising democracy online: A civic commons in cyberspace. *IPPR/Citizens Online Papers*.

Brisola, A. C., & Leal, T. (2012). Internet e Participação: uma análise do Portal Gabinete Digital. *Enagramas, 6*(1), 1–15.

Chadwick, A. (2006). *Internet politics: States, citizens, and new communication technologies*. Oxford: Oxford University Press.

Coleman, S., & Blumler, J. G. (2009). *The internet and democratic citizenship: Theory, practice and policy* (p. ix, 220). Cambridge: Cambridge University Press.

Coleman, S., & Gøtze, J. (2001). *Bowling together: Online public engagement in policy deliberation*. Retrieved January. Hansard Society. London, 2002, http://www.hansardsociety.org.uk/wp-content/uploads/2012/10/Bowling-Together-Online-Public-Engagement-in-Policy-Deliberation-2001.pdf.

Coleman, S., & Price, V. (2012). Democracy, distance, and reach: The new media landscape. In S. Coleman & P. M. Shane (Eds.), *Connecting democracy: Online consultation and the flow of political communication*. Cambridge, MA/London: MIT Press.

Coleman, S., & Shane, P. M. (2012). *Connecting democracy: Online consultation and the flow of political communication.* Cambridge, MA/London: MIT Press.

Coleman, S., Morrison, D. E., & Svennevig, M. (2008). New media and political efficacy. *International Journal of Communication, 2,* 771–791.

da Silva Resende, C. A., & da Mata, J. F. (2012). *Gabinete digital: uma experiência de democracia na rede.* Conference OIDP, Porto Alegre, Brazil.

Dahlberg, L. (2001). Democracy via cyberspace: Mapping the rhetorics and practices of three prominent camps. *New Media and Society, 3*(2), 157–177.

De Cindio, F. & Schuler, D. (2007). *Deliberation and community networks: A strong link waiting to be forged.* Proceedings of CIRN conference: Communities and Action (pp. 1–11), Prato.

Dryzek, J. S. (1993). Policy analysis and planning: From science to argument. In F. Fischer & J. Forester (Eds.), *The argumentative turn in policy analysis and planning* (p. viii, 327). Durham: Duke University Press.

Fischer, F. (2000). *Citizens, experts and the environment: The politics of local knowledge.* Durham: Duke University Press.

Fischer, F. (2003). *Reframing public policy: Discursive politics and deliberative practices* (p. xi, 266). Oxford/New York: Oxford University Press.

Fischer, F., & Hajer, M. (1999). *Living with nature: Environmental politics as cultural discourse.* Oxford: Oxford University Press.

Fishkin, J. S. (2009). *When the people speak: Deliberative democracy and public consultation.* Oxford/New York: Oxford University Press.

Fountain, J. E. (2003). Prospects for improving the regulatory process using e-rulemaking. *Communications of the ACM, 46*(1), 43–44. doi:10.1145/602421.602445.

Freelon, D. G. (2010). Analyzing online political discussion using three models of democratic communication. *New Media and Society, 12*(7), 1172–1190. doi:10.1177/1461444809357927.

Goodin, R. E. (2006). Deliberative impacts: The macro-political uptake of mini-publics. *Politics and Society, 34*(2), 219–244. doi:10.1177/0032329206288152.

Habermas, J. (1996). *Between facts and norms: Contributions to a discourse theory of law and democracy.* Cambridge, MA: MIT Press.

Hajer, M., & Wagenaar, H. (2003). *Deliberative policy analysis: Understanding governance in the network society.* Cambridge: Cambridge University Press.

Innes, J. E., & Booher, D. E. (2003). Collaborative policymaking: Governance through dialogue. In M. A. Hajer & H. Wagenaar (Eds.), *Deliberative policy analysis: Understanding governance in the network society* (pp. 33–60). Cambridge: Cambridge University Press Cambridge.

Innes, J. E., & Booher, D. E. (2005). Reframing public participation: Strategies for the 21st century. *Planning Theory and Practice, 5*(4), 33–60.

Jaeger, P. T., & Bertot, J. C. (2010). Transparency and technological change: Ensuring equal and sustained public access to government information. *Government Information Quarterly, 27*(4), 371–376. doi:10.1016/j.giq.2010.05.003.

Jann, W., & Wegrich, K. (2007). Theories of the policy cycle. In F. Fischer, G. J. Miller, & M. S. Sidney (Eds.), *Handbook of public policy analysis: Theory, politics and methods.* Boca Raton: CRC Press.

Jensen, J. L. (2003). Virtual democratic dialogue ? Bringing together citizens and politicians. *Information Polity, 8,* 29–47.

Kingdon, J. W. (2002). *Agendas, alternatives, and public policies (Longman classics edition).* New York: Longman.

Lasswell, H. D. (2003). The policy orientation. In H. D. Lasswell, D. Lerner, & S. Braman (Eds.), *Communication researchers and policy-making.* Cambridge, MA/London: MIT Press/Stanford University Press.

Macintosh, A. (2003). *Promise and problems of E-democracy: Challenges of online citizen engagement.* OECD Publishing. doi:10.1787/9789264019492-en. Leeds, UK.

Norris, P. (2000). *A virtuous circle: Political communications in postindustrial societies* (p. xvii, 398). Cambridge: Cambridge University Press.

Norris, P. (2001). *Digital divide: Civic engagement, information poverty, and the Internet worldwide* (Vol. 40). Cambridge: Cambridge University Press. http://www.cambridge.org/ar/academic/subjects/politics-international-relations/comparative-politics/digital-divide-civic-engagement-information-poverty-and-internet-worldwide.

Noveck, B. S. (2010, November 2). *Wiki government: How technology can make government better, democracy stronger, and citizens more powerful* (p. xxii, 224). Brookings Institution Press; Reprint edition (November 2, 2010), ISBN-10: 0815705107. Washington, DC: Brookings Institution Press. http://www.amazon.com/Wiki-Government-Technology-Democracy-Stronger/dp/0815705107/ref=sr_1_1?ie=UTF8&qid=1423535280&sr=8-1&keywords=wiki+government

O'Reilly, T. (2005). *What is Web 2.0? Design patterns and business models for the next generation of software*. Author's webpage.

Parkinson, J. (2004). Hearing voices: Negotiating representation claims in public deliberation. *The British Journal of Politics and International Relations, 6*(3), 370–388. http://onlinelibrary.wiley.com/doi/10.1111/j.1467-856X.2004.00145.x/abstract.

Plant, R. (2004). Online communities. *Technology in Society, 26*(1), 51–65. doi:10.1016/j.techsoc.2003.10.005.

Postigo, H. (2011). Questioning the web 2.0 discourse: Social roles, production, values, and the case of the human rights portal. *The Information Society, 27*(3), 181–193.

Sack, W. (2005). Discourse architecture and very large-scale conversation. In R. Latham & S. Sassen (Eds.), *Digital formations: IT and new architectures in the global realm*. Princeton: Princeton University Press.

Salganik, M. J., & Levy, K. E. C. (2012). Wiki surveys: Open and quantifiable social data collection*. http://arxiv.org/abs/1202.0500, pp. 1–29.

Savazoni, R., & Cohn, S. (2009). *Cultura digital.br* (R. Savazoni & S. Cohn, Eds.) (p. 312). Rio de Janeiro: Beco do Azougue.

Schneider, A. L., & Ingram, H. M. (1997). *Policy design for democracy*. Lawrence: University Press of Kansas.

Segurado, R. (2011). Política da Internet: a regulamentação do ciberespaço. *Confibercom. Revistausp, 90*, doi:http://dx.doi.org/10.11606/issn.2316-9036.v0i90p43-57.

Shane, P. M. (2004). *Democracy online: The prospects for political renewal through the internet*. New York: Routledge.

Shane, P. M. (2012). Online consultation and political communication in the era of obama: An introduction. In S. Coleman & P. M. Shane (Eds.), *Connecting democracy: Online consultation and the flow of political communication*. Cambridge, MA/London: MIT Press. (Kindle.). Kindle edition.

Sidney, M. S. (2007). Policy formulation: Design and tools. In F. Fischer, G. J. Miller, & M. S. Sidney (Eds.), *Handbook of public policy analysis: Theory, politics and methods*. Boca Raton: CRC Press.

Steibel, F. (2012). Ferramentas Web 2.0 e o design de consultas públicas online: o caso do Marco Civil Regulatório. *Compos, GT Comunicação e Democracia*, Juiz de Fora, Brazil.

Steibel, F., & Beltramelli, F. (2012). Policy, research and online public consultations in Brazil and Uruguay. In B. Girard & E. A. y Lara (Eds.), *Impact 2.0: New mechanism for linking research and policy*. Montevideo: Fundación Comuca.

Street, J. (1992). *Politics & technology*. London: MacMillan.

Sunstein, C. R. (2007). *Republic.com 2.0*. Princeton: Princeton University Press.

Tambouris, E., & Tarabanis, K. (2007). *DEMO-net: D14.3 The role of Web 2.0 technologies in eParticipation*. DEMO-net consortium, Leeds, UK.

Taylor, P., & Preece, J. (2010). Determining and measuring success Sociability and usability in online communities: Determining and measuring success. *Behaviour and Information Technology, 20*(5), 37–41. doi:10.1080/0144929011008468.

Van der Heijden, H. (2003). Factors influencing the usage of websites: The case of a generic portal in The Netherlands. *Information and Management, 40*(6), 541–549. doi:10.1016/S0378-7206(02)00079-4.

Winner, L. (1980, Winter). Do artifacts have politics? (In D. MacKenzie & J. Wajcman, Eds. *Modern technology: Problem or opportunity?*). *Daedalus, 109*(1), 121–136. http://innovate.ucsb.edu/wp-content/uploads/2010/02/Winner-Do-Artifacts-Have-Politics-1980.pdf

Wright, S. (2006). Government-run online discussion fora: Moderation, censorship and the shadow of control. *The British Journal of Politics and International Relations, 8*(4), 550–568.

Wright, S., & Street, J. (2007). Democracy, deliberation and design: The case of online discussion forums. *New Media and Society, 9*(5), 849–869. doi:10.1177/1461444807081230.

# ICTs and Opinion Expression: An Empirical Study of New-Generation Migrant Workers in Shanghai

Baohua Zhou

This chapter focuses on the opinion expression of new-generation migrant workers and empirically examine its relationship with ICTs in China. It differentiates two kinds of problems that the migrant workers encounter, labour rights problems and personal emotion problems, as well as three types of expressive channels, interpersonal networks, new media, and institutional channels. By analysing the data from a questionnaire survey conducted in Shanghai of China ($N = 869$), we find that when faced with realistic problems, the intention of expression among new-generation migrant workers is relatively high on the whole. In terms of expressive channels, they would like to express more via interpersonal network, followed by new media channels, and least via institutional channels. The online news and online interaction have been found to be significantly related to the expressive intention through new media, although they have no direct impact on offline expression. We conclude this chapter by discussing the academic and practical implications of this study.

## 1 Introduction

How should we make sense of Information and Communication Technologies' (ICTs) impact on society? An important approach is to empirically examine whether ICTs are empowering powerless groups. Empowerment is an enduring theme in Information and Communication Technology for Development (ICT4D) studies which refers to "the process through which individuals perceive that they control situations" (Rogers and Singhai 2003: p. 67). Scholars believe that ICTs can help

---

B. Zhou (✉)
School of Journalism, Fudan University, Shanghai, China
e-mail: zhoubaohua@yeah.net

© The Authors(s) 2015
A. Chib et al. (eds.), *Impact of Information Society Research in the Global South*,
DOI 10.1007/978-981-287-381-1_14

powerless, disabled, or marginalised groups in society to gain more opportunities to resolve personal and collective problems, to participate in decisions that affect their lives, and finally to earn control of their own lives (Elijah and Ogunlade 2006; Green 2008). By conducting empirical studies among women, farmers, poor, and other grass roots in the developing world, scholars have explored issues such as how the development and access to low-cost ICTs enhances access to accurate and reliable information for the poor (Elijah and Ogunlade 2006); how acquiring ICTs skills increases women's confidence, general knowledge, as well as knowledge of their rights (Green 2008); how ICTs can assist women in addressing the chronic issues of widespread poverty (Elijah and Ogunlade 2006); and how engagement with ICTs empowers grass roots to facilitate their civic engagement (Kwon and Nam 2009; Leung 2009), among others. Even so, a less studied issue concerning the empowerment role of ICTs is whether new technologies could help the marginalised group to voice themselves when they are faced with actual problems in their working and living situations.

The current study will address this question by examining the status of opinion expression of new-generation migrant workers in China and its relationship to ICTs. New-generation migrant workers, referring to those young rural–urban migrant workers aged under 30 who differ from their parents in their lack of farmer experience, low intention to return to home villages, and increasing reliance on ICTs, are typically marginalised people in post-reform China. This happens because they are poor, lack equal social welfare, are excluded by urban society, and are without significant power over their lives despite of their contributions to China's urbanisation. Compared to other social strata, new-generation migrant workers have less opportunities and skills to express their opinions in a traditional media environment, because they are inaccessible to such marginalised groups. Then, will new ICTs empower new-generation migrant workers by facilitating their intention to express their opinions via new media channels? Will expression via new media channels further encourage them to express via other channels such as interpersonal network or official institutions? These questions deserve careful empirical studies. Here, we treat facilitating opinion expression as one important dimension of empowerment of ICTs for migrant workers in China.

This chapter begins with literature review on the realistic needs of migrant workers, the expressive behaviours in the Chinese context, and the role of the Internet in facilitating opinion expression. Based on the discussions, we develop a set of specific hypotheses and research questions. Next, we quantitatively test our theoretical propositions with the data from a questionnaire survey we conducted in Shanghai of China. We conclude this chapter by discussing the implications of the research findings for academic research, governmental innovation, and corporation co-operations.

## 2 Literature Review

### 2.1 *Realistic Needs[1] of Migrant Workers*

Although the expression of opinion is conceived as the most basic form of political participation, prior empirical studies have showed that, in general, ordinary Chinese people are not active in expressing their opinions, especially in the public space, mostly due to the strict political control and low political efficacy (Pan et al. 2010; Chan and Zhou 2011). While scholars also argue that when people encounter realistic needs to resolve existential problems such as pollution in their living environment, their intention to engage in expressive behaviours would be higher (Yang 2009; Zhou 2011). So when studying the expression of migrant workers in China, it is better to examine the expressive intentions when they are faced with realistic needs or existential problems.

To focus on the realistic needs of migrant workers is also responding to the call for attention to the concept of "needs" in the ICTD research or "southern" communication studies. As Qiu (2010) put it, it is imperative for "southern" communication studies to tackle issues of needs (versus wants). "Needs" are basic resources or fundamental problem-resolving demands to help ordinary people especially marginalised groups to survive in the living world, to address issues such as job seeking, housing, education, health care, and social networking, while "wants" refer to those perceptions promoted by advertising and marketing campaigns, imposed by peer group pressure, and internalised as one's personal wants for "modern life" and "urban style", such as conspicuous consumption (Fitzgerald 1977). Following this conceptual distinction, Qiu (2009) argues that the most fundamental force to drive the information have less to adopt, and the use of new media technologies is the bottom-up needs of these marginalised people to deal with the existential issues created by the transformations of Chinese society, exacerbated by existing structural inequalities. The rise of working-class network society in China is, in this sense, a grounded transformation consisting of daily struggles by members of the have-less people to use working-class ICTs as micro-solutions to meet their needs at the grass-roots level. Hence, it is important to explore whether and to what extent migrant workers will voice themselves when they are faced with realistic problems.

What then are the main needs or realistic problems of new-generation migrant workers in the contemporary China? Prior empirical investigations, albeit limited in number, have showed two lines of them: one is the labour rights protection and the other is personal emotion problems (Chinese National Labour Union 2010; Wang and Chen 2012). The labour right protection problems refer to those problems that they are faced during the working process, including excessive working time, delay in paying, and working-related injury or illness. The personal emotion problems

---

[1] By "realistic needs", we emphasise that these needs are not "pseudo-needs" constructed by other social forms, such as the government or mass media, but stated by migrant workers themselves.

include emotional confusion, loneliness, and monotony, which are the most serious problems facing migrant workers when they live in cities without their traditional emotional support from their families and relatives. Based on the discussion, we propose the first research question:

RQ1: What's the pattern of expressive intention among new-generation migrant workers when they are faced with realistic problems?

Since the labour rights protection problems are more basic and crucial for migrant workers to stay in the cities compared with personal emotion problems, we hypothesise that:

H1: The expressive intention will be higher when migrant workers are faced with labour rights problems than with personal emotion problems.

## 2.2 Three Spaces of Opinion Expression in the Chinese Context

Citizens can express their opinions in different social spaces or via different channels. In the context of transformational Chinese society, scholars have conceptualised three distinct spaces of expression based on activity contexts, corresponding venues, and felt inhibitory forces: face-to-face interpersonal private space, institutional space, and hybrid new media space, referring to the Internet and mobile phones (Pan et al. 2010; Chan and Zhou 2011). The idea of expressive space comes from what Zhou (2000: p. 604) refers to as the "discourse universe". In a study of how Chinese journalists engage in duplicity with official ideology, Zhou (2000) identifies two distinct universes of discourse: (1) the public discourse made up of discursive activities in the mass media, public meetings, and other institutionalised settings and (2) the private discourse comprised of information communicated through personal networks. In the Chinese context, the former universe is inclined towards the official sphere while the latter is nonofficial in nature (Zhou 2008). According to Habermas (2002), the distinction between the private and public spheres is also marked by a set of communicative conditions. People's experiences are first interpreted "privately" within the networks of interaction found in families, circles of friends, neighbours, colleagues, and acquaintances before being articulated and expressed in the public realm. In other words, the expression of opinion in private space is much more convenient and requires the least resources. For migrant workers, when they are faced with realistic problems in their work or life, it is natural to expect that they will first express their opinions in their personal networks. Although they may receive strategic or emotional support from their family members or close friends, they will not get substantive solutions from a discussion with them since most of the power and resources for problem-resolving have been strictly controlled by institutions such as governments or employers.

Ironically, although theoretically it is more efficient to express opinions to official institutions for help, people seldom do so (Chan and Zhou 2011). Besides the more strict restriction and higher engagement cost, another important explanation is the lack of effective responses from the institutions. The official institutions such as the Labour Union are more like channels for propaganda, education, and social networking than for workers' rights protection, organisation, and participation (Hong 2003). As for the government's reaction, most of the complaints from the public never get responded. Wang (1997) found that only 1 % of the complaints sent by mails were replied by the governments. The situation is even worse for migrant workers since they are among the highly oppressed and marginalised groups in the stratified Chinese society. The lack of economic, social, and cultural capital has made them almost invisible in the real-life public sphere (Qiu 2009; Zhao 2008).

Under such circumstances, the Internet becomes a new expressive space for marginalised groups to voice themselves (Mitra 2001, 2004). Thanks to the technical advantages, the Internet, especially Web 2.0 featuring the user-generated content (UGC), has for the first time provided a platform for the marginalised people to express themselves and participate in politics (Mehra et al. 2004; Zheng 2008). There is no question that the Chinese Internet is subject to tight regulations which bar negative references to the top leadership, the legitimacy of the Chinese Communist Party (CCP), and other politically sensitive issues (Sohmen 2001). Nevertheless, it is fair to say that cyberspace is much more liberal than the state-controlled mass media and has raised opinion expression to an unprecedented level (Goldman 2005; Zheng 2008; Yang 2009). With the rapid expansion of the Internet across various social strata, it also gained a striking penetration among the group of new-generation migrant workers. Rather than "information have-nots", new-generation migrant workers have actively adopted and used ICTs, being capable of interacting with digital media, integrating them into their modes of communication and ways of life, and even domesticating ICTs to meet their needs and resolve their existential problems (Peng 2008; Lin and Tong 2008; Qiu 2009). In this sense, they are "information have-less". An earlier survey conducted in Shanghai found that 75.4 % of new-generation migrant workers are Internet users, among whom 74.5 % are reading bulletin board system (BBS), 65 % are posting messages on BBS or online forums, 76.4 % are reading blogs (including microblogs), and 65 % of them are writing their personal blogs or microblogs (Zhou and Lu 2011). This shows that in addition to being a convenient tool for migrant workers to access news online, the Internet has also become a site for their potential opinion expression, although there are no systematic empirical studies on the opinion expression among Chinese new-generation migrant workers until now, especially those studies focusing on the situation when they encounter realistic problems. More importantly, it deserves the examination of the emerging role of the Internet for the migrant workers' expression in the structure of three distinctive expressive spaces in China. Since the new media space was less restrictive than the institutional space but more restrictive than the interpersonal space, we propose the following hypothesis:

H2: When migrant workers are faced with realistic problems, they will be more likely to express their opinions via new media channels than institutional channels (H2a), though their expressive intentions in personal networks are still highest (H2b).

It should be noted that this study focused on expressive intentions rather than actual expressive behaviours, given that Chinese people in general seldom express themselves especially in the public space. Expressive intention is argued to be the most robust predictable variable for actual expressive behaviours (Fishbein and Ajzen 1975); hence, we adopted expressive intention measures as a proxy for real behaviour.

## 2.3 Factors Influencing Opinion Expression and the Role of the Internet

The extant literature based primarily on the evidence from developed democracies and upper-middle class in Chinese society suggests four basic sets of predictors of citizens' expressive engagement in public life: (1) the socio-demographic variables that index individuals' possession of material and, indirectly, social as well as expressive competence resources, (2) the psychological variables that index individuals' interest in public life and their attitudes towards involvement, (3) social interaction and social networking variables that index individuals' possession of social capital and placement in the social ecology that facilitate their engagement, and (4) media use and other communicative activities that index individuals' interactions with their information environment, which can be viewed both as a constituting part of individuals' engagement and a key venue to prepare them for engagement in other forms (e.g. Delli Carpini 2004; Jacobs et al. 2009; Verba et al. 1995; Lei 2011; Shen et al. 2009; Shyu 2009).

We argue that these factors are also important to influence the expressive intentions of new-generation migrant workers. First, we expect individual migrant workers to show higher levels of expressive engagement if they possess more material resources. As scholars have noticed (e.g. Qiu 2009; Zhou and Lu 2011), the issue of internal stratification or variation among migrant workers has been ignored in the extant literature. Although migrant workers could be treated and analysed as a group, when they are compared with other groups with higher socioeconomic status (SES), they are not a single homogeneous group in terms of their socioeconomic positioning and ICT connectivity. Instead, there is a significant internal variation and in-group stratification in terms of gender, education, income, profession, and ownership of employers.[2] How these internal variations influence their expressive

---

[2]The ownership of employers includes three models: state-owned, private-owned, and overseas investment (investment from Hong Kong, Macau, and Taiwan included).

intentions will be empirically examined in this study. Second, we expect those migrant workers perceiving greater seriousness of the problems they encounter and having higher levels of political efficacy to be more likely to express their opinions. Third, social networking and social capital are important resources for migrant workers to engage in expressive behaviours. Based on case studies, scholars (e.g. Guo et al. 2011) have argued that whether having everyday social network influences the action models of migrant workers when they are faced with serious problems, in the "atomized" situation where the primary social network is missing, they tend to resort to personal solutions such as suicide; the existence of social network (e.g. the hometown fellow organisations) will facilitate them to organise and take collective actions to protect their own rights. Therefore, whether having supporting social networks plays a critical role in influencing migrant workers' expressive intentions, carrying out broader and more frequent social interactions with others will positively relate to their opinion expression. Fourth, as an important information resource, news media use is expected to facilitate expressive behaviours of migrant workers.

The Internet has been viewed in a similar way. Scholars have argued that the Internet may contribute to citizens' expressive engagement because of its abundant and diverse information, open and democratic discussion, and "horizontal communication" (Polak 2005; Hung 2006; Shirk 2007). Empirically, in a meta-analysis of 38 studies with 166 effect estimates, Boulianne (2009) reports that online news use had significant positive effects on citizenship engagement. In China, the Internet has also been found to be positively related to engagement in general. For example, Lei (2011) showed using the 2007 World Value Survey Data that compared with exclusive traditional media users and non-media users, Chinese netizens were more likely to be opinionated, more likely to be "politicised"—that is, simultaneously embracing the norms of democracy and being critical of the current political conditions and the party-state—and more likely to have experiences in collective action. Chan and Zhou (2011) show Internet news use had significantly positive influence on opinion expression in the new media space, though it had no significant effect on offline expression.

Following these theoretical arguments, we expect that the Internet also has potential to facilitate expressive intentions among new-generation migrant workers via various processes. First, as a source for abundant and diverse information, the Internet can facilitate steady and fuller information flows which serves as the basic resources for migrant workers to express their opinions. For example, when they are faced with labour protection problems, they can easily find useful information online which can help them to express their opinions with more knowledge and confidence. Second, the Internet can serve as an openly accessible virtual space for migrant workers to interact with others and organise to voice their own opinions. Scholars have argued that the Internet becomes an important space for migrant workers to interact with each other and to maintain and extend their social networks (Law and Chu 2008; Lin and Tong 2008), which can make expression much easier, be it labour issues or emotional issues. Third, both the information acquisition and interpersonal interactions via the Internet may facilitate their offline expression. Online expression is actually an exercise for marginalised people to engage in public affairs, and this kind of exercise might even ease the fear about expression in public sphere. In other words, the Internet has the potential to encourage migrant workers to speak out when they are faced with realistic problems. Thus we will test the following hypothesis:

H3: The Internet use (e.g. online news use and communicative use) is positively related to the levels of expressive intention of new-generation migrant workers.

In addition, we tested if each of the following factors, higher socioeconomic status, perceived problem seriousness, political efficacy, social interactions, and news media use, is positively related to the levels of expressive intention of new-generation migrant workers.

## 3 Method

### 3.1 Survey

The data came from a survey of new-generation migrants conducted in Shanghai. To improve the representativeness of the sampling, this study sampled based on two variables—industry (manufacturing, construction, and service) and ownership of their workplace (state-owned, private-owned, and overseas investment including those from Hong Kong, Macau, and Taiwan). Firstly 13 companies[3] were selected as survey locations. At each selected location, we conducted either census survey (if there were 100 or less than 100 migrant workers in the selected company) or systematic random sampling survey (if there were more than 100 migrant workers in the selected company) on migrant workers aged from 16 to 30. By conducting self-administered questionnaire survey, we collected 1,000 questionnaires among which 869 were fully completed, which amounted to a response rate of 86.9 %.

### 3.2 Measures

Dependent variable: *Expressive intentions*. The respondents were asked to indicate how likely (1 = absolutely will not, 5 = absolutely will) they will express their opinions or appeals in each of the three types of channel (personal network, new media, and institutional) when they are faced with labour rights problems or

---

[3] Among them, seven are manufacturing companies, four are service, and two are construction ones. Three are state-owned, six are private, and four are overseas investment. They were chosen based on personal network. Although it is difficult to assess the accurate representativeness of this sample because of the lack of a census of new-generation migrant workers in Shanghai, the sample did show a sound representativeness by this stratified sampling design. Please refer to the demographics section for reference.

personal emotion problems. Separate factor analyses of 13 items measuring the respondents' expressive intentions under each problem domain were conducted, and each resulted in the following three factors as expected: expressive intentions via personal network (including discussion with family members and friends, workmates, or fellow townsmen), new media channels (including posting comments on the online forums or communities, on the blogs, on the QQ/MSN group, talking with friends on QQ/MSN, and sending or forwarding comments via SMS), and institutional channels (reporting to traditional media, employers, the Labour Union or the Youth League, government departments, and social organisations). An index of expressive intentions via each channel was subsequently created by averaging the scores among the items contained within this factor to represent the level of opinion expression concerning each problem (Cronbach's alpha ranged from .82 to .92).

Independent variables include:

**Demographics** We measured the following demographical variables: gender (male = 59.1 %), age ($M = 23.2$, $SD = 3.5$), education (45.7 % of the respondents graduated from high school, followed by junior high school with 25.9 %), average monthly income (36.8 % ranged from 1,501 to 2,000 RMB and 24.1 % ranged from 1,001 to 1,500 RMB), industry (manufacturing = 47.2 %, construction = 25.7 %, service = 27.1 %), and ownership of their workplace (state-owned = 24.6 %, private-owned = 35.3 %, and overseas investment = 30.1 %).

**Perceived Seriousness** Five items were used to measure the perceived seriousness of the labour rights problems and personal emotional problems the migrant workers came across (1 = almost no, 5 = very serious). The former included excessive working time, delay in paying, and working-related injury or illness ($\alpha = .73$). The latter included emotional confusion, loneliness, and monotony ($\alpha = .76$).

**Political Efficacy** Respondents were asked to what extent they agreed to the following four statements (1 = strongly disagree, 5 = strongly agree): "I have clear understanding of the problems that should be settled by the governmental policy"; "I'm able to make constructive suggestions on the policies and decisions of the government"; "every citizen like me can influence the policies and decisions of the government"; and "the government responds to the suggestions of citizens properly". Factor analysis showed only one factor indicating that new-generation migrant workers had no clear awareness of discriminating two aspects of political efficacy—internal efficacy and external efficacy—which are often distinguished in the related literature. Therefore, we averaged the scores among the items to represent the political efficacy ($\alpha = .81$, $M = 2.69$, $SD = .75$).

**Social Network** Eight questions were asked about how often (1 = almost not, 5 = very often) do new-generation migrant workers have contact with the following four groups in their work and life: fellow townsmen, migrant workers from other towns, local Shanghainese, and leaders or bosses of the workplace. Factor analysis was then conducted and resulted in two factors as expected: the homogeneous network (contacting with fellow townsmen and workmates; $\alpha = .77$, $M = 3.47$, $SD = .76$) and heterogeneous network (contacting with local people and leaders/bosses; $\alpha = .82$, $M = 2.23$, $SD = .85$).

**Traditional Media Use** Respondents were asked how closely (1 = almost no, 5 = very closely) they paid attention to international, national, local, and hometown news, respectively. The four items for TV were averaged into an index ($\alpha = .74$) and the four for newspaper were averaged into another ($\alpha = .70$).

**New Media Use** Online news use was measured by averaging the scores of frequency (1 = never, 5 = always when connected) of respondents' use of online domestic news, international news, local Shanghai news, and hometown news ($\alpha = .82$). Online interaction was measured by averaging their frequency of using QQ, other chat tools, and SMS ($\alpha = .54$). When analysing the expressive intentions via personal network and institutional channels, we averaged the two variables of online news use and online interaction to avoid the multicollinearity because of their high correlation ($r = .80$, $p < .001$). It should be mentioned that we did not include the ownership of cellphone into our model as 96.0 % of our samples own cellphones.

## 4 Results

### 4.1 Perceived Problem Seriousness

Firstly, we would like to report the basic situation of the problems that new-generation migrants come across in Shanghai. The survey shows (Table 1) that 78.4 % of new-generation migrant workers consider themselves to have encountered "a little" labour rights protect problems (among which the most serious one is excessive working time) and 91.4 % of them think that they have "a little" personal emotion problems (the most serious one is monotony). These results illustrate that the two types of issues defined here are common and realistic problems facing this group. Paired sample $t$-test shows that personal emotion problems are significantly higher than labour rights problems ($t = 13.0$, $p < .001$).

Table 1 Basic situation of the problems that new-generation migrant workers come across

|  | Mean (SD) | Percentage of at least having "a little" (%) |
|---|---|---|
| Labour rights problems ($N = 862$) | 1.96 (.85) | 78.4 |
| Delay in paying | 1.51 (.85) | 32.2 |
| Working-related injury/illness | 1.87 (1.01) | 51.6 |
| Excessive working time | 2.49 (1.26) | 73.4 |
| Personal emotion problems ($N = 858$) | 2.41 (.84) | 91.4 |
| Emotional confusion | 2.29 (.98) | 74.1 |
| Loneliness | 2.33 (1.01) | 76.3 |
| Monotony | 2.62 (1.05) | 84.6 |

## 4.2 Expressive Intentions

We were interested in new-generation migrant workers' expressive intentions when they came across problems above (*RQ1*). Table 2 shows that when they come across labour rights problems, 90.5 % will express via personal networks, 77.8 %

Table 2 Expressive intentions of new-generation migrant workers

|  | Labour rights problems |  | Personal emotion problems |  |
| --- | --- | --- | --- | --- |
|  | Mean (SD) | Percentage of at least "might do occasionally" (%) | Mean (SD) | Percentage of at least "might do occasionally" (%) |
| *Personal networks* | 3.39 (.92) | 90.5 | 2.98 (1.00) | 79.3 |
| Discussion with family members or friends | 3.31 (1.19) | 76.5 | 3.08 (1.27) | 67.7 |
| Discussion with workmates | 3.48 (.98) | 87.1 | 2.91 (1.10) | 67.8 |
| Discussion with fellow townsmen | 3.37 (1.04) | 82.9 | 2.95 (1.13) | 68.2 |
| *New media* | 2.48 (.97) | 77.8 | 2.25 (.92) | 72.0 |
| Posting on online forums or communities | 2.35 (1.16) | 41.6 | 1.90 (1.02) | 25.6 |
| Comment on the blogs | 2.22 (1.14) | 36.0 | 1.92 (1.05) | 25.8 |
| Talking to friends via QQ/MSN | 2.54 (1.25) | 49.0 | 2.42 (1.23) | 48.1 |
| Posting on QQ/MSN group | 2.31 (1.20) | 40.0 | 2.17 (1.19) | 35.3 |
| Sending or forwarding comments via SMS | 2.94 (1.23) | 63.3 | 2.82 (1.25) | 60.6 |
| *Institutional channels* | 2.28 (.95) | 66.2 | 1.73 (.82) | 35.1 |
| Reporting to traditional media | 2.02 (1.08) | 29.1 | 1.70 (.92) | 18.2 |
| Communicating with the employers | 2.58 (1.17) | 53.0 | 1.80 (.97) | 23.1 |
| Reporting to the Labour Union or the Youth League | 2.28 (1.14) | 40.3 | 1.72 (.89) | 19.2 |
| Reporting to related government department | 2.21 (1.13) | 37.1 | 1.66 (.92) | 17.3 |
| Reporting to other social organisations | 2.23 (1.11) | 39.3 | 1.75 (.96) | 20.4 |

Note: *N* ranges between 830 and 861. This table is compiled by the author

would like to express via new media, and 66.2 % choose to express through institutional channels. At the same time, when they are faced with personal emotion problems, 79.3 % will express via personal networks, 72.0 % would like to express via new media, and 35.1 % choose to express through institutional channels. This indicates that the expressive intentions of new-generation migrant workers in Shanghai are relatively high when they come across realistic problems in work or life. Paired sample tests show that their expressive intentions when they are faced with labour rights problems are significantly higher than when they are faced with personal emotion problems ($t = 13.0$, $p < .001$ [personal networks], $t = 8.3$, $p < .001$ [new media], $t = 18.0$, $p < .001$ [institutional channels]), although the perceived seriousness of the former is lower than the latter. This could be explained by two reasons: firstly, labour rights problems are more related to basic needs of migrant workers (e.g. health, the income to support a family); secondly, compared with personal emotion problems, labour rights problems are more public and easier to arouse sympathy and resonance; thus they are more proper to be discussed in the public sphere. H1 is thus supported.

In terms of expression channels, results show that the new-generation migrant workers are more willing to express through personal networks regardless of the type of problem they are facing. Expressive intentions through new media come next (the differences between these two channels are significant, $t = 24.1$, $p < .001$ [labour rights problems], $t = 19.0$, $p < .001$ [life and emotion problems]). Institutional channels take the last place (the differences between institutional channels and new media channels are also significant, $t = 6.3$, $p < .001$ [labour rights problems], $t = 17.6$, $p < .001$ [life and emotion problems]). It illustrates that when new-generation migrant workers come across realistic problems, they tend to turn to primary group and personal network for support. At the same time, new media is becoming a vital expression space which includes not only the interpersonal contact tools for long-distance interaction like QQ and cellphones, but also online forums, blogs, and QQ groups which are more public and have the potential to transform private expression into public expression. In contrast, new-generation migrant workers do not like to express opinions or ask for help through institutional channels, although many of their problems need to be settled by their company, Labour Union, and even the government. This pattern supports H2.

## 4.3 Influencing Factors of Expressive Intentions

We apply OLS regressions to analyse the influence of independent variables on the expressive intentions of new-generation migrant workers (Table 3). First of all, personal background has some impact on the expressive intentions. The major influences come from three variables—gender (females are more likely to express personal emotion problems through personal networks than males while males are more inclined to express through institutional organisations), age (the elder are more willing to express personal emotion problems through personal networks),

**Table 3** OLS regression: predicting the expressive intention of new-generation migrant workers

|  | Personal networks | | New media | | Institutional channels | |
|---|---|---|---|---|---|---|
|  | Labour rights | Personal emotion | Labour rights | Personal emotion | Labour rights | Personal emotion |
| *Personal background* | | | | | | |
| Gender (male = 1) | −.074 | −.127** | −.025 | .009 | .155** | .143** |
| Age | .085 | .098* | .069 | .021 | .012 | .032 |
| Education | .022 | .034 | .024 | .042 | −.055 | .004 |
| Income | .054 | .041 | −.024 | −.043 | .023 | −.074 |
| Private-owned (state-owned =0) | −.008 | −.018 | −.049 | −.079 | −.028 | −.020 |
| Overseas investment (state-owned =0) | .057 | .026 | .067 | −.021 | .110 | −.086 |
| Manufacturing (service = 0) | −.007 | −.026 | −.034 | −.143** | .157** | .000 |
| Construction (service = 0) | .140* | .112 | −.047 | −.012 | .058 | −.083 |
| *Psychological factors* | | | | | | |
| Perceived seriousness | −.001 | .089* | .062 | .091* | .034 | .125** |
| Political efficacy | .068 | .137** | .015 | .100* | .125** | .146** |
| *Social network* | | | | | | |
| Contact with fellow townsmen or workmates | .278*** | .245*** | .041 | .053 | .032 | −.052 |
| Contact with leaders or bosses | −.087 | .003 | .058 | .098 | .142** | .180*** |
| *Traditional media* | | | | | | |
| Newspaper news | −.014 | .010 | .008 | −.023 | .067 | .171*** |
| TV news | .136** | .068 | −.090 | −.153** | .176*** | .021 |
| *New media* | | | | | | |
| Online news | .085 | −.068 | .228*** | .217*** | .009 | .017 |
| Online interaction | .207*** | .158** | | | | |
| Total $R^2$ (%) | 14.3*** | 12.4*** | 19.9*** | 23.5*** | 16.7*** | 16.1*** |
| N | 576 | 574 | 446 | 444 | 573 | 570 |

Note: Figures in this table are standardised regression coefficients. This table is compiled by the author
*$p < .05$; **$p < .01$; ***$p < .001$

and industry (compared with workers of service industry, manufacturing industry workers express less personal emotion problems in the new media space while they assert more demands via the institutional channels about labour rights problems; construction industry workers discuss labour rights problems more often in the private space).

Secondly, psychological factors have significant impact on expressive intentions. Nevertheless, the impact of perceived seriousness differs from issue to issue—only significant with the personal emotion problems, while the perceived seriousness of labour rights problems has no significant influence on expression. The influence of political efficacy is also more powerful for personal emotion problems, though it has significant influence on the expressive intention through institutional channels when they are faced with labour rights problems ($\beta = .125$, p $< .01$).

Thirdly, the social network shows significant influence while mainly for expressive intentions through offline spaces. The more frequent the interaction with fellow townsmen or workmates, the stronger the intention to express through personal networks ($\beta = .278, p < .001$ [labour rights], $\beta = .245, p < .001$ [life and emotion]). More frequent interaction with local Shanghainese or leaders and bosses promotes their intention to express through institutional channels ($\beta = .142, p < .001$[labour rights], $\beta = .180, p < .001$ [life and emotion]).

Fourthly, TV news has stronger influence than newspaper—the latter only has significant influence on expression through institutional channels about personal emotion problems. It has significant influence on the expressive intention through both personal networks and institutional channels when they are faced with labour rights problems. This may be due to the higher frequency of TV news exposure ($M = 2.44$, SD $= 1.63$, while for newspaper news, $M = 1.18$, SD $= 1.69$) and that watching TV is more like a kind of collective behaviour (this survey shows that 61.9 % of migrant workers watch TV together) thus setting creating a space for them to communicate with each other, although more TV news use constrains their intentions to express personal emotion problems via new media.

Last but not least, when all other variables are controlled, online news and online interaction are still the major influencing factors of expressive intention via new media ($\beta$ ranges from .158 to .228 and significantly at between .01 and .001). This illustrates that more Internet news browsing and more active Internet interaction encourage stronger intention to express their appeals via new media. Yet, consistent with previous studies (Chan, and Zhou 2011; Pan et al. 2010), no direct influence of the Internet use on offline expression is found in this study. H3 is thus partially supported.

## 5 Conclusion

The current study empirically investigates the intention of expression when faced with realistic problems among new-generation migrant workers in Shanghai. The main findings are as follows.

First, when faced with realistic problems, the intention of expression among new-generation migrant workers in Shanghai is relatively high on the whole. Second, consistent with the general public, the expressive channels of this group decrease progressively in a way of "personal networks–new media–institutional channels". Third, internal diversity is found in the expression of new-generation migrant workers in terms of gender, age, and industry where they work. Fourth, the influence of psychological factors varies with issues—it only has significant influence on the intention of expression concerning personal emotion problems; psychological factors have little influence on expression of labour rights problems, except for the impact of political efficacy on expression via institutional channels. Fifth, social network is found to be a vital factor in the matter of offline expression in the way that homogeneous network (i.e. contact with fellow townsmen and workmates) facilitates expression through personal network while heterogeneous network (i.e. contact with local people and employers) facilitates expression via institutional channels. Sixth, exposure to TV news increases the likelihood of offline expression on labour rights problems. Lastly, online news and online interaction have significant impact on expression through new media but have no direct impact on offline expression.

Based on the above empirical findings, we could better understand the impacts and implications of the new media for the new-generation migrant workers in China. First, new media channels, including multiple online platforms and mobile phones, have become an important space for the migrant workers to voice themselves when they are faced with actual problems. Since expressive intention is believed to be the strongest predictor of real expressive behaviours according to the classic theory of reasoned action (Fishbein and Ajzen 1975), the higher expressive intention through new media channels could be transferred into their actual opinion expression through these platforms, which may have some potential for the marginalised group to change their personal as well as collective lives. In this sense, we could say that the Internet has the potential to empower the migrant workers in China in terms of making them and their lives more visible. Second, online news exposure and online interaction have been proven to have positive impacts on the expressive intentions among migrant workers, which show the significance of these two kinds of online behaviours. Our study also cautions that there are no direct linkages between ICTs use and offline expressive intentions, which suggests that more concrete actions should be taken to promote the online expressions to offline actions. In other words, the online activities should be combined with offline actions to improve the motivation for the migrant workers to voice themselves.

This study has implications for academia, policy, and practitioners. In terms of theoretical contributions, it is the first empirical study to examine the status of opinion expression among new-generation migrant workers and how it is influenced by various factors especially Internet use. By focusing on a specific social group, this study has confirmed the spatial structure of "personal networks–new media–institutional channels" for people to express opinions when they are faced with personal or social problems. It has also explicated the different influential patterns for two kinds of realistic needs—resolving labour rights problems and personal emotion problems—which suggest that they require different resources to facilitate expressive intentions. It has also shown that the Internet could empower migrant workers at least in two ways—as an expressive space itself and as a positive factor to encourage online expressions. Thus, this study endeavours to operationalise "empowerment" from the angle of opinion expression in the Chinese context.

For policymakers and social activists, they may obtain inspiration from these findings to consider designing effective campaigns via online platforms to facilitate the active opinion expression among Chinese new-generation migrant workers. For example, nowadays the Chinese government is launching a new campaign to promote "e-government". How can the migrant workers benefit from this campaign? Can they interact with governmental labour offices to protect their rights via "e-government" social media such as *Weibo* or *WeChat*? Can their voices be heard by the government? Our study suggests that the government should make efforts to build online platforms for the migrant workers to express their opinions when they are faced with problems and develop an efficient system to receive and respond to these feedback. For the practitioners (i.e. online corporations), highly active opinion expression is a necessary foundation for the energy of the online community. Given that the number of new-generation migrant workers in China is huge, to build an online community for them to engage in expressive behaviours is very valuable. It is not only the social accountability of big companies but also a new opportunity for them to develop a booming online market.

This study is a quantitative study to show the basic pattern of migrant workers' expressive intentions and their influential factors. For the future studies, we suggest more empirical work, especially qualitative ones, could be done to further explicate the mechanism of how the Internet facilitates the migrant workers' expressive intentions.

**Acknowledgements** This chapter is supported by the SIRCAII project, Shanghai Social Science Foundation Project (#2012BXW004) and Shanghai Pujiang Talents Program (#13PJC020). The author wishes to thank Dr. Jack Linchuan Qiu for his valuable collaborations and suggestions during the whole research, and thank Arul Chib, Julian May, Roxana Barrantes, Ang Peng Hwa, and Roger Harris for their insightful comments on the earlier manuscript. The author also thanks two RAs Shuning Lu and Miao Xiao for their help in the fieldwork.

**Open Access** This chapter is distributed under the terms of the Creative Commons Attribution Noncommercial License, which permits any noncommercial use, distribution, and reproduction in any medium, provided the original author(s) and source are credited.

# References

Boulianne, S. (2009). Does Internet use affect engagement? A meta-analysis of research. *Political Communication, 26*(2), 193–211.
Chan, J. M., & Zhou, B. (2011). Expressive behaviors across discursive spaces and issue types. *Asian Journal of Communication, 21*(2), 150–166.

Chinese National Labor Union. (2010, June 21). A research report on new generation migrant workers. *Workers' Daily*.
Delli Carpini, M. X. (2004). Mediating democratic engagement: The impact of communications on citizens' involvement in political and civic life. In L. L. Kaid (Ed.), *Handbook of political communication research* (pp. 395–434). Mahwah: Lawrence Erlbaum Associates.
Elijah, A. O., & Ogunlade, I. (2006). Analysis of the uses of information and communication technology for gender employment and sustainable poverty alleviation in Nigeria. *International Journal of Education and Development Using ICT, 2*(3), 45–69.
Fishbein, M., & Ajzen, I. (1975). *Belief, attitude, intention and behavior: An introduction to theory and research*. Reading: Addison-Wesley.
Fitzgerald, R. (1977). *Human needs and politics*. Sydney: Pergamon Press.
Goldman, M. (2005). *From comrade to citizen: The struggle for political rights in China*. Cambridge, MA: Harvard University Press.
Green, J. H. (2008). Measuring women's empowerment: Development of a model. *International Journal of Media and Cultural Politics, 4*(3), 369–389.
Guo, Y., Shen, Y., Pan, Y., & Lu, H. (2011). The struggle of migrant workers and the transformational labor-capital relationship in contemporary China. *21th Century Review, 4*, 4–14.
Habermas, J. (2002). Civil society and the political public sphere. In C. Calhoun, J. Gerteis, J. Moody, & S. Pfaff (Eds.), *Contemporary sociological theory* (pp. 358–376). Oxford: Blackwell.
Hong, Y. (2003). On political efficacy in China. *Academic Exploration, 7*, 42–45 (In Chinese).
Hu, R. (2008). Social capital and political participation in an urban city. *Jounal of Sociology, 5*, 142–159 (in Chinese).
Hung, C. F. (2006). The politics of cyber participation in the PRC: The implications of contingency for the awareness of citizens' rights. *Issues and Studies, 42*(4), 137–173.
Jacobs, L. R., Cook, F. L., & Delli Carpini, M. X. (2009). *Talking together: Public deliberation and political participation in America*. Chicago: University of Chicago Press.
Kwon, K., & Nam, Y. (2009). *Instrumental utilization of ICTs in mobilization processes of political collective actions: In the context of grassroots protest of Korea 2008*. Paper presented at the annual meeting of the International Communication Association (pp. 1–31), Chicago.
Law, P. L., & Chu, W. C. R. (2008). ICTs and migrant workers in contemporary China. *Knowledge Technology and Policy, 21*(2), 43–45.
Lei, Y. W. (2011). The political consequences of the rise of the Internet: Political beliefs and practices of Chinese netizens. *Political Communication, 28*(3), 291–322.
Leung, L. (2009). User-generated content on the internet: An examination of gratifications, civic engagement and psychological empowerment. *New Media and Society, 11*(8), 1327–1347.
Lin, A., & Tong, A. (2008). Mobile cultures of migrant workers in Southern China: Informal literacies in the negotiation of (new) social relations of the new working women. *Knowledge Technology and Policy, 21*(2), 73–81.
Mehra, B., Merkel, C., & Bishop, A. P. (2004). The internet for empowerment of minority and marginalized users. *New Media and Society, 6*(6), 781–802.
Mitra, A. (2001). Marginal voices in cyberspace. *New Media and Society, 3*(1), 29–48.
Mitra, A. (2004). Voices of the marginalized on the internet: Examples from a website for women of South Asia. *Journal of Communication, 54*(3), 492–510.
Pan, Z., Jing, G., Yan, W., & Zheng, J. (2010, June). *Understanding expressive engagement in urban China: Differentiating domains, settings, and media effects*. Paper presented at the annual conference of the International Communication Association, Singapore.
Peng, Y. (2008). Internet use of migrant workers in the Pearl River Delta. *Knowledge Technology and Policy, 21*(2), 47–54.
Polak, R. K. (2005). The Internet and political participation: Exploring the explanatory links. *European Journal of Communication, 20*(4), 435–459.
Qiu, J. L. (2009). *Working-class network society: Communication technology and the information have-less in urban China*. Cambridge, MA: MIT Press.

Qiu, J. L. (2010). Southern imagination: Class, network, and communication. *Chinese Journal of Communication Research, 12*, 51–69.

Rogers, E., & Singhai, A. (2003). Empowerment and communication: Lessons learned from organizing for social change. *Communication Yearbook, 27*, 67–86.

Shen, F., Wang, N., Guo, Z., & Guo, L. (2009). Online network size, efficacy, and opinion expression: Assessing the impacts of Internet use in China. *International Journal of Public Opinion Research, 21*(4), 451–476.

Shirk, S. (2007). *China: Fragile superpower*. Oxford: Oxford University Press.

Shyu, H. (2009). Psychological resources of political participation: Comparing Hong Kong, Taiwan, and Mainland China. *Journal of International Cooperation Studies, 17*(2), 25–47.

Sohmen, P. (2001). Taming the dragon: China's efforts to regulate the Internet. *Stanford Journal of East Asian Affairs, 1*, 17–26.

Verba, S., Schlozman, K. L., & Brady, H. E. (1995). *Voice and equality: Civic voluntarism in American politics*. Cambridge, MA: Harvard University Press.

Wang, J. (1997). *On government power*. Beijing: China Fangzheng Publishing (In Chinese).

Wang, F., & Chen, Y. (2012). From potential users to actual users: Use of e-government service by Chinese migrant farmer workers. *Government Information Quarterly, 29*, 98–111.

Yang, G. (2009). *The power of the Internet in China: Citizen activism online*. New York: Columbia University Press.

Zhao, Y. (2008). *Communication in China: Political economy, power, and conflict*. Lanham: Rowman & Littlefield.

Zheng, Y. (2008). *Technological empowerment: The Internet, state, and society in China*. Stanford: Stanford University Press.

Zhou, H. (2000). Working with a dying ideology: Dissonance and its reduction in Chinese journalism. *Journalism Studies, 1*(4), 599–616. doi:10.1080/146167000441321.

Zhou, H. (2008). SMS in China: A major carrier of the nonofficial discourse universe. *The Information Society, 24*(3), 182–190. doi:10.1080/01972240802020101.

Zhou, B. (2011). Media exposure, civic participation and political efficacy: An empirical study on "Xiamen PX Event". *Open Times, 5*, 123–141 (In Chinese).

Zhou, B., & Lu, S. (2011). An empirical study of ICTs and new generation migrant workers in Shanghai. *Journalism Quarterly, 2* (In Chinese).

# Impact of Research or Research on Impact: More Than a Matter of Semantics and Sequence

**Julian May and Roxana Barrantes**

## 1 Introduction

The rich data and analysis contained in this volume permit a number of lessons to be drawn, some of which resolve the questions identified by the authors and others which may inspire new inquiries. Looking at the variety of topics, disciplines and areas of research, a palpable conclusion is that ICT now pervades the lives of people throughout the world. This is unsurprising since ICTs are general purpose technologies, widely adopted, used and adapted by people of all ages worldwide and who are from different socioeconomic backgrounds. By focusing on the global south, however, one outstanding characteristic of the research relates to the different pace of use and appropriation within these countries. This is apparent in both the theoretical chapters and the evidence gathered from the empirical ones. Whatever the pace, it is apparent that ICT is having an impact on most economic sectors and social processes. This impact may be positive or negative and is not independent of the context into which ICT is inserted or the existing socioeconomic dynamics of that context.

In the concluding chapter to an earlier edited volume, Harris and Chib (2012) propose three possible areas through which research may have influence: capacity development, socioeconomic benefits and policy impact. We add a fourth and suggest that the lessons drawn from the research contained in this book also

---

J. May, Ph.D. (✉)
Institute for Social Development, University of the Western Cape, Bellville, Western Cape, South Africa
e-mail: jmay@uwc.ac.za

R. Barrantes, Ph.D.
Instituto de Estudios Peruanos, Lima, Peru
e-mail: roxbarrantes@iep.org.pe

© The Author(s) 2015
A. Chib et al. (eds.), *Impact of Information Society Research in the Global South*,
DOI 10.1007/978-981-287-381-1_15

contribute towards theory and, in some instances, reveal areas in which further theory building is required. In this chapter we will focus on policy and theory impacts.

In the realm of policy influence, more than simply building appropriate linkages to policymakers, the media and development practitioners is required. The timing and content of these linkages must be considered and better understanding is needed of the politics and process of policy. Without this, significant resources may be invested in research but without proper communication, and engagement with multiple stakeholders, lessons arising from the gathered evidence may be lost, limited to circles with no real influence into either policy or theory, or misused. This is true of any topic and the case of ICT is no exception.

In the case of theory, it is apparent that the studies in this volume fall into two groups. There are those which find that ICT amplifies existing structures and dynamics as is proposed by Agre (2002) and more recently Toyama (2011), and thus, existing theory may need only extension. There are those in which ICT reconfigures existing structures and dynamics as proposed by Dutton (2004) and supported by interventionists such as Brewer et al. (2005). In these cases, existing theory may be inadequate, may be out of date or does not exist.

## 2 Research on Impact

The objective for many of the chapters in this volume is identifying the evidence concerning the impact of ICT on society and its transformation. The volume begins to collect and draw together evidence about the impact that ICTs are having and, as such, contributes new findings to the already vast body of knowledge that informs about how ICTs are contributing to society. In this respect, it is generally recognized that there are both social and economic dimensions of change and, further, that consideration should be given to both actual and perceived changes in deprivation. This is a substantial question that requires a more sustained research effort than the one undertaken by the SIRCA II research cycle. As the chapter by Flor reminds us, ICT can have an impact on sectors (such as agriculture, health or education) and on development themes (such as poverty, gender or governance). The mechanisms through which this impact takes place are complex, as are the sectors and themes in which the impact is to be felt. In addition, information and communications is a fast-growing economic sector in its own right and directly contributes towards the growth of a quaternary sector of knowledge-based services.

Attribution of change to ICT, whether positive or not, is thus complex and links of causation may flow in different directions. As with any attribution attempt, it is necessary to separate outputs, outcomes and impacts and to decide which is feasible to measure, as well as which is necessary before conclusions can be reached concerning the contribution of the intervention. Thought must also be given as to what is meant by 'causation' and whether it can be convincingly demonstrated. There are also lags between ICT interventions and their impact, and these lags in turn may result in confounding influences that further test the ability to determine causation and the mechanisms through which change is brought about. Finally, as Flor observes, there are unintended consequences to which any intervention is prone. All this should be expected for as policy analyst Ray Pawson (2006:35) comments, interventions are 'complex systems thrust amidst complex systems'. Further, in these systems they compete with prior, existing and new interventions for impact.

The field of ICTD possibly has an advantage over other forms of research in development studies as it is relatively specific about the link to be investigated between the intervention and its intended outcome. By keeping people communicated and informed, the underlying expectation is that society will be strengthened in some way and individuals will not only be better off but will feel better as well. Taking advantage of this, an important step that is consciously made in several of the chapters attempting to trace the link from an ICT activity to change is the deconstruction of how such links manifest.

Questioning how an ICT activity manifests, several authors in this volume have proposed options. As an example, Dodel and Ramirez emphasize a triad: access, use and appropriation. Thus, for ICT to bring benefits to individuals, households, communities and society in general, ICT must be accessible, used and appropriated – meaning that users both adopt and adapt ICT to meet their needs. This then can increase well-being (following a mainly economics perspective) or augment capabilities and freedoms (following a Sen-oriented framework). Also in this volume, Diga and May speak of access, ownership and usage. Elsewhere Heeks (2010) has proposed a similar triad of readiness, availability and uptake. In each case ICT cannot simply be present in order to have an impact and must satisfy additional criteria of usage.

For the sake of brevity, we will focus our discussion of research on impact in the following domains: education, gender and society, poverty and political participation. Regarding education, useful evidence is provided from Peru, where achievement is very low representing a deep developmental problem. Through participatory action research (PAR) the team set out to adapt pedagogical tools to online platforms, working directly with school teachers. These were initially disconcerted when confronted with the technologies, but their confidence improved along their computational skills. An important lesson that comes out of the study is that a key interaction that should be taken into consideration when designing educational materials is the one between teachers and software developers.

Two studies inform our conclusions on gender and society, both from Asia. Being in the forefront of IS development is an aspiration shared by many countries, the Philippines being not exception. We learned that legislation may not keep up when there is a clash between how ICTs are used for a living, i.e. cybersex, and society's efforts to fight cybercrime. Understood as affective labour, cybersex challenges moral-based perceptions about crime and poses the question as to why it should be penalized, especially in a context in which women have few decent employment alternatives.

We have evidence from Indonesia showing how effective social media can be to boost women's entrepreneurship abilities and outcomes because of the flexibilities it offers. In turn, this percolates positively into women's position in the family and household. Women may feel empowered due to their social media use.

Diga and May's review of the literature suggests that evidence is leaning towards ICT on average reducing poverty rather than exacerbating it. This work shows that ICT access among poor people is now high in many countries and is boosted by shared ownership of handsets and in some instance SIM cards. Numerous applications now exist for mobile phones which are tailored towards the needs of the poor, with ways of transferring money conveniently and at low cost being one of the most frequent uses. They comment on the high willingness of poor people to commit resources to ICT which can run at around 25 % of income and they show that the poor spend a greater share of their income on ICT than the nonpoor. This could be a significant drain on household resources if this usage does not result in income generation or cost reductions elsewhere. This is not necessarily impoverishment since information and communication may bring about greater freedoms in other dimensions of well-being. Alternatively it could mean improvements in the quality of life of some within a household at the cost of others and may have particular gendered consequences. More direct measurement of well-being would help resolve this, perhaps through indicators such as those used for happiness studies might be an option.

Other chapters have traced the link between ICT and well-being. The case study in the Ivory Coast reveals that mobile phones are no panacea since they are used by people already engaged in specific social and power relations, as is the case with any other ICT. We learn that women take advantage of the mobile for reasons similar to those heard of in other places: security, reduced coordination costs, avoided transportation costs and so forth. But we also learn that some people may inhibit use due to their traditional beliefs, some of which contribute towards a fear of the technology.

Regarding political influence, chapters in the volume have revealed that ICT can affect both politics and policies and, equally, that ICT can affect governance and not simply government. These terms are often used interchangeably in the literature, including in this volume, but the distinction is important. Policies might not be an issue, their proper implementation may well be. Supporting this view, ICT is often blamed, by repressive governments, or thanked, by community organizations, for enabling social contact for political mobilization. Two research projects, one from Brazil and the other from China, show this. The Brazilian study focuses on political consultations processes aided by ICT, which is a promising tool to encourage participation and strengthen democratic processes as a by-product. Different results are found depending on which platform is used, the requisites that each one demands and the functionalities allowed, particularly taking into consideration the stage in the consultation process in which the Web 2.0 tool is used. For instance, in the initial stages of agenda setting, platforms allowing for comments and social interaction may be recommended.

In China, where migrant workers face a whole set of hurdles to legally work outside their place of origin, through ICT, the intention of expression of young migrant workers is considered high, beginning with interpersonal communication in social media, up to institutional channels, which are less personal. This finding is important since young people find a way of expression, through ICT, in an otherwise closed society, facing a difficult predicament because of migration. What is apparent from these case studies is that it is not appropriate to view either ICT or (D)evelopment as being apolitical and instead ICT is embedded in society/polity.

## 3 Impact of Research

Several authors emphasize that impact of research is a process which begins when the research project is designed. This has resonance with the chapter by Harris and could be separated into the intent to influence, the means to influence, the opportunity to do, the absorptive capacity of those being influenced, their willingness to be influenced, their ability to influence others in power and the extent to which influence can endure into the creation or reform of policy and the implementation of policy. In considering this, several authors note that when all relevant stakeholders are engaged from the onset, research and communication are better linked, and the results informed by policy considerations and as a result are rapidly taken into consideration in policymaking. This demands different attitudes and skills from researchers as well as a new way to schedule research activities. Not only mastering research methodologies and being at the forefront of research methods are in order, but the ability to communicate and convey ideas in lay terms. Further, these processes of moving research findings into policy do not occur in a vacuum but in the institutional framework in each country or subnational partition. These may be functioning, well established and integrated, but can be partial, politicized and disconnected.

This is further complicated by the modes and protocols of communication that are appropriate to the sectors being investigated. For example, the chapters show how work on educational issues and professionals is different than from that which involves interacting with health specialists. In this volume, this issue is illustrated by the discussion on how ICT affects the different logics of health care: choice or care. The case teaches us the importance of an open mind when designing and implementing interventions and the necessity to regard the cultural dimensions of the people involved, that is, those whose well-being is supposed to increase by the intervention.

Wider issues concerning policy impact are commented on by several of the authors who problematize the question. These authors question what policy can be influenced by research, when and under what circumstances as well as the extent to which such influence is desirable. Furthermore, governments are by no means the most important sites of decision-making in ICT in which a relatively small number of large multinational companies provide services globally. As such, the priorities of different stakeholders will differ, and those of the private service providers accountable to international shareholders are sure to differ from those of governments accountable to their citizens and local businesses.

Addressing this is tricky. As is pointed out by Ordóñez in this volume, if research is to be useful, it must be timely. This means that appropriate research questions may have to be determined in advance of their need by policymakers. Policymakers are under pressure to deliver and cannot necessarily wait for the gestation and delivery of rigorous research. Thus, time must be allowed for data collection, analysis and interpretation and the development of appropriate communication strategies that feed into the policy cycle of government. The point of intervention into the policy cycle is by no means clear and will vary both by the policy under review and according to the structures of governance and over time. Further, over the medium to long term, there are distinct policymaking episodes when governments may be more open to research.

Addressing these issues is not without challenge. Funding is not necessarily available for as yet unfelt needs, while researchers are not necessarily integrated into governance structures and may not know the appropriate points through which to enter the policy dialogue. In addition, most researchers must also manage their own academic cycles and obligations, and these may conflict with those of the government. As a result it should also not be taken as a given that policymakers are the only or prime audience of researchers. There are other users of research which researchers may wish to prioritize, including their peers and the users of ICT themselves. An interesting pathway for research influence is implicit in the discussion by Steibel and Estevez on ICT and public participation. Thus, researchers who wish to be activists in taking forward their findings could use such processes to bring their findings to the attention of both citizens and policymakers.

Once these issues have been resolved, the results must be wisely communicated but in a manner that is nonetheless accessible to the audience, whether decision makers or beneficiaries. Pathways to influence risk becoming oversimplified to marketing exercises and the development of packages which make research results appear palatable. There is also the concern that research findings may be presented as being more conclusive than the data actually permit or that only research questions which lend themselves to apparently conclusive results are deemed worthy of policy influence.

These challenges are well documented elsewhere (e.g. Boaz and Pawson 2005) and it is important that donors of research funding realize that longer research cycles may be needed in order to deal with these complexities. These should specifically include a component which focuses on building pathways to submit results into debates over policy and its implementation. In some cases, this is already recognized and a few grant-making bodies now require that reporting on research output and impact extends well beyond the period that has been funded. The United Kingdom's Economic and Social Research Council (ESRC) is an example.

Turning now to the impact of research on theory building, ICT4D is a relatively new topic and various contributions in the book address the novelties associated. At the same time, the chapters have demonstrated opportunities to use existing theory in application to ICT with a diverse theoretical base drawn from Habermas, Spinoza, Bourdieu and Putnam. Theory development also offers a way of resolving some of the difficulty of attribution mentioned above. By developing and testing theories of change that consider how, why and under what circumstance an anticipated impact of ICT may take place, attribution can at least be deduced if not measured.

The chapters in the book deal with a variety of approaches of development, from which several concepts are used – and these are not necessarily interchangeable. From the mainstream MDG discussion, and the need to build theories of change along the unintended consequences of ICT interventions, to the discussion around the inextricably linked relationships between ICT and poverty, the first part is mainly theoretical, dwelling with relevant concepts and methodological approaches to include ICT as a part of answers to development challenges. In addition, much is made of putting the D back into ICTD. But 'D' has always been a contested notion. ICTD is equally of relevance in wealthy countries in which there is significant poverty, and development is a concern in wealthy countries which are consuming unsustainable quantities of the world's non-renewable assets. The ICTD debate also should begin to separate out what is development from what is development studies, the latter being a field of critical inquiry concerning the contemporary dynamics of social, economic, political and population change that makes use of interdisciplinary conceptual frameworks and methods to identify development destinations that are possible and desirable, for whom and at what cost. In addition, theory building in this respect should distinguish between the study of ICT for developing countries and the study of ICT in developing countries.

The notion that communication can bring about developmental change is not new (e.g. Stevenson 1988). The notion that differential access to information may impact on economic development is somewhat more recent (Stiglitz 2002 provides a useful overview). As Wilbur Schramm observes in a retrospective of the field, 'communication (is) a relationship built around the exchange of information' (Schramm 1983:5). However, the chapters in this volume reveal that communication through ICT has a number of unique characteristics which may shape these relationships. These include its capacity to pool, deconstruct, amplify, reconfigure, redistribute, store and curate information. In addition, the pivotal actors in this relationship who link ICT to beneficiaries may be individuals, institutions, corporates or governments or combinations of these. Further, ICT can be used in research to inform opinion or make decisions, comment and respond; in promotion/advocacy networking and organization and forms of identity or group formation; other forms of activism; sharing of information, opinions and preferences; and showing support, consensus or disagreement, scolding, abuse and flaming. On the other side, ICT can also be used for rule compliance and rule enforcement, surveillance, misinformation and propaganda, scolding and exposure; to give sermons; and to advance views of small connected groups over the other, most probably those of elites.

Understanding how these characteristics, dynamics and actors interact and the systems within which they operate is likely to be an area both for the application of existing approaches such as the actor-network theory and for the development of new concepts such as the ICT ecosystem. Of relevance here is the notion ICT renewing previous flagging communication relations that is used specifically around political communication by Fabro. This could apply to other dimensions as well such as intellectual curiosity, reading or art and suggests a role that ICT could play for greater engagement and agency promotion.

Closely related to this, a possible area of theory extension concerns the technical dimension of ICT and this might be a shortcoming of the current volume. This does not refer to better understanding of the technical component, but rather to understand the implications of different technical choices. These choices (and level of knowledge about the choice being made) may well shape later options and create a path dependency. Thus, there may be many arenas within which policy debate and conflict take place which typically include those that are political or bureaucratic. It may be necessary to recognize the importance of the technical arena in which design decisions are taken. As is shown in this volume, these design choices may influence the outcomes of interaction. One gap in our understanding that could address this gap concerns the origins of the ICT champions who make such decisions. This would include how such champions come about, how they survive or flourish and which champions do not succeed, including the reasons and the implications of this.

Digital poverty is an important theoretical advance discussed in Diga and May. However, as is the case with other alternatives proposed for the conceptualization of poverty, digital poverty does not necessarily map neatly onto more conventional forms of well-being. As an example, 15 % of the nonpoor measured in a money-metric approach are digitally poor. The reasons for this require further development and can be related to existing studies of subjective and structural poverty that have similar results.

Cruz and Sajo draw attention to the potentially subversive role of the Internet which simultaneously provides opportunities for exploitation and survival through cybersex. They also remind us of the potentially dark side of digital capitalism. Not all ICT is good, and equally not all deemed bad is necessarily bad for all. Some part of this consideration is ICT and financialization of capitalism and the consequences of this for development. A useful concept that they introduce is that of 'affective labour' and they raise the possible labour issues that might be associated with digital capitalism. The consequences of employment changes as a result of ICT, especially for those that are poor and marginalized, are an area of future investigation and should include considerations of the circumstances in which ICTs reduce or increase control over labour and perhaps provide freedom from drudgery or hazardous work.

Finally, future areas of research could make use of the typology of research proposed by Ordóñez to organize the ambitious task just provided. These are:

- Conceptual research which is blue-sky in nature
- Planning research, what to be done, in what sequence and where, establishing proof of mechanisms
- Instrumental research, testing new ideas, demonstrating proof of concept
- Action research
- Impact assessment/M&E, attribution

Any single research project may fall into multiple categories, and a study grounded in action research principles and methodologies may well have a conceptual component in which new ideas are explored or old ones reassessed.

**Open Access** This chapter is distributed under the terms of the Creative Commons Attribution Noncommercial License, which permits any noncommercial use, distribution, and reproduction in any medium, provided the original author(s) and source are credited.

# References

Agre, P. E. (2002). Real-time politics: The internet and the political process. *The Information Society, 18*(5), 311–331.

Boaz, A., & Pawson, R. (2005). The perilous road from evidence to policy: Five journeys compared. *Journal of Social Policy, 34*(2), 175–194.

Brewer, E., Demmer, M., Du, B., Ho, M., Kam, M., Nedevschi, S., & Fall, K. (2005). The case for technology in developing regions. *Computer, 38*(6), 25–38.

Dutton, W. H. (2004). *Social transformation in an information society: Rethinking access to you and the world.* Paris: UNESCO. Retrieved from http://citeseerx.ist.psu.edu/viewdoc/download?doi=10.1.1.137.6534&rep=rep1&type=pdf. Accessed 16 Jan 2014.

Harris, R., & Chib, A. (2012). Finding a path to influencing policy. In A. Chib & R. W. Harris (Eds.), *Linking research to practice: Strengthening ICT for development research capacity in Asia.* Singapore: Institute of Southeast Asian Studies.

Heeks, R. (2010). Do information and communication technologies (ICTs) contribute to development? *Journal of International Development, 22*(5), 625–640.

Pawson, R. (2006). *Evidence-based policy: A realist perspective.* London: Sage.

Schramm, W. (1983). The unique perspective of communication: A retrospective view. *Journal of Communication, 33*(3), 6–17.

Stevenson, R. L. (1988). *Communication, development, and the third world: The global politics of information.* New York: Longman.

Stiglitz, J. E. (2002). Information and the change in the paradigm in economics. *The American Economic Review, 92*(3), 460–501.

Toyama, K. (2011, February). *Technology as amplifier in international development.* In Proceedings of the 2011 iConference (pp. 75–82). Seattle, USA: ACM.

Printed by Printforce, the Netherlands